Trespassing
on
Einstein's
Lawn

Trespassing on Einstein's Lawn

A Father, a Daughter,
the Meaning of Nothing, and
the Beginning of Everything

Amanda Gefter

Bantam Books | New York

Published in the United States by Bantam Books, an imprint of The Random House
Publishing Group, a division of Random House LLC,
a Penguin Random House Company, New York.

BANTAM BOOKS and the HOUSE colophon are registered trademarks
of Random House LLC.

LIBRARY OF CONGRESS CATALOGING-IN-PUBLICATION DATA
Gefter, Amanda.
Trespassing on Einstein's lawn : a father, a daughter, the meaning of nothing, and the
beginning of everything / Amanda Gefter.
pages cm
Includes bibliographical references.
ISBN 978-0-345-53143-8 (hardback : acid-free paper)—ISBN 978-0-345-53963-2
(eBook) 1. Physics—Philosophy. 2. Beginning. 3. Quantum theory. 4. Gefter, Amanda—
Anecdotes. I. Title.
QC6.G34 2014
530.01—dc23
2013013874

Printed in the United States of America on acid-free paper

www.bantamdell.com

987654321 246897531 123456789

First Edition

Book design by Susan Turner

For my dad, who gave me the universe

"We used to think that the world exists 'out there'
independent of us, we the observer safely hidden behind
a one-foot thick slab of plate glass, not getting involved,
only observing. However, we've concluded in the meantime
that that isn't the way the world works. In fact we
have to smash the glass, reach in."

JOHN ARCHIBALD WHEELER

Contents

A Note to the Reader

The book you're about to read contains cutting-edge physics packaged in a personal memoir, which spans the last seventeen years of my life. As such, it is inevitably subject to the failures of human memory, which I always hear neuroscientists claiming to be pathetically unreliable. Nevertheless, in reconstructing scenes and dialogue, I have done my best to keep everything as accurate as possible—by consulting my own notes and photographs, talking with others who were on the scene, and, most important, asking my mother, who somehow manages to remember my life in far greater detail than I ever do. My conversations with physicists have all been transcribed directly from recordings, though edited for ease of reading and length. In some cases, I've combined multiple interviews with the same physicist into a single conversation. When necessary, I've adjusted the chronology of scenes to allow me to present the physics in a logical, meaningful way. I've spent seventeen years wandering a tortuous, circuitous road trying to piece together a deep understanding of physics and the nature of reality; I figured I'd try to relay what I've learned in a slightly more straightforward book. Of course, I could have opted for perfect accuracy, but then there'd be far too many scenes of me watching bad TV, quietly reading, or sleeping for hours on end. Plus, it would take me way more than seventeen years to write and it would take you way more than seventeen years to read, and I think by the end we'd all agree it probably wasn't the best option. The logician Kurt Gödel proved that any form of self-reference is plagued by uncertainty, and I can't think of a better example than a memoir. Still, I have worked to produce a book that rings deeply true. We are in search of ultimate reality, after all.

—AG

Trespassing
on
Einstein's
Lawn

Crashing the Ultimate Reality Party

It's hard to know where to begin. What even counts as a beginning? I could say my story begins in a Chinese restaurant, circa 1995, when my father asked me a question about nothing. More likely it begins circa 14 billion years ago, when the so-called universe was allegedly born, broiling and thick with existence. Then again, I've come to suspect that *that* story is only beginning right now. I realize how weird that must sound. Trust me, it gets weirder.

As for my story, it probably begins the day I lied and said I was a journalist. Not that I knew at the time that it was a beginning. There's no way I could have known how far the whole thing would go. That I'd soon be hanging out with the world's most brilliant physicists. That I'd turn a minor deception into an entire career. I could never have guessed that I'd be getting emails from Stephen Hawking, lunching with Nobel laureates, or stalking a man in a Panama hat. I never once imagined driving through the desert with my father to Los Alamos, or poring over fragile manuscripts in search of clues to a cosmic riddle. If I had stopped to think about it, I couldn't have foreseen that one little lie, one impulsive decision to go somewhere I didn't belong, would launch an all-consuming hunt for ultimate reality.

But the strangest part is that I no longer believe any of these things

is the beginning. Because after everything that's happened, after everything I've learned, I've come to see that this story begins with you.
With you opening a book, hearing the soft crack of a spine, the whisper
of a turning page. Don't get me wrong—I'd love to say that this is my
story. My universe. My book. But after everything I've been through,
I'm pretty certain that it's yours.

I was working in a magazine office when the lie was born. That was the
idea, anyway—"working" in an "office." In reality I was stuffing envelopes in the dusty one-bedroom apartment of a guy named Rick. The
idea was that I worked for *Manhattan* magazine. The reality was that I
worked for *Manhattan Bride*.

Manhattan covered New York's socialite charity-event circuit, but
the magazine was bordering on extinction when I first took the job, and
it was laid to rest shortly after.* Rick's newly launched glossy bridal
magazine, on the other hand, was alive and well. So even though I spent
most days fielding calls from florists and cake decorators, and one long
afternoon scowling in an obscenely puffy wedding gown, I continued to
tell people that I worked for *Manhattan* magazine. It sounded better.

I was there in the office, wondering if I could use the rubber-band
ball to fling myself back to Brooklyn, when I spotted the article in *The
New York Times.* John Archibald Wheeler, leading light of theoretical
physics, poet laureate of existence, had just turned ninety and physicists from around the world were heading to Princeton to celebrate.
"This weekend," the article read, "the Really Big Questions that Dr.
Wheeler loves will be on the table when prominent scientists gather at
a conference center here in his honor for a symposium modestly titled
'Science and Ultimate Reality.'"

As it happened, I was burning to ask Wheeler one particular Really
Big Question. If only I were a "prominent scientist." I slumped back in
my seat and gazed absentmindedly at an old *Manhattan* cover hanging
on the wall.

---

* Since then, a new *Manhattan Magazine* has come into existence, but it has nothing to do with
Rick.

And then it hit me.

I waited until Rick left to get lunch, then picked up the phone, called the people in charge of publicity for the conference, and told them, in the most professional voice I could muster, that I was a journalist calling from *Manhattan* magazine and I was interested in covering the event. "Oh, of course, we would love you to come," they said.

"Great," I said. "Put me down plus one."

I was utterly certain that these kind public relations people had never heard of *Manhattan* magazine. Most people in New York, let alone the rest of the world, had never heard of any such publication, but when I told people I worked for *Manhattan* magazine they always said, "Oh, of course!" *Manhattan* magazine is just a name that everyone thinks they know. Only they don't. And that, I realized, was my ticket to Science and Ultimate Reality.

I was equally certain that these same PR people assumed that my "plus one" would be a fellow journalist or a photographer there to shoot pictures as I covered my big story. I picked up the phone and called my father. "Clear your schedule for this weekend. We're going to Princeton."

My sudden urge to crash a physics conference with my father can be traced to a conversation seven years earlier.

I was fifteen at the time, and my father had taken me out for dinner at our favorite Chinese restaurant near our home in a small suburb just west of Philadelphia. Usually we ate there with my mother and older brother, but this time it was just the two of us. I was pushing a cashew around my plate with a chopstick when he looked at me intently and asked, "How would you define nothing?"

It was a strange dinner-table question, to be sure, but not entirely out of character for my father, who, thanks to his days as an intellectual hippie Buddhist back in the sixties, was prone to posing Zen-koan-like questions.

I had discovered that side of him the day I came across his college yearbook, flipping pages only to discover a photo of my father sitting shirtless in a lotus pose reading a copy of Alan Watts's *This Is It—*

a hilarious sight considering that these days he was a radiologist at the University of Pennsylvania, where he not only wore a shirt every day but often sported a well-coordinated tie, too. He had made a name for himself by explaining how a whole array of lung diseases were caused by a single kind of fungus, and by inventing the disposable nipple marker—a sort of pastie that you stick on someone's nipple when they're getting a chest X-ray so the radiologists don't mistake the nipple's shadow for a tumor. But behind all the fungus and nipples, that groovy lotus-posing dude was still in there waiting for a chance to speak up. When he did, he would offer unlikely morsels of parental guidance, like, "There's something about reality you need to know. I know it seems like there's you and then there's the rest of the world outside you. You feel that separation, but it's all an illusion. Inside, outside—it's all just one thing."

As a dogmatically skeptical teenager, I had my own Zen-like practice of zoning out when adults offered me advice, but when it came to my father I listened—maybe because when he spoke it sounded less like an authoritarian command and more like the confession of a secret. *It's all an illusion.* Now here he was speaking in that same quietly intense tone, leaning in so as not to let the other diners overhear, asking me how I'd define nothing.

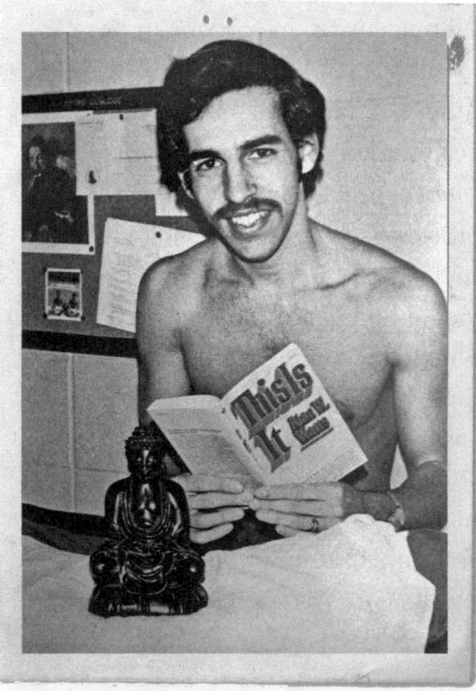

My dad's Haverford College
yearbook photo, 1970
Peter Gorski

I wondered if he was asking me about nothing because he suspected I was entertaining some kind of nihilistic streak. I was a contemplative but restless kid, the kind that parents describe as "hard to handle." In truth I think I was just bored and not cut out for the suburbs. An aspiring writer with a learner's permit, I had read Jack Kerouac and I was itching to hit the road. To make matters worse, I had discovered philosophy. When you're fifteen, boredom plus suburbia plus existentialism equals trouble. I couldn't imagine Sisyphus happy, and frankly, I didn't bother to try. Kurt Cobain had offed himself and I didn't believe in math. I had read somewhere that between the numbers 1 and 2 there were an infinite number of numbers, and I just kept thinking, how do you ever get to 2? My mother, a math teacher, would valiantly attempt to tutor me in geometry, but I'd refuse on principle. "Sure, I'll find that area," I'd say, "just as soon as you explain to me how you get to 2." She'd throw up her hands in defeat and storm off, leaving me to fail the class as a conscientious objector. It was Zeno's angst, in retrospect, but no one told me that then.

"How would I define nothing? I guess I'd define it as the absence of something. The absence of everything. Why?"

"I've been thinking about it for years," he said, "this question of how you can get something from nothing. It just seemed so impossible, but I figured we must be thinking about nothing the wrong way. And then the other day I was at the mechanic waiting for my car to be fixed and it just hit me! I finally understood it."

"You understood nothing?"

He nodded excitedly. "I thought, what if you had a state that was infinite, unbounded, and perfectly the same everywhere?"

I shrugged. "I'm guessing it would be nothing?"

"Right! Think about it—a 'thing' is defined by its boundaries. By what differentiates it from something else. That's why when you draw something, it's enough to draw its outline. Its edges. The edges define the 'thing.' But if you have a completely homogeneous state with no edges, and it's infinite so there's nothing else to differentiate it from . . . it would contain no 'things.' It would be nothing!"

I spooned some more rice onto my plate. "Okay . . ."

My father continued, his excitement mounting. "Usually people think that to get to nothing, you have to remove everything. But if nothing is defined as an infinite, unbounded homogeneous state, you don't have to remove anything to get to it—you just have to put everything into a specific configuration. Think about it this way. You take a blender to the world—you blend up every object, every chair and table and fortune cookie in this place, you blend it all until everything is just atoms and then you keep blending the atoms until any remaining structure is gone, until everything in the universe looks exactly the same, and this completely undifferentiated stuff is spread out infinitely without bound. Everything will have disappeared into sameness. Everything becomes nothing. But in some sense it's still everything, because everything you started with is still in there. Nothing is just everything in a different configuration."

"Okay, that's pretty cool," I said. "Something and nothing aren't really opposites, they're just different patterns of the same thing."

"Exactly!" My father beamed. "And if that's true, then it seems much more plausible that you can get something from nothing. Because, in a way, the something is always there. It's like if you build a sandcastle at the beach and then knock it down—where does the castle go? The castle's 'thingness' was defined by its form, by the boundaries that differentiated it from the rest of the beach. When you knock it down, the castle disappears back into the homogeneity of the beach. The castle and the beach, the something and the nothing, are just two different patterns."

The idea was beginning to click. In my existentialist musings I had thought a bit about nothing—not the transcendent, oneness brand of nothing that my father was drawn to but the Heideggerian variety, laced with indifference and dread. A nothing that was an absence, not only of stuff but of meaning, a vast and impenetrable darkness, like the void I'd find behind my eyelids at night. It was a concept that easily gave way to vertigo, a word that was a paradox by its mere existence. By its name it was a thing, yet it was no thing, and somehow it was the very thing that defined the world. Inasmuch as anything existed, it existed in opposition to nothing, but nothing was a noun doomed to self-destruct, an idea that came complete with its own negation, poised as the limit not only of reality but of knowledge and of language. Heidegger said that the question "what is nothing?" was the most fundamental of all philosophical questions, and yet "no one," wrote Henning Genz, "has ever given us an answer to what exactly defines nothing, other than by characterizing it simply by negatives." Only that's exactly what my father was trying to do. To define nothing not in terms of what it isn't, but in terms of what it is. A state of infinite, unbounded homogeneity.

"I like it," I told him.

He smiled.

And then this happened.

My father looked at me—his fifteen-year-old daughter—and in all seriousness asked, "Do you think that could explain how the universe began?"

I opened my mouth to speak, then paused, mouth open, searching for the right words, any words, to convey my mounting concern for his sanity. Had he gotten into the pot I had hidden under my mattress? "You're asking me how the universe began?"

"Well, before the universe there was nothing. So to get a universe, nothing has to become something. For years I've been thinking they must be two different states of the same underlying thing—the same underlying reality—otherwise there'd be no way to transform one into the other. But how could nothing be a state of anything? Only now I realize that it's a state of infinite, unbounded homogeneity. If you start from that, the problem of the origin of the universe becomes thinkable, at least. Tractable, maybe."

I had been on board when I thought we were playing a philosophical game of semantic Jenga, but now he was bringing the universe into it?

"Isn't this, like, *physics*?" I asked.

He nodded.

"I'm not even taking physics. I opted out of physics and took meteorology with the other underachievers. And I can't even tell you how a hurricane begins because I slept through the class."

He motioned to the waitress for the check. "Well, I think we should figure it out."

We should figure it out. It wasn't the kind of thing a parent says to a child. It was the kind of thing a person says to another person. I was intrigued. The whole thing sounded crazy, but crazy was infinitely better than boring. Besides, if there was one thing I knew, it was that my father was brilliant. Everyone knew my father was brilliant. He played it down with his sweet exterior and goofy sense of humor. You'd be forgiven for not seeing his brilliance right away, since he was always making wrong turns, zoning out midsentence, or, according to family legend, forgetting his pants. But just past that polite, absentminded demeanor was a bold, creative, insightful mind, and people who spoke to him for even a few minutes walked away knowing they had encountered something extraordinary. If you had to choose one guy to lead you off a cliff with his crazy idea, my father was that guy. For the first time in what felt like years, I smiled.

"Okay. How?"

He shrugged. "We'll do a little research."

So we began to read. If there was a book about physics or cosmology, our noses were in it. We read about the big bang, inflation, relativity, quantum theory, galaxy formation, particle physics, thermodynamics, extra dimensions, black holes, the microwave background. We discussed the ideas late into the night until my mother yelled at us to go to bed. Each piece of knowledge gained brought with it a hundred new questions, and reading became an endless scavenger hunt. We pored over enormous stacks of books in a feverish attempt to learn what was known about how the universe began, how something came from nothing. It became our own secret world.

Soon we needed to dedicate an entire room of the house just to physics books. Luckily, we had a spare—a small bedroom I had burned down in a freak accident involving a trick birthday candle. We cleaned out the ashes, painted the walls, and built shelves. The books multiplied at an exponential rate, climbing the walls up to the ceiling and eventually blanketing the floor.

My father had me convinced that the nothing before the universe was a state of infinite, unbounded homogeneity, a featureless, uniform sameness that stretched on and on for eternity. Or at least until the universe was born. Which of course begged the million-dollar question: Why would the nothing ever change? How could something defined by relentless sameness ever become different? Why would anything, like a universe, ever happen at all?

It was an infuriating dilemma. On one hand, by suggesting that something and nothing were just different configurations of the same thing, an answer to the question of how you get something from nothing seemed possible. On the other, the perfect uniformity of the homogeneous state seemed to rule out the possibility of change altogether.

The more we discussed it, the more annoyed I became at having to repeat the phrase "infinite, unbounded homogeneous state." I tried referring to it simply as "nothing," but the grammatical ambiguity inevitably led us into a kind of philosopher's rendition of "Who's on First." "Seriously," I told my father. "If I have to say 'homogeneous state' one more time, I'm going to kill myself with a physics book."

"We'll come up with a shorthand," my father said. "How about just 'H-state'?"

I considered it for a moment. "H-state. I can live with that."

To figure out why the H-state would ever change, we needed to know why cosmologists believed the big bang happened. Exactly what kind of physical process could get a whole universe to suddenly burst into existence from nothing?

In the growing piles of books we found intriguing suggestions, but no answers. The problem was that cosmology didn't start with nothing. It started with everything—with an expanding universe full of matter and radiation—and worked backward, running the clock 14 billion years in reverse and watching as the universe contracted, galaxies crowding closer together until the entire observable universe occupied a single point, the origin whence something relatively big presumably went bang, an infinitely hot, infinitely dense, infinitely twisted cosmic seed. A singularity.

It was tempting to think of a singularity as small, but, as my father and I quickly learned, that's a rookie mistake. It only seems small because you picture it as a point in space, as if you're looking at it from the outside. But the singularity has no outside. It's not a point in space because it *is* space. It's the universe, it's everything. We're *in* the point. Besides, a point isn't small—it's sizeless. I had learned that in geometry class, in spite of my protests. You can just as well think of a point as infinitely big. *The big bang happened everywhere,* I scribbled in my notebook. *Even in the suburbs.*

Watching cosmic evolution play out in reverse, you see everything turn into nothing at the singularity. The answer to why the H-state would change was hidden there. It was hidden well. The whole expanding universe was described by the equations of general relativity,

Einstein's theory of space and time and gravity, but the singularity was the one place where the equations couldn't go. If general relativity provides a map of the universe, the singularity is the uncharted spot that the cartographers aren't sure how to draw. *Here be dragons.*

Quantum dragons, most likely. The singularity suggested that general relativity would eventually give way to a more fundamental theory, but physicists already knew that. Einstein's theory wasn't compatible with quantum mechanics, the theory that describes the behavior of matter at extremely small scales. In their day-to-day lives physicists could ignore the problem by keeping the two theories separate, using general relativity to describe how big things such as planets and galaxies distort spacetime and using quantum mechanics to describe the strange dice game subatomic particles play. But at the end of the day, the separation can't hold up. Spacetime and matter talk to each other all the time. As Wheeler put it, "Matter tells space how to curve. Space tells matter how to move." As helpful as it may have been to pretend otherwise, the two theories described one and the same universe. Something had to give.

The singularity on the cosmic map wasn't a thing but a message: spacetime, at least as Einstein dreamed it, can't be reality's bottom layer. Something lies beneath it, something more fundamental from which spacetime as we know it emerges, something that will be revealed only by a theory that unites general relativity with quantum mechanics: a theory of quantum gravity.

Need quantum gravity to understand singularities, I jotted down in my notebook. *To understand nothing.* It was kind of funny, I thought. You needed the theory of everything before you could have a theory of nothing.

It suddenly dawned on me that if singularities were nothing but placeholders on the map, then the big bang was nothing but a placeholder, too. It was an important theory backed by powerful evidence—but it wasn't the whole story. It couldn't be.

So we kept on reading. Eventually we stumbled on some articles by John Wheeler. I was instantly drawn to his writing—it wasn't like any physics writing I had seen. It was more like poetry: intellectually daring and provocative, full of whimsical yet powerful turns of phrase. Wheeler emphasized that spacetime couldn't be reality's ultimate ingredient, be-cause at its highest resolution quantum mechanics and general relativ-ity conspire to destroy it, warping its geometry until it isn't geometry anymore. And in a bizarre twist, he suspected that it might not be pos-sible to understand how the big bang happened—how the nothing turned into something—without considering the role of observers. "Can one only hope some day to understand 'genesis' via a proper apprecia-tion of the role of the 'observer'?" he wrote. "Is the architecture of exis-tence such that only through 'observership' does the universe have a way to come into being?" The idea sounded totally outlandish, but I knew that Wheeler was considered a genius, on par with some of the greatest physicists of all time. There had to be something to it. Still, we couldn't get past the most obvious question: if observers are necessary ingredients for existence, where do the observers come from? I was tempted to dismiss it as a nonstarter, but the idea was so strange that I couldn't quite let it go. *Wheeler says observers play a role in the big bang,* I wrote in my notebook. *Find out what the hell that could possibly mean.*

Eventually, going out and partying with my friends seemed less excit-ing than venturing through the universe with my father. On nights that I did go out, I would come home at 3:00 A.M. to find my dad still awake, reading, and we would sit at the kitchen table, snack on cereal, and talk about physics until dawn.

I was loving every minute of it—which was weird, considering I had never had much interest in science. In fact, when I really thought about it, there were only two instances in which I had ever considered science at all. The first was when I was seven years old and someone gave me a set of children's science books. I was only interested in one of them, the volume about air. For months I carried it with me, fasci-nated by the idea that what looked like nothing was actually something—something intricate and vital. The second was in tenth-grade chemistry class. My teacher, Mr. McAfoos, was one of those rare high school teachers who managed to be passionate yet cool—even the most cynical of his students (namely, me) found his enthusiasm bear-able. On the day he taught us about the structure of atoms he jumped up on top of his desk and did a little dance to illustrate the dynamics of electron energy levels, a twist that got a little bit softer now as the electron sank into a lazier level. But the part of the lesson that caught my attention was when he noted that more than 99 percent of the atom is just empty space. Not empty like air, which is made of atoms, but empty like nothing at all. "This desk," he told us, grinning and pounding his fist on the wood to emphasize its deceitful solidity, "is mostly made of *nothing*." For weeks I couldn't get that out of my head. This time I was amazed by the thought that what looked like something was mostly nothing. That behind the world was another world— behind the barely visible, something more invisible still.

Something about nothing had captured my imagination—maybe because it was the least likely thing to catch one's imagination, in the same way that "the most ordinary thing in the world" would, by definition, be extraordinary. How strange to find out that nothing had captured my dad's imagination, too. That all those times I'd looked over to see him lost in thought, he was probably thinking deeply about nothing.

But those two episodes aside, I had never cared about science— because until my cosmic adventures with my father, I had no idea what science was. No one ever tells you. You sit down in a classroom and the teacher starts hurling facts your way and you're supposed to memorize them and regurgitate them and you have no idea why. They present the whole thing as if it was a done deal, a list of facts that together constitute a kind of instruction manual for nature. But the instruction manual hasn't been written yet. Einstein said, "This huge world stands before us like a great eternal riddle." Why couldn't any of my teachers have told me that? "Listen," they could have said, "no one has any idea what the hell is going on. We wake up in this world and we don't know why we're here or how anything works. I mean, look around. Look how bizarre it all is! What the hell is all this stuff? Reality is a huge mystery, and you have a choice to make. You can run from it, you can placate yourself with fairy tales, you can just pretend everything's normal, *or* you can stare that mystery in the eye and try to solve it. If you are one of the brave ones to choose the latter, welcome to science. Science is the quest to solve the eternal riddle. We haven't done it yet, but we've uncovered some pretty cool clues. The purpose of this class is for you to learn what clues we've already got, so that you can take those and head out into the world to find more. And who knows? Maybe you'll be the one who finally solves the riddle." If just one of my teachers had said that, I wouldn't have taken meteorology.

Luckily, my father had let me in on science's secret. So while my teachers at school made me feel like just another unremarkable kid, at night I went home to a covert world in which I had been chosen for the ultimate quest, one that required intensive training, one in which nothing less than the universe itself was at stake.

"Someday, when we find the answer to the universe, we should write our own book," my father said one night as we were scouring stacks of newly acquired cosmology books. "We're reading all these books in search of answers, but maybe the book we're looking for hasn't been written yet. Maybe *we* need to write it."

"A physics book?"

"You've always wanted to be a writer."

"Well, yeah," I said, "but I want to write poetry and short stories."

"What could be more poetic than the answer to the universe?"

I couldn't help grinning at the prospect. I wasn't sure whether my father really believed we'd find the answer to the universe someday, but his optimism was contagious. Not long before, my whole world had been steeped in insignificance and apathy. Now every atom in the universe was a mystery, every word a clue. In the blink of an eye my father had turned my world into a treasure hunt, and now, as if that weren't good enough, he had revealed the twist: we'd have to draw the treasure map for ourselves.

"When we write our book, I'm going to clear out every last one of these," he said, motioning to the hundreds of books in the room with the sweep of his arm, "and replace them with ours. We'll have a whole library just for a single book."

The intellectual excitement I felt learning physics with my father made high school feel all the more mind-numbing by comparison. I grew so bored with the place that I racked up enough credits to graduate a year early. My big plan was to move to New York City to become a writer while continuing to hunt reality with my father on the side. I applied to the New School, enamored with their alternative liberal arts program that didn't require students to take any math. My mother, who had adored the experience of touring sprawling green New England campuses when my brother had visited potential colleges several years before, was anxious to accompany me to New York to visit the New School. For the grand campus tour, we squeezed into a small elevator with a handful of other prospective students who sported a colorful array of dyed hair, tattoos, and piercings. The entirety of the tour consisted of four sentences flatly mumbled by our androgynous guide: "This is the first floor. This is the second floor. This is the third floor. Any questions?" My mother started to cry. I was sold.

Back at home, I didn't bother attending graduation. Finishing high school didn't strike me as enough of an achievement to warrant a polyester gown. Still, my parents threw a little party at our house to celebrate. In the midst of the festivities, my father pulled me aside and handed me a blue folder.

While everyone talked and laughed in the next room, I sat down on the stairs and opened the folder.

Your first years were so silent.
Waiting, waiting for the words.

I grinned. He had written a kind of Beat-style poem, a nod to my literary tastes at the time, and it chronicled my childhood: the time I learned to read, the night I ran away from home, all the books and ideas that made life seem worthwhile.

And Kerouac, and On the Road
The rhythm, the rhythm of the words
And Ginsberg and "Howl" and "Kaddish"
Chanting the rhythm
And Kesey and Burroughs, Fitzgerald and Proust
The words, the words
And Lou Reed and the Velvet Underground
The rhythm, the rhythm, rhythm and the words

And Zen and existentialism, Walden *and Thoreau*
The meaning, the reason, the meaning of the words
Soon New York City, the New School
The village and Washington Square
The world's a big blank journal
Waiting for your words
Let all hear the rhythm, the rhythm of your words.

A few days after my seventeenth birthday I packed up my things and moved from suburbia to New York's East Village. At the New School, students don't "declare a major," they "choose a path." I chose two: philosophy and creative writing. I was interested in ideas: in Plato's forms, Spinoza's God, Wittgenstein's things of which we cannot speak. I was interested in how one can take these ideas and weave them into story, how to use narrative to bring meaning to the universe.

The Gefter family
circa 1998: Warren,
Brian, Marlene, and me
Harry Bergelson

At the New School, my philosophy classes were inspiring, but my writing classes were saturated with a postmodern sociopolitical agenda that was too liberal even for me. The secular Jewish household in which I'd grown up had been ultraliberal, but we still held to some basic standards—like "facts" and "spelling." When a professor handed one of my stories back to me with a big red circle around the word *women* and a note instructing me to spell it as *womyn* so as not to perpetuate the misogynistic patriarchy of the English language, I decided I had had enough. I transferred to the Gallatin School of Individualized Study at New York University, located just a few blocks away. When I graduated, I resigned myself to a life as a starving writer, but to keep me in snacks I took a somewhat dubious job at the little-known but fortunately named *Manhattan* magazine.

When the weekend of the Science and Ultimate Reality symposium rolled around, I took an early train from New York to Princeton. My father met me at the station. Together we drove to the conference center, gearing up to ask Wheeler our burning question.

We strolled confidently into the lobby. Having never been to a physics conference, I didn't know what to expect. Still, with the all-star lineup of speakers, I had assumed there would be some kind of audience. A crowd of ordinary, subgenius people. Throngs, perhaps, of fools gaping at the physicists milling about with their danishes and coffee. But no. There were only two.

We just stood there, two dumbfounded deer caught in Mensa-grade headlights. We were clearly the only outsiders in a room full of the world's leading physicists and legit journalists qualified to cover the event.

"*Journalists*," I muttered to my father. "Remember, we're journalists."

He nodded. He looked handsome and proper in his navy suit. Looking at him now, it occurred to me that he didn't exactly stand out in this homogeneous sea of white, middle-aged men. Surely his most incongruous feature was the jittery twenty-one-year-old girl standing next to him, eyeing him suspiciously.

"Don't we need some kind of badges?" he whispered.

"Badges! Yes. I'll get the badges. You stay here."

I knew if questioned I could stick to my *Manhattan* magazine story at the press table, but I had no idea how I'd explain him. *This journalist? Why, yes, it is weird how he looks exactly like me. Old enough to be my father? You think?*

I headed over to the check-in table, peeking at name tags as I went, finally attaching faces to what had for us become household names. I scanned the table for my badge. There it was: *Amanda Gefter*, Manhattan *Magazine*. Next to it, a blank for my plus one. As I leaned over to grab them, I accidentally bumped shoulders with the man next to me. "Sorry," I said, glancing up. I blushed and sprinted giddily back to my father. "Oh my God!" I squealed. "I just touched Brian Greene!"

As we took our seats in the conference room we looked around in awe, jabbing each other with our elbows and whispering things like, "Holy shit! That's Alan Guth!" and "Max Tegmark is sitting right in front of us!" We were awestruck, starstruck, dumbstruck. These people had played the lead characters in our conversations for years, and now we were sitting among them. I nudged my father and pointed. There, taking a seat in the front row, was the man whom everyone was here to celebrate: John Archibald Wheeler.

Physicist, philosopher, poet, prophet, legend. Even at ninety, Wheeler had a boyish face. Sweet-looking, with a mischievous gleam in his eye. As a younger man, Wheeler had studied quantum physics in Copenhagen under Niels Bohr, and taught the first course on general relativity at Princeton, where he strolled the tree-lined streets discussing the nature of reality with Einstein. With Bohr, he helped work out the physics of nuclear fission, and he went on to work on the development of the atomic bomb in the Manhattan Project and the hydrogen bomb after that. He coined the terms *black hole* and *wormhole*. He led countless students to profound discoveries—students including Richard Feynman, Hugh Everett, Jacob Bekenstein, and Kip Thorne. He emphasized the importance of "ideas for ideas," fearless in the face of mystery.

Four quintessential Wheeler questions served as the inspiration for the symposium: Why the quantum? It from bit? A participatory universe? And how come existence? My father and I were sure that answering them would be key to solving the mystery.

Why the quantum? The thing about quantum mechanics was that it offered up a picture of reality that didn't seem to jibe with the world as we know it: a picture in which effects had no causes, observers determined the observed, and everywhere you looked there were boxes full of simultaneously dead and alive cats. Then again, maybe quantum mechanics didn't offer up any picture of reality at all. Maybe it just took the ones we had and blurred them beyond recognition. The theory allowed physicists to make extraordinarily accurate predictions, but the predictions alone provided no clues as to what the whole thing meant. Wheeler had lived through quantum theory's founding and had been there in the thick of it as physicists such as Bohr, Feynman, Everett, and Einstein tried desperately to make sense of the bizarre facts unfolding before them. With no principles to stand on, physicists' most successful theory just sort of hovered there, looking arbitrarily weird, a weirdness to which many surrendered, waving white flags, muttering, "Shut up and calculate." But Wheeler refused to surrender. He knew that particles' seemingly arbitrary behaviors had to be some kind of clue. The weirdness had to be telling us something.

It from bit was Wheeler's slogan for the idea that the physical universe is made not of matter but of information. In quantum theory, making an observation is equivalent to asking a yes-or-no question. Is the particle here or elsewhere? Is the cat dead or alive? Wheeler suggested that the very posing of the question *creates* a bit of information, and that such bits were the fundamental building blocks of reality. "The universe and all that it contains ('it') may arise from the myriad yes-no choices of measurement (the 'bits')," Wheeler wrote. "Information may not be just what we *learn* about the world. It may be what *makes* the world." It was a pretty weird idea, given our intuition that the fundamental building blocks of matter are just smaller pieces of matter, like particles. Of course, as Mr. McAfoos had taught me, particles are 99 percent empty. Still, you'd hope that this meager 1 percent would somehow be enough to fortify a world. But according to Wheeler, even the 1 percent was made of nothing sturdier than an observer-elicited yes or no. *A house is made of bricks but the bricks are made of information?* I wrote in my note-book. How could it be that when we look closely enough at the physical world we find something that's not physical at all, as if this whole universe were a kind of virtual reality? Then again, what's the difference? What could "physical" even mean?

A *participatory universe?* If measurements built the universe bit by bit, as Wheeler suspected, then observers were somehow implicated in the creation of reality—a radical picture that, if true, would mean that ours was a participatory universe. As the physicist Paul Davies wrote, "Wheeler seeks to . . . turn the conventional explanatory relationship: matter → information → observers on its head, and place observership at the base of the explanatory chain: observers → information → matter." That was the notion that had struck a chord in my father and me—could it somehow be that observers turn nothing into something? The idea seemed impossible from the start, because where would the observers come from? What would even count as an observer? Surely it didn't have to be conscious or human . . . but what?

Finally, *how come existence?* That was the big one. Why is there something rather than nothing? It was the question that had taunted my father for years, the one that had set our journey in motion and led us to crash this conference in search of an answer. How come existence? How come, indeed.

"I had the good fortune of having my first and only heart attack last January," Wheeler told the captivated audience when he finally took the stage, speaking slowly and quietly, his quavering voice betraying not only his age but his urgency. "I call it good fortune because it taught me that there's a limited amount of time left and I better concentrate on one thing: How come existence? How come the quantum? Maybe those questions sound too philosophical, but maybe philosophy is too important to be left to the philosophers."

When the session ended, a swarm of physicists descended on Wheeler, who remained seated in the front row, smiling and nodding as one physicist after the next sat down to speak with him. We waited patiently in our seats, biding our time. Eventually our moment arrived. The crowd was thinning, so we made our way down to the front row. This was it. This was why we had come.

John Archibald Wheeler (left) and my father (right) at the Science and
Ultimate Reality Conference, 2002
A. Gefter

We bent forward, and each of us shook his hand. "I'm Warren Gefter; this is my daughter, Amanda. She's here covering the conference for *Manhattan* magazine. We're so thrilled to meet you," my father said. Wheeler nodded, but it appeared as if he couldn't hear and was too polite to say so. My father leaned in closer and spoke louder. "We have a question we've always wanted to ask you," he said, over-enunciating every word. "If observers create reality, where do the observers come from?"

Wheeler smiled. "From physics. From the universe. I like to say"—he paused, trying to find the words—"that the universe is a self-excited circuit."

My father nodded appreciatively, then pensively. "So it's all from nothing?"

Again Wheeler appeared not to hear, so my father asked once more, louder still. "So it's all from nothing?"

Wheeler nodded and spoke slowly. "There is a principle that says the boundary of a boundary is zero."

Just then some physicists turned to speak with him, so we thanked him, told him what an honor it was, smiled, and walked away.

The meeting had let out for the day, so we decided to head outside and wander around Princeton. The crisp springtime air seemed electric. Walking down the street, manically chattering about the people we had seen and the ideas we had heard, we felt we were now a part of something, even if we hadn't been invited.

"We spoke to Wheeler!" my father said, looking dazed and grinning in disbelief.

"Hell, yeah, we did!" We slapped a high five.

We found our way to Mercer Street, the quiet road where Einstein had lived during his years in Princeton, where he and Wheeler had walked, discussing the great cosmic mysteries. It suddenly struck me as really amusing that Einstein had lived in New Jersey. *New Jersey?* It's like spotting Shakespeare eating a burger at Wendy's, or finding out that Plato wasn't really Greek but Canadian.

We found Einstein's former house—112 Mercer—and stood side by side in front of it, gazing ahead. We stood in awe, but the house stood in modesty. It was quaint and unassuming, painted small-town, ironic, anonymous white. But it was undergoing some sort of renovation and the front steps were cordoned off by yellow tape, like a murder scene.

My father pointed to the tape. "Maybe that guy he was always dropping off the roof in his thought experiments finally met his doom."

I knew that Einstein had specifically asked Princeton to keep the home as an ordinary residence rather than turning it into some kind of landmark or museum. "This house will never become a place of pilgrimage where the pilgrims come to worship the bones of the saint," he had said. But I didn't feel too bad. With all that yellow tape, we could barely worship the front porch. Besides, the pathologist who did Ein-

Trespassing at Einstein's Princeton home W. Gefter

stein's autopsy had stolen his brain. Compared to that, standing in reverence on his front lawn hardly seemed like a violation.

My father had once shown me an old beat-up hardback book of Einstein's essays on relativity that his father had given to him when he was just a kid. He told me how he tried to read it when he was only ten or eleven years old, how he would pretend to understand it. He kept the book on a shelf in his bedroom throughout his childhood; he would stare at it, flip through its pages, and wish he could untangle its meaning because he was sure it held the promise of some shining truth. Now, staring at the house, the March sun glistening off the white paint, I felt as if I had stepped out of my own head and into something vastly larger than myself. I looked at my father and had the sense that he was at last on the path he had always privately desired to tread. And me? I just wanted to follow.

We watched the house intently, as if Einstein were going to come strolling out the front door at any minute, sticking out his tongue and yelling at us to get the hell off his lawn.

Is that house really made of information? I wondered. Is that information made by me? By us? Is anything what it seems? Is any of this real?

I knew that the world around me had to be more than it seemed. Physics clearly bolstered that view—after all, a table was mostly empty space, and if you zoomed in on the empty space, it dissolved into something else, something unknown. Look closely enough at anything, and everything we know seems to fall away, leaving behind . . . what? Some basic ingredient of reality? Something as intangible as information? I didn't need science to tell me that appearances were deceiving— I knew it in my bones. I had known it since the day my father told me it was all an illusion. I knew it because without it, I wasn't going to make it. I couldn't bear the notion that reality stopped at wedding cakes and rubber-band balls and suburban homes. If the world in front of my eyes was the be-all and end-all of existence, I thought, then count me out. I needed mystery. I needed to know that there was more to the story.

I suspected my father needed that, too. Though he would never admit it, it was becoming increasingly obvious to me that his quiet, professional, suburban life wasn't cutting it. His inner rebel—that shirtless hippie sitting lotus-style and meditating on the nature of self and existence, the one who had been silenced by career pressures and the arbitrary demands of adulthood—was looking for a chance to emerge, and he glimpsed that chance in physics. And in me.

We sat down on the sidewalk in front of the house.

"A self-excited circuit . . . ," I mumbled.

My father nodded. "The boundary of a boundary is zero. . . ."

"Who is he, Yoda?" I asked. "The man speaks in riddles. What do you think it means?"

My father smiled. "I have no freakin' idea."

So there we were, in Princeton, New Jersey, in a vast and expanding universe, exhilarated, trespassing, half real, realizing that what once had been a hobby had now become a mission.

We sat for a few moments steeped in the heavy silence. Then a car pulled up, and we ran.

2

The Perfect Alibi

When I returned to New York, I had an epiphany.

I was staring up at the Science and Ultimate Reality press pass that hung on the wall above my computer as a souvenir of the conference and, with *"Manhattan* Magazine" printed proudly beneath my name, as a kind of inside joke. The badge contained an image of a pixilated sphere with a 0 or a 1 marked in each pixel. I was pretty sure it was one of Wheeler's drawings—but what did it mean? I quickly sketched it in my notebook with a reminder to myself to find out.

It was amazing how powerful that thing was. A piece of laminated paper on a string could give you access to ultimate reality's inner circle? It was like the golden ticket to cosmology's chocolate factory— find one and you win the chance to attend every lecture, talk with every physicist, even enjoy the lunches and banquets.

If we wanted to crash the ultimate reality party in the hopes of finding some answers, a press pass was clearly the way to do it. Still, I was pretty sure that my little scam wasn't going to cut it for long. Eventually someone was bound to look up *Manhattan* magazine and realize that it had nothing whatsoever to do with physics and that the magazine's ontological status was . . . well, anyone peering inside that box was going to find a dead cat.

If only there was some other way to get press passes.

A cartoon lightbulb lit up overhead.

I called my father. "I'm going to be a journalist."

"Okay . . ."

"Think about it! If we want to figure out the nature of reality, we need access to the best physicists around, the cutting-edge data, the meetings, the journals, everything. And if you're press, it's all yours! When we have a question about cosmology, we won't have to dig through twenty books to find the answer; we'll just ask a cosmologist! It's the perfect alibi!"

"That's a great idea," he said. "Maybe you could get some kind of internship. Or do you have to get a degree first?"

"No, no," I said, "you're missing my point. I'm going to become a journalist *today*."

"Sorry, what?"

"Yeah, I'm going to call *Scientific American* and ask if I can write something for them about the symposium. Then we'll be golden."

"Listen," he said, "I don't want to burst your bubble, and I think you could make a great journalist someday, but you can't just call *Scientific American*."

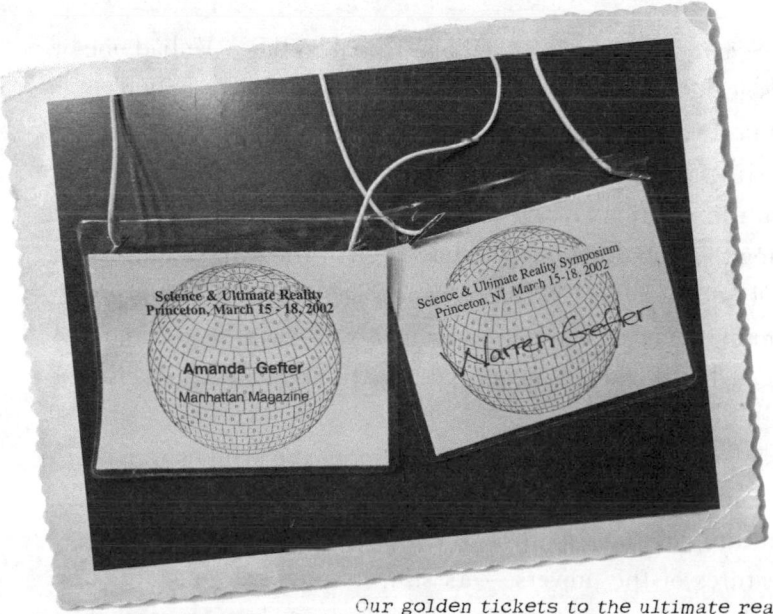

Our golden tickets to the ultimate reality party

W. Gefter

"Oh, yeah?" I said. "Watch me."

I knew I sounded a little nuts. After all, I hadn't gone to journalism school. Hell, I had never taken a physics class. But who cares, I thought. I would just have to learn on the job. Besides, I didn't need a degree and an internship and job experience—that would be like going to culinary school just to open a restaurant as a mob front. I wasn't try-ing to win a Pulitzer; I was just trying to scam some press passes.

I hung up the phone and dialed the number for *Scientific Ameri-can*'s news editor. Voicemail. When I heard the beep, I cleared my throat and tried my best to muster up a voice that would sound like that of a professional colleague, not a twenty-one-year-old girl. "Oh, hey, Phil, this is Amanda Gefter over here at *Manhattan* magazine, I'm just calling because I attended yesterday's symposium down at Prince-ton in honor of John Wheeler, and I wanted to check in with you to see if you need any coverage of it. We don't really run science stories here at *Manhattan*, but physics is sort of a side hobby of mine. Anyway, there was some good stuff, so feel free to give me a call."

I left my number and hung up the phone. If we were going to do this, we were going to do it right, and that meant starting at the top.

Phil from *Scientific American* called back the next day. "We had one of our editors at the symposium," he told me, "so we've pretty much got it covered. But if you come up with some interesting angle, email me."

Interesting angle? I could do that. I sat down and looked over my notes from the meeting. Trying to pitch an article about any one talk wasn't going to work—anyone who had been at the symposium could do that. Of course Wheeler's strange message—*the universe is a self-excited circuit, the boundary of a boundary is zero*—was an angle, but I had no idea what the hell it meant. I'd have to find something else for now.

That's when I noticed a theme. Throughout the symposium, there had been a giant elephant in the room: the anthropic principle.

The anthropic principle invoked our own existence to account for certain features of the universe—its size, its physical constants, the existence of stars and galaxies. Had those features been even slightly different, we wouldn't be here to wonder about them. At its worst, the anthropic principle was an empty tautology: we exist, therefore the universe is the kind of place that allows us to exist. At its best, it offered a way for physicists to explain why many cosmic features have such extraordinarily unlikely values—unlikely, but perfectly ripe for life.

I sat down at my computer and composed an email to Phil suggesting a small news item entitled, "Physicists Can't Avoid the A-Word." *At the Science and Ultimate Reality conference, I wrote, physicist Andy Albrecht began his presentation by assuring the audience, "I'm not going to use the A-word."*

Anthropic had become a four-letter word because it veered uncomfortably close to religion, I explained—as if the universe, somehow, were built just for us. That is, unless our universe isn't the only one. It's like the Earth. Our homey planet is positioned at the perfect distance from the Sun to host liquid water; any closer and the water would turn to gas, any farther and it would freeze. If the Earth were the only planet—a lone rock adrift in the solar system—its water-bearing position would seem awfully miraculous. But with seven other planets wandering around out there, it's hardly a miracle that we find ourselves on the one that's suitable for life—we're here because it's the only place we can be. The same kind of anthropic selection bias can explain the life-bearing features of the universe, all for the low, low price of a few trillion extra universes. Not that it's the kind of explanation anyone wants. Physicists want to explain the universe through logical and mathematical necessity. They want the world to be the way it is because that's the *only* way a world could be. But according to the anthropic principle, anything goes.

Still, everyone's efforts to avoid the A-word seemed a little odd, given the occasion. *After all, I wrote, it was Wheeler who famously described the universe as "participatory," and once asked, "On what else can a comprehensible universe be built but the demand for comprehensibility?"* Wheeler didn't believe that the universe was designed for us, nor that ours was a small island in a vast multiverse. He believed that the universe was right for observers because, somehow, observers *create* the universe.

I noted that the $10,000 prize for the Young Researchers Competition in Physics was awarded at the conference to Fotini Markopoulou, a loop quantum gravity researcher at the Perimeter Institute for Theoretical Physics in Canada. In her winning paper she argued that cosmology must describe the universe as seen by observers who are stuck inside it. *In the end, it seems, the more we pursue the deepest mysteries of the cosmos,* I wrote, *the closer we come to ourselves.*

I clicked "Send."

Phil emailed back right away. He explained that *Scientific American* had a similar piece in the works about the anthropic principle, but they were interested in Fotini Markopoulou. "What do you know about loop quantum gravity?" he asked.

What did I know about loop quantum gravity? Approximately . . . nothing. I called my father and read him the email.

"Oh, well," he said. "You tried."

"Tried?" I said. "This isn't over. This is our one shot!"

"Okay, but—"

"We have one night to learn loop quantum gravity."

"You're joking," he said. "Why one night?"

"If I don't write back tomorrow, it will seem like I took the time to look this stuff up. It has to look like I know it off the top of my head. We don't have much time—start reading and call me back in a few hours!"

I absorbed what I was reading as best I could. Loop quantum gravity was an attempt to unify general relativity and quantum mechanics— the seemingly correct yet mutually exclusive pillars of modern physics. Such unification is the key to the origin of the universe, that placeholder on the map where nothing turns into something, where the H-state becomes the world. *Need quantum gravity to understand singularities,* I had written in my notebook. *To understand nothing.* Loop quantum gravity's approach was to zoom in on space, peering down to nature's smallest scale to see what dragons lurk there.

That nature even *has* a smallest scale was pretty hard to grasp. I couldn't wrap my mind around the notion that if I were to zoom in on some modest stretch of space, magnifying it and peering into increasingly smaller depths, I would eventually reach a place further from me in scale than the entire observable universe—and yet, somehow, right here on the tip of my finger. A universe larger than the universe, sitting in the palm of my hand. Only you can't keep zooming forever. At a millionth of a billionth of a billionth of a billionth of a centimeter, you hit the bottom of reality. Sorry, folks, you've reached the end—the teeny tiny edge of the universe.

Space ends there, at the so-called Planck scale, because that's where quantum mechanics and general relativity join forces to bend spacetime until it breaks. The sheer density of gravity produces a sea of black holes, which Wheeler dubbed "spacetime foam."

It was a counterintuitive notion—usually when you're dealing with small things, gravity is negligible. Gravity acts on mass, and you need a lot of it before you notice its pull. Even at the human scale, gravity is pretty insignificant. A refrigerator magnet can overpower the gravitational force of the entire planet just to lift a paper clip. At the scale of protons and electrons, gravity barely exists at all.

But keep zooming in and, strangely, things start to turn around. The laws of quantum mechanics contain a loophole that allows large fluctuations of energy to burst forth from the vacuum, provided they don't stick around too long. At increasingly shorter time scales, energy blinks in and out of existence in the form of fleeting, or "virtual," particles. The more localized the virtual particle, the greater its momentum, and the greater its momentum, the larger its energy. Thanks to $E = mc^2$, more energy means more mass. So as you look at smaller and smaller distances, virtual particles grow increasingly massive until, at the Planck scale, gravity grows as powerful as the other forces. An energy in its own right, gravity's crescendo generates a runaway feedback disaster of the same variety that can collapse a 10^{32}-pound star into a black hole. At distances smaller than the Planck scale, gravitational feedback turns pathological. The universe turns on itself, cannibalizes. The fabric of reality bursts at the seams. Before melds into after, here conquers there, distance and duration give way to confusion and chaos, space and time dissolve and disappear. Equations sizzle and spark, mathematics falls apart into nothingness. In short, everything goes to hell. It's the end of the world.

Loop quantum gravity is a model of space at the Planck scale, just before gravity rips it to pieces. Lee Smolin, another Perimeter Institute physicist and a founder of the theory, had realized that the situation would stabilize if space, like matter, has a kind of atomic structure. That way, when you zoom in on a region of space—smaller, smaller, smaller still—you eventually hit a roadblock: a chunk of space that can't be further divided, a spatial "atom" that's as small as small gets. As long as space's atoms are no smaller than the Planck scale, Smolin said, gravity would remain in check. Its energy could only grow so large; its destruction could only wreak so much havoc.

As I paged through Smolin's book on the subject, one section caught my attention. The universe, he wrote, has to be considered a closed system. "It is true that the universe is as beautiful as it is intricately structured. But it cannot have been made by anything that exists outside it, for by definition the universe is all there is, and there can be nothing outside it. And, by definition, neither can there have been anything before the universe that caused it, for if anything existed it must have been part of the universe. So the first principle of cosmology must be 'There is nothing outside the universe.'" I couldn't help thinking that the same principle would have to apply to my father's H-state, since nothing, by definition, was infinite and unbounded. It had no outside.

My father gave up reading around 4:00 A.M., but I made it through the night and in the morning wrote the best email I could manage on no sleep, offering a breezy, off-the-cuff explanation of loop quantum gravity.

But something was nagging me. At the Wheeler symposium, Markopoulou's talk had been about more than atomic geometry—it had been about the meaning of observers in a quantum universe. "A physical cosmological theory must refer to observers inside the universe," she had said, echoing Smolin's first principle. Rather than a single description of the universe from the outside, the best we can get is a vast array of partial internal views. Quantum gravity, she said, ought to be a set of rules for translating between them. Markopoulou's internal cosmology, built of observers' limited perspectives, reminded me of Wheeler's suspicion that observers play some part in the creation of the universe. Did her loop quantum world have something to do with Wheeler's self-excited circuit? I knew if I could get the *Scientific American* article, I'd have my chance to find out.

I was sitting at my computer at *Manhattan Bride* when the email from Phil popped up in my in-box. I glanced over my shoulder and made sure Rick was engrossed in the layouts he was fiddling with on his computer screen before opening it: *Hi, Amanda. Thanks for this. Would you be interested in writing a profile of Fotini Markopoulou?*

I called my father, huddling over the phone, my hand covering my mouth so Rick couldn't hear. "I'm writing for *Scientific American*," I whispered. "We're in."

I made a plan to meet Fotini Markopoulou a few months later when she would be visiting New York. The following morning, I walked into the office and quit.

"It's not personal," I told Rick. "It's just that I want to write about physics."

"Maybe I could find a way for you to do that here," he said.

I blinked. "For a bridal magazine?"

An hour later I was back on the subway headed for Brooklyn. As the train made its way downtown toward Wall Street, it dawned on me that quitting my job might have been impulsive. One article was not going to pay the bills, and neither was my quest for ultimate reality, no matter how impassioned. But by the time we hit the East River, I was sure I had made the right call. In my bones I knew that this was the start of something big. That there was a leap to be taken, that adventures would ensue. That if we were going to do this thing, we needed to take it all the way.

My excitement descended into nervousness when I called my parents. "I know it wasn't the most fiscally responsible decision," I said sheepishly. "But it felt right."

"You've got to trust your gut," my father said. "Money is useful. But this thing is *important*."

I heard my mother sigh. "You better hope you can marry a doctor."

Without a day job to worry about, I dedicated all of my time to reading and thinking about physics. Unfortunately, to my friends and to the boyfriend I was living with at the time, "thinking about physics" looked awfully similar to "doing nothing." When I explained myself, everyone nodded politely, but their questions betrayed a deep conviction that I had officially lost my mind. "It's just that, well, do you think it might be difficult to devote your life to physics when you've never taken a single physics class?" they'd ask.

"Nah," I'd say. "It'll be a piece of cake."

To keep me in cash, I worked a few nights a week in my brother's Manhattan nightclub as a coat-check girl. It was a good gig. Amidst the deafening thump of hip-hop, in my miniskirt and high heels, I'd sit on the floor, lean back against a mountain of designer coats, and silently read about the universe.

When the weather warmed, I spent my afternoons sitting outside on the stoop with Cassidy, my black Labrador retriever, soaking in the sunlight and thinking about reality in preparation for my meeting with Fotini Markopoulou.

I knew that physicists needed a theory of quantum gravity because general relativity and quantum mechanics couldn't manage to peaceably coexist in a single universe. But what exactly made them so hopelessly incompatible? Everywhere I looked I found technicalities—the world of relativity is continuous and the quantum world discrete; relativity regards positions in spacetime as well defined, while quantum theory renders them fuzzy. They were obstacles, sure, but they struck me as mere couple's squabbles, not deep, unbridgeable rifts. It was like relativity preferred chocolate and quantum theory vanilla—not like relativity was a Protestant and quantum theory was a duck.

Relativity, at its core, was about what space and time mean to different observers. It began with a simple question that obsessed a sixteen-year-old Albert Einstein. What would a light beam look like if you were flying at the same speed alongside it? Would it appear to stand still, the way a car in the next lane looks like it's not moving if you're cruising alongside it at exactly the same speed? James Clerk Maxwell's equations of electromagnetism demanded that electromagnetic waves, otherwise known as light, always travel at 186,000 miles per second. Einstein immediately saw the problem. For an observer traveling at 186,000 miles per second, light's speed would drop to zero. And what then? Would electromagnetism cease to exist? Would the universe crumble?

Einstein realized that for the universe to hold together, and for the laws of electromagnetism to apply equally to every observer, there couldn't be any frame of reference in which light stood still. And while it seemed impossible, he knew there was only one way to make that rule hold: observers must always measure light to be moving at a constant 186,000 miles per second, *regardless of how fast the observers themselves are moving relative to the light.* No matter how fast you run you can never catch a light beam. Even if you were speeding alongside it, it would still stream away from you at a relentless 186,000 miles per second, a horizon that recedes just as quickly as you approach.

It's easy to miss how insane a statement that really is. Speed is a measure of how much space something crosses in a particular amount of time. For light to always move at the same speed regardless of how fast an observer is moving when measuring it, *space and time themselves* have to vary from observer to observer to make up the differences. The total distance in space and time *combined* remains the same for everyone—a unified four-dimensional spacetime that observers slice up in different ways, choosing some coordinates to call "space" and others to call "time," according to their individual points of view.

Einstein knew that there had to be some way of translating between all the different points of view, some prescription for figuring out how the same spacetime appears to different observers, since presumably there's only one universe. When he found it, he named it special relativity—"special" because it only provided translations for observers in uniform motion. It did not contain the rules for translating between a uniformly moving, or inertial, observer and an accelerated one—suggesting that a guy cruising down Bedford Avenue at a constant speed would find himself in a different universe than the guy who's picking up speed in the car next to him, even though they're both in Brooklyn.

For Einstein, this was nothing short of tragedy. He believed deeply that the true nature of the universe shouldn't depend on an arbitrary choice of coordinates. That reality was unified and singular, transcending our fragmented perspectives. That there was some way in which the world *really was,* in and of itself, regardless of who was looking or how they were moving as they looked. He was desperate to peel back the layers of false appearances and get at the truth that lay beneath. That meant finding a way to translate between what an inertial observer sees and what an accelerated observer sees. It was a mission that led him to create his masterpiece: general relativity.

I thought back to the night when general relativity had finally made sense to me. I'd been in high school and it was late at night; I was hunkered down at the kitchen table with my father. It was one of those perfect gestalt moments when something in the brain clicks and nothing is ever the same.

I had read all the usual explanations. That space and time are sewn together by the constant speed of light into a four-dimensional spacetime. That mass or energy in this spacetime causes its metric properties to warp, laying a landscape of slopes and valleys that we call the gravitational field. That what appears to us as the force of gravity is really space's hidden geometry.

Okay, so gravity isn't a force, it's the curvature of spacetime. Everyone said that as if it were to be followed by the sound of my kneecaps hitting the floor. But I didn't see what the big deal was. Sure, "the curvature of spacetime" sounded mysterious and arcane, but so did "the gravitational force." It was like replacing one phantom with another.

"Think about it this way," my father said, flipping to a fresh sheet of paper in one of those yellow legal pads he was always writing in. "This is a diagram of the universe," he said, drawing an L-shaped coordinate system, labeling the vertical axis *time* and the horizontal *space.* "This area in here," he said, sweeping his hand over the empty yellow expanse framed by the two axes, "is four-dimensional spacetime. Let's

say I'm moving through spacetime at a constant speed. So here's me."
He traced out a straight line, cutting a diagonal across the paper. "And
you're moving at some different but also constant speed, so here's you."
He traced a second straight line, drawn at a slightly different angle.
"But we're both looking at the same world. It might look different to
each of us, we'll each measure different distances and times, what
looks like space to you might look like time to me . . . but ultimately it's
just one world described by two different points of view, right? So special relativity gives you the equations that allow you to rotate my path
here in spacetime until it aligns with your path. That's a Lorentz transformation. You can move one until it lines up perfectly with the other—
that tells you that we're looking at one and the same world."

"Okay," I said, curious to see where this was headed.

He flipped up the yellow paper to reveal a fresh sheet, quickly
scratching out new coordinate axes. "Okay. Here's me again," he said,
drawing another straight line at an angle. "But this time, you are moving at a changing speed, you're accelerating. An accelerated path in
spacetime is a curved line, right, because you keep crossing more space
in less time," he said, drawing a curve that swept up toward the right-
hand corner of the page, then slowly arced back downward. "Now,
imagine rotating your curve until it matches up perfectly with my
straight line."

I thought for a minute. "It's impossible," I said. "A curve can never
match up with a straight line."

"But it is possible," he said. "Einstein knew it *had* to be possible,
because there's still only one universe. If you can't get that curve to
match up with that line, it means that you and I see entirely different
worlds just because I'm moving at a constant speed and you're accelerating."

"Einstein figured out how you can make a curve match up with a
straight line?"

"Yup." My father looked at me, grinning. "Bend the paper."

Suddenly it all became illuminated in what I can only describe as
a religious experience. Somewhere a choir was singing "Hallelujah."
Bend the paper! If you wrinkle the paper in just the right way, you can
turn a curve into a line. The wrinkles are gravity. They connect the

world. General relativity is at once incredibly profound and incredibly simple, a classic case of thinking outside the box. "That Einstein was some kind of freakin' genius, huh?" I said.

"The bending of the paper—of spacetime—is called a diffeomorphism transformation," my father said. "You have to be able to curve spacetime to ensure that everyone sees the same reality. Here in the lower-dimensional world of space and time, we see the curvature as gravity."

General relativity is about how we can stitch reality back together when it's broken into pieces by our various points of view. We *can* translate between the perspectives of inertial and accelerated observers—we just need gravity. An inertial frame with a gravitational field is indistinguishable from an accelerated frame without a gravitational field. That means there's nothing fundamentally unique about acceleration, and that all observers, regardless of their state of motion, are on equal footing. The universe looks radically different from one perspective to the next, but at the end of the day there's just one ultimate reality.

Quantum theory was a little more complicated. All the physics books I read told me not to feel discouraged if my brain blew a fuse while trying to comprehend it. *If quantum theory seems bat-shit crazy,* they'd say, *don't worry. It is.* Then they'd try to make me feel better by quoting a handful of brilliant quantum physicists talking about how no one understands quantum physics, and in case that wasn't enough to console me, they'd toss in some objections from Einstein to boot.

But I was sick of being patted on the head and told not to worry about understanding it. The quantum works in mysterious ways? Really, *science*?

After reading enough so-called explanations of the theory, it was clear to me that any hope I had for understanding quantum mechanics hinged on a single experiment: the double slit. It went something like this.

Physicists shine a laser at a screen with two parallel slits. The light travels through the slits and hits a photographic plate on the other side, so you can see where it lands. If light was made of particles—and Einstein had already proven that it was—you'd expect to see two blobs of light opposite each slit. But you don't. Instead you see a series of light and dark vertical stripes, like a barcode.

Physicists realized that they could make sense of the barcode if the photons en masse were traveling as a wave, which splits apart at the screen, travels through both slits, then recombines on the other side. When it recombines, it's partially out of phase. In spots where the two waves are in phase—crests line up with crests and the troughs line up with troughs—they reinforce each other, producing a bright strip of light on the screen. Where they are out of phase—crests align with troughs and troughs with crests—they cancel out, leaving only stripes of darkness.

Okay, that was kind of strange, but it was nothing compared to what happens next. The physicists repeat the experiment, turning down the intensity of the laser until it shoots a single photon at a time. They fire off one photon and see a single bit of light register on the plate—just like you'd expect. They fire another, and another dot appears. They fire another and another and another and slowly but surely a pattern begins to emerge. One by one, the individual photons build up the same interference pattern of light and dark stripes.

The books all concluded that the interference showed that light was "both a particle and a wave"—the so-called wave-particle duality—but when you measure it, light is *always* a particle. A single photon will invariably show up as a single spot of light. It's only when you map out the probability for the particle to land at any given spot on the plate that you find a wave.

The wave that describes the quantum particle is a mathematical wave, a wavefunction. Whereas physical waves carry energy in their amplitudes, mathematical wavefunctions carry probability. Square the amplitude at any point along a wavefunction mapped out in position space and you get the probability of finding the particle there. Make enough measurements and the dots of light will map out the whole distribution.

As far as I could tell, it wasn't so weird that a particle's probability distribution could be mapped as a wave. What *was* weird was the interference pattern that shows up photon by single photon. The probability distribution mapped out by the bright and dark stripes is not the distribution encoded in the wavefunction of a single photon—it's the distribution that results from the interference of *two* wavefunctions. It's as if the single photon travels through both slits and its wavefunction splits in two. When the pieces recombine on the other side, they are out of phase and they interfere with each other to form a new wavefunction. The individual photons then land only at positions that are allowed by the probability distribution encoded in the combined wavefunction. Hence the stripes.

If you close the second slit and run the experiment again, shooting one photon at a time at the screen, the interference pattern disappears. Now the spots that show up on the plate are only those allowed by the wavefunction of a single photon, which is always in phase with itself. It's only when both slits are open that the stripes appear.

Finally, the books all told me, there's one more version of the experiment that physicists run in an attempt to catch the photon traveling through both slits at once. They keep both slits open but this time they rig them with detectors that will trigger when a photon passes through. Then they turn on the laser and shoot off one photon at a time. Two open slits always show interference. But not this time. This time there's just the distribution of a single photon's wavefunction on the plate. It's as if the photon knows it's being watched.

Okay, I thought. This was what they had warned me about: the smell of my neurons sizzling. It *knows* when it's being watched?

Obviously the photon doesn't *know* anything. But how do you explain what's going on? Is the photon really in two places at once when no one's watching and in one when someone is? What does it mean for us to be watching? Why should our watching make any difference at all?

Double slit boils down to this, I wrote in my notebook. *Why do the probability distributions of single photons trace out an interference pattern, as if the photons are traveling two paths simultaneously? And why does the interference pattern disappear when you try to measure which path the photon takes?*

Different physicists viewed the situation differently. Feynman, for instance, said that when we're not watching, the particle really does take multiple paths. Bohr, on the other hand, argued that if we're not watching, we have no right to say anything about the particle at all. Until we measure it, Bohr said, the particle has no position. Until we measure it, it's not even a particle. It's not anything. But if particles aren't particles until you measure them, what exactly is the interference pattern interference of? Stark stripes of counterfactuals? A scattered pile of mere could-have-beens that never quite were?

Clearly *something* happens when we make a measurement—observe which path the photon takes and the interference pattern disappears. But quantum theory itself doesn't describe any such thing. It never breathes a word about measurement at all. According to the theory, *everything* is described by wavefunctions: the photon, the slits, the detectors, the photographic plate, even the physicist performing the experiment. According to the theory, when the photon passes through a detector-rigged slit, its wavefunction superposes and interferes with the wavefunction of the detector; no single event registers. The combined system of photon-plus-detector is now described by their combined wavefunction, hovering in a simultaneous state of yes-the-photon-took-this-path and no-it-didn't. According to the theory, when the physicist checks the detector's readout, his wavefunction superposes with the combined photon-plus-detector wavefunction, a mangled heap of probability, a haze of parallel would-be realities: the-physicist-sees-that-detector-A-registered-a-photon and the-physicist-sees-that-detector-A-did-not.

The universe, according to quantum theory, is just superpositions piled atop superpositions, yet we never see even one. Sure, we see their remnants in paradoxical stripes of interference. But I'd never once found myself in Manhattan *and* Brooklyn, or checked someone's coat only to find it hanging on multiple hangers. If the world was really so quantum, where were all the simultaneously alive-and-dead cats?

Physicists called it the measurement problem: the wavefunction encodes a host of possible states and yet we only ever measure one. What happens in the course of a measurement that collapses the wavefunction's probability distribution down to a single outcome? Given the many positions allowed by the photon's probability distribution, how does it choose one? The choice appears to be truly random—an effect with no cause. Was the universe at bottom truly random? Einstein didn't buy it, but the universe didn't seem to care.

Bohr argued that quantum phenomena, like particles, have real properties only after they are measured; it makes no sense to even ask about their pre-measurement state. There's no mysterious collapse, he said, because there's nothing to collapse. Bohr didn't believe that observers magically influence the outcomes of experiments or create reality through their minds—it was just that a measurement outcome was objectively relative to the frame of reference of a measuring device, be it a detector or a photographic plate or a human eye.

That's not to say he didn't realize how seriously weird the whole thing was, requiring, as he wrote, "a radical revision of our attitude toward the problem of physical reality." But in some sense the fact that properties were relative to observers wasn't that different from Einstein's relativity, a fact that Bohr happily pointed out after Einstein had insisted that quantum theory couldn't be a complete description of reality. "I like to think that the moon is there even if I'm not looking at it," Einstein had said. In response, Bohr wrote that quantum theory "may be paralleled with the fundamental modification of all ideas regarding the absolute character of physical phenomena brought about by the general theory of relativity." In other words, *Sure, quantum theory fucks with reality, but you started it.*

Then again, there was something distinctly weirder about quantum mechanics than relativity. At least in relativity there was some basic reality—the unified four-dimensional spacetime—that simply *looked* different relative to different observers, and Einstein had kindly offered up tools such as Lorentz and diffeomorphism transformations to translate between different points of view. But what was the basic reality in quantum theory? It was as if there was no reality at all until someone made a measurement.

Of course, if that was true, you couldn't have an observer to make the measurement in the first place. The observer's got to live in some kind of reality. That was the problem with Bohr's view. If measurement is the arbiter of reality, then the measuring device has to sit outside reality—which, even within the bizarro universe of quantum mechanics, is downright impossible. Besides, any measuring device, human or otherwise, is ultimately made up of subatomic particles, so drawing some kind of ontological line between the two was just plain schizophrenic.

The assertion that a particle doesn't have any "real" attributes until someone measures them becomes particularly weird when you realize that certain attributes can't be measured at the same time. Which means that certain attributes can't *exist* at the same time. Take position and momentum. There's no conceivable experiment that can measure both a particle's position and momentum to perfect accuracy. If you want to accurately measure position, you need a rigidly fixed measuring device that won't move when the particle hits it; otherwise its movement will smear out the position measurement. But if you want to accurately measure momentum, your device had better move easily when hit, so that its recoil can register the amount of momentum imparted by the particle.

No matter how you set it up, the two measurements are mutually exclusive. The more accurately you know position, the less accurately you know momentum. And it's not merely a practical matter. It's not just that you can't measure both at once. It's that the particle doesn't *have* both at once. The uncertainty relation between position and momentum is built into the mathematical structure of the theory. A particle's position wavefunction and its momentum wavefunction are Fourier transforms of each other—two equally true but mutually exclusive ways of looking at the same thing. Choose to look at one and you obliterate the other. The probability distributions encoded in the wavefunctions reflect this mutual exclusivity. If you were to assume that the particle had both attributes at the same time, your probability distribution wouldn't match up with experiment. In other words, you can pretend the whole thing is merely a pragmatic problem, a reflection of the limits of measurement rather than the limits of reality, but you'll get the wrong answer.

So here was the situation. A particle can't have a well-defined position and momentum, yet an observer can measure either one with perfect accuracy and is free to choose which one to measure. The moral of the story was clear: there's no normal reality lurking behind the quantum scene, no objective Einsteinian world that sits idly by regard-less of who's looking. There's just the stuff we measure. The whole thing reeked of paradox, but as Feynman said, "The 'paradox' is only a conflict between reality and your feeling of what reality 'ought to be.'"

It was clear to me that in our hunt for ultimate reality, my father and I needed to be prepared for the ground to give way beneath our feet. Reality according to quantum theory was not the run-of-the-mill, steady-mooned world we thought we knew. But it was also clear that Bohr and his followers didn't have the last word on the theory's interpretation—because the distinction between observer and ob-served would never hold up. If that presumably false dividing line marked the very birthplace of reality, it was going to be crucial to figure out what happens to reality when it blurs.

It was also clear that we needed to give careful consideration to the meaning and role of "observers" in general. Both relativity and quan-tum theory had changed the role observers played in physics—not ob-servers as in humans or conscious creatures, but observers as in points of view. Relativity taught us that we can't talk about space or time without first specifying a frame of reference. Independent of observ-ers, those terms lose all meaning, since one observer's time is another's space. Quantum mechanics taught us that we can't talk about proper-ties of matter without first specifying what we're measuring—its posi-tion, for example, or its momentum. At the heart of both theories was a single epiphany: perspective matters. For some as-yet-unknown rea-son, points of view determine not only how we *see* things, but how things *are*.

That was what they had in common, anyway. So what was really at the heart of their incompatibility? Why couldn't those two crazy kids make it work?

Summer had descended on New York when I finally met Fotini Mar-kopoulou. We had agreed to meet in the lounge of the Tribeca Grand Hotel. I figured we could find a quiet spot to talk, and that it would be air-conditioned—a luxury I had grown to appreciate. My Brooklyn apartment didn't have any such luxury, and I had taken to reading and writing in the bathtub to stay cool.

I arrived with plenty of time to spare and snagged a table in the corner. It was too early for drinks or dinner, so the room was practically empty, only a few people scattered throughout the large space, chatting, reading magazines, and sipping icy beverages, having ducked into the darkened lounge to escape the relentless heat.

Markopoulou sauntered into the room in a long, flowing skirt and sandals. She was prettier than I had remembered from the Wheeler conference, with striking Greek features and long, shiny black hair. She seemed younger, too. She had a decade on me, but I'm not sure anyone would know it, and in her early thirties she was practically a newborn in her field. All said, she was not how you might imagine a physicist. When I told people I was into physics, they always seemed a little too surprised, and I imagined that Markopoulou knew what that was like. I smiled to myself, knowing that anyone glancing over at the two of us would assume that we were talking about boys or fashion, not the microscopic structure of spacetime. Not that I didn't enjoy talking about boys and fashion. But today it was loop quantum gravity.

I stood up to greet Markopoulou, shook her hand, and told her how great it was to finally meet in person. If she was thrown by my age, she didn't show it. She slid into the banquette alongside me and we ordered some cold drinks. After some obligatory small talk, I launched into a barrage of questions. I was sure that she would be able to tell what a rookie I was, but I didn't care. I was too excited by the prospect of learning physics straight from the mouth of a physicist. Who knew if I'd ever get the chance again?

Markopoulou explained to me the notorious obstacles involved in uniting general relativity with quantum mechanics. It was Wheeler who first took seriously the need for such unification and made the bold leap of applying quantum theory to the universe as a whole. You would think there would be no need for such a feat, since quantum theory is about tiny things, not about universes. But, as even Bohr himself acknowledged, there's no clear boundary separating the quantum world from the classical world, no state line marked by a billboard that reads "Welcome to the non-quantum realm." Yes, quantum mechanics requires a separation between the quantum system and its environment, observed and observer, inside and out. But the theory never tells us where to place the dividing line. The line is a moving target; it can be drawn anywhere and shifted to ever-bigger scales. If reality is quantum, then reality is quantum. It doesn't reach some scale and stop—it's quantum mechanics all the way up.

Of course, in ordinary quantum mechanics you could at least *pretend* to draw a distinction between observer and observed, arbitrarily slicing the universe in two, calling one side the classical measuring device and the other the quantum system. But when it came to the universe as a whole, you couldn't even fake the procedure. The universe, by definition, is the whole of spacetime, the complete set of everything that exists. It has no outside. No outside, no observers.

Quantum cosmology was born when Wheeler had to kill some time between flights. It was 1965 and he had a layover in North Carolina. He asked his friend and fellow physicist Bryce DeWitt, who happened to live nearby, to keep him company for a few hours at the airport. It was there that they wrote down an equation, which Wheeler called the Einstein-Schrödinger equation, everyone else came to call the Wheeler-DeWitt equation, and DeWitt himself eventually called "that damned equation."

That damned equation was meant to solve a problem that had plagued earlier attempts to quantize general relativity. In quantum mechanics, time is always external to the system; clocks live in that murky classical realm—the "environment"—where observers reside. Wavefunctions describe the physical system at an instant of time; the wavefunction then evolves *in* time according to the Schrödinger equation. When it comes to spacetime, though, there's no such thing as spacetime *at an instant,* because spacetime contains *all* instants. And you can't have spacetime evolve *in* time, because it *is* time. The only way forward seemed to be this: break four-dimensional spacetime into three dimensions of space and one of time, then describe the spatial portion as a wavefunction that can evolve relative to the dimension you called "time."

In this procedure, however, something crucial gets lost. The key feature of general relativity, known as general covariance, is that there's no preferred way to slice up spacetime. All reference frames are relative to other reference frames, none more fundamental than the next. Different observers can slice up spacetime in different ways. So when we decide to quantize only the three dimensions of space, we have to choose certain coordinates to call "space" and others to call "time." But whose space? Whose time? Making any kind of choice would suggest that one observer had a truer view of reality than all others. But that can't be so. That was Einstein's whole point: *the laws of physics must be the same for everyone.*

Wheeler and DeWitt saw a way out. As long as the quantum space evolved according to their damned equation—a kind of Schrödinger equation for spacetime—general covariance would be restored, all observers would be created equal, the laws of physics would be the same for everyone, and all would be right in the quantum universe. But there was a snag in the plan. The equation required that the total energy of the universe be precisely zero.

In itself, that wasn't so strange—if the universe really came from nothing, it would have to have a total energy of zero. But quantum mechanics is never so certain. Just as position and momentum are bound together by uncertainty—the more precisely you know one, the less you know the other—so, too, are time and energy. As soon as you've specified a quantum universe's energy with exact precision, you'd better say goodbye to time.

Wheeler and DeWitt had successfully rescued the attempts to quantize spacetime, but at a cost: they ended up with a quantum universe that was frozen in time, stuck in a single, eternal instant. It was a universe in limbo—no giant clock hovering on the outskirts of reality, ticking away each second after absolute second so that we might live in a world in which time actually means something, in which anything ever changes at all.

When you think about it, it ought to have been obvious from the start that there's no possible way to have both general covariance and a universe that evolves in time—the two ideas are mutually exclusive, because for the universe as a whole to evolve in time, it must be evolving relative to a frame of reference that is outside the universe. That frame is now a preferred frame, and you've violated general relativity. It's one or the other—you can't have an evolving universe and eat it, too.

As Markopoulou talked, it occurred to me that the very notion of "the universe as a whole" might be similarly doomed. Could you talk about "the universe as a whole" without talking about it from an impossible reference frame outside the universe?

The problem of Wheeler and DeWitt's frozen universe is intimately tied to the measurement problem in quantum mechanics. Quantum systems seem to hover in a ghostly state of almost-existence until an observer or measuring apparatus makes a measurement, thereby collapsing the wavefunction of possibilities into a single actuality. But if the quantum system is the universe itself, who can collapse the wavefunction? Again the problem comes down to the fact that no one can step outside the bounds of the universe, turn around, and look back. "That's a whole sticky thing," Markopoulou said. "Who looks at the universe?" The cosmos is a half-dead, half-alive cat. An almost, but never an is.

Markopoulou explained that she had set out to address the problem of quantum cosmology without falling into the trap set by that damned equation, heeding Smolin's slogan that "the first principle of cosmology must be 'There is nothing outside the universe.'" No clocks, no observers. No God's-eye point of view. How strange, I thought, that the universe is the only object with an inside but no outside. It reminded me of a line from a Borges poem: *Obverse without a reverse, one-sided coin, the side of things* . . . The universe is a one-sided coin. Not quite an object, but an impossible object, like Escher's staircase or Penrose's triangle. Quantum cosmology is a science of impossible objects.

Still, Markopoulou believed there was a way forward, and it meant embracing a radically new view of things. "Any satisfactory theory of quantum cosmology has to refer to observations that can be made by observers inside the universe," she said. "No Wheeler-DeWitt equation, no wavefunction of the universe." By "observers," she explained, she meant not humans or conscious creatures but simply reference frames, possible points of view. And a quantum cosmology that refers only to the reference frames of internal observers requires us to change one thing that seems fundamentally unchangeable: logic.

You'd think logic is logic is logic, eternal and unbreakable. But if that was true, ordinary logic wouldn't need a name. It has one: Boolean. Codified in the countless "if P, then Q" statements that philosophy students around the world were memorizing as we spoke, Boolean logic is a binary logic, the logic of yes or no, 0 or 1, true or false, black or white.

But quantum cosmology needs shades of gray, Markopoulou explained, thanks to a simple yet profoundly important fact: the speed of light is finite. Whenever we observe something, light has to travel from the object to our eyes, and it doesn't happen instantaneously. It takes 186,000 miles per second. Sunlight takes eight minutes to reach the Earth—looking up at the Sun is like hopping into an eight-minute time machine. Look up at the stars and you're looking back thousands of years; grab a telescope and you can see billions of years into the past. But the point is this: there are stars whose light hasn't had enough time since the big bang to reach us yet. Wait long enough and some of it will. But with a finite speed of light, there will always be portions of the universe that we can't see.

Markopoulou explained that the slice of universe I *can* see is called my light cone—a sphere of space that grows with time, so that if you drew it in the spacetime coordinates on my father's yellow legal pad, you'd see nested spheres, swelling in diameter as they move upward along the time axis, tracing a cone. If an event is in my past light cone, I can see it; if it's not, I can't. I knew my light cone had to be pretty big, given the nearly 14 billion years of travel time that light has enjoyed since the beginning of the universe. But it still felt a little claustrophobic.

"Let's talk about an event, say a supernova explosion," Markopoulou said. "It can have two possible values: yes or no. It happened or it didn't. That way of thinking about observables follows Boolean logic. But let's ask, is there a supernova explosion according to this particular observer? Now there are the following possibilities. If the supernova is in his past, we can say yes. Another possibility is that it's not in his past, but if he waits long enough he's going to see it. So it's 'yes, but later.' Another possibility is that the supernova is so far away from him that he'll never see it, so it's no. The fact that the supernova occurred doesn't matter, because the question was, did it occur *according to this observer?* So whereas before, in the old way of thinking, there were just two possible values, yes and no, now there's a whole range of possibilities." This new kind of non-Boolean logic was called intuitionistic logic, she said—a name that made me chuckle to myself, given how counterintuitive it was. It had existed as its own kind of logic game among mathematicians, but Markopoulou was among the first to apply it to cosmology.

I started to understand the gist of what she had done that had so impressed the judges of the young researchers' competition at the Wheeler conference. She had attached tiny light cones to the atomic lattice of a quantum space, let the light cone structure determine how the network could evolve in time, applied the rules of intuitionistic logic in a mathematical form called a Heyting algebra, laid down some rules for transforming from one observer's perspective to another, and voilà—a theory of quantum cosmology that doesn't require observers or clocks to lurk beyond the bounds of space and time. It was a different kind of quantum cosmology, to be sure. It wasn't a quantum description of *the* universe; it was a quantum description of *each individual observer's* universe.

The conversation seemed to fly by, but when I checked my watch, several hours had passed. I felt guilty for taking up so much of her time, but it was my first one-on-one conversation with a physicist, and for all I knew my last, so I was hell-bent on learning everything I could. I was just glad she didn't attempt to quantum-tunnel her way out of there to escape my incessant questions. I wrapped up by asking her what she had thought of the Wheeler symposium.

"I've never been to such an open-minded meeting," she said. "People just stood up and said what they really thought. That basically never happens. And it was all because of Johnny Wheeler. Not only did he stick to the big issues, the important problems, he also must have been very accepting of people. Oftentimes we're not so good at encouraging people to go out on a limb. The culture is to always say that something looks wrong. This is science—it's a bunch of immature boys who want to look smart."

We laughed, then made our way out into the blazing sunlight, said goodbye, and walked off in opposite directions. I headed down into SoHo to catch the train back to Brooklyn. I couldn't wait to get in the tub and start writing.

As I walked toward the subway station, my brain was buzzing. I had already learned that both relativity and quantum mechanics were trying to tell us the same thing. we run into trouble when we try to describe physics from an impossible God's-eye view, a view from nowhere. We have to specify a reference frame, an observer. But now I finally understood the real tension between the two theories. The whole mess could be summed up with one question: where's the observer?

In general relativity, observers have to be inside the system—since "the system" is all of spacetime and the theory accounts for every quirk that arises as a result of their differing perspectives. It's a closed, self-contained whole. Quantum mechanics, on the other hand, is about open systems—the observer has to be outside the system in order to make a measurement, to transform possibility into reality. If you want to unite them into a single theory, you have to figure out where the observer goes: inside or out? Quantum gravity was going to have to negotiate a perilous catch-22: stick the observer outside the universe and you violate general covariance; keep the observer inside and the universe's wavefunction can never be collapsed.

It seemed obvious that the first option was barely an option at all: you can't have an observer standing outside space and time. So the question seemed to be, how do you do quantum mechanics in a closed system? Then again, maybe quantum mechanics wasn't the culprit. Maybe quantum cosmology was trying to tell us that there *is* no closed system. After all, in Markopoulou's model, the universe was just a collection of open systems, each defined by its own observer. But if there's no single closed system, what happens to reality? To the universe as a whole?

Did it even make sense to talk about the universe as a whole, or did that very notion require an impossible God's-eye view? Maybe we had all gone wrong in thinking of the universe as a thing, a noun, an object replete with all the properties that objects normally have, like an outside. But if the universe isn't a "thing," what is it? A loose assortment of points of view? And if that's the case, points of view *of what*?

The question reminded me of a late-night conversation with my father back when I was in high school. We were talking about curved spacetime. "Wait a second," I had said. "If spacetime is all there is, how can it be curved? It would have to be curved relative to something outside; it would have to be embedded in a higher-dimensional space." I had smiled proudly, sure that at fifteen years old I had just spotted the long-overlooked flaw in Einstein's theory.

My father had laughed. "The math allows you to talk about intrinsic curvature, so you can measure curvature from within the spacetime. It doesn't have to refer to anything outside it. Curvature is just the warping of the metric."

Now I wondered if something similar could save the universe. Was there some way to continue talking about the universe while only referring to it from the inside? Markopoulou seemed to think so, but it came at a serious price. It meant tossing aside ordinary Boolean logic and replacing it with a kind of logic that depended on the observer. It meant redefining what we mean by "true." It meant stripping physics of the ability to make absolute statements about ultimate reality. Propositions were no longer true or false. They were true or false *according to some particular observer*.

I laughed to myself, thinking about how much some of my old classmates at the New School, who were constantly spouting some kind of postmodern bullshit, would have loved to hear that truth was relative to observers. Of course, what Markopoulou was saying didn't fit their agenda at all—truth and falsity were relative to observers as dictated by the geometry of their reference frames and the objective laws of physics. It wasn't like you could will a star into exploding or render the Earth flat with sheer multicultural tolerance. "Observers" didn't mean people, and "observer-dependent" didn't mean subjective. But I could imagine how easily it could all be misconstrued.

After talking with Markopoulou, one thing was clear: the old way of thinking about cosmology wasn't going to cut it. We couldn't keep pretending that we can describe the universe from the outside, from an impossible God's-eye view. There had to be some accounting for what individual observers, shackled to their light cones, could see from the inside.

As I walked through the sweltering New York streets, the heat rising up from the pavement, I felt as though I had been let in on a secret, as though the city I had left behind when I entered the lobby of the Tribeca Grand was not the same city in front of me now. I watched all the people hurry past me on the sidewalk, rushing through their day like it was any other, as if concepts like "true" and "false" and "space" and "time" meant anything at all. That's right, I thought, keep sipping your Frappuccinos and gazing at the world through your Boolean-colored glasses. Binary logic is bliss.

A man being walked by several dogs bumped me as he passed, a knot of leashes lurching in his hand. It's okay, I told myself when he didn't bother to apologize, we don't even live in the same universe. For someone like me who was a bit crowd-phobic, it was an oddly comforting thought. Lonely, maybe, but beautiful: we each have our own universe. We just don't notice because there's so much overlap.

I was so lost in thought that I didn't notice I had walked past the subway station until I was nearly at the next one. Before descending into the ungodly heat of the subway platform, I fished my cell phone out of my bag and called my dad at his office.

"It was amazing," I told him. "I'll send you the whole interview once I've transcribed it. But it's got me thinking—maybe it doesn't make sense to think about the nothingness from the outside, since there is no outside. By definition it's infinite and unbounded. Maybe we should be thinking about what it looks like from the inside, since that's the only possible perspective." I paused for a moment. "Do you think light cones could be the boundaries that turn nothing into something?"

3

Smile!

My mother says that she remembers exactly what she was doing the moment she heard John F. Kennedy had been shot. My father remembers exactly where he was when Neil Armstrong stepped onto the Moon. Me? I will never forget the day the WMAP data were released. It was the twelfth of February in 2003, I was in my Brooklyn apartment, and I was on the phone with Poison Control.

Cassidy, my black Lab, had managed to nose her way into the pantry, where she promptly ate an entire package of roach traps, box and all. When I saw the mangled remains of her snack, I panicked and called Animal Poison Control. The woman on the other end of the phone calmly asked me what brand the traps had been, then took a few minutes to search for the ingredients. As she typed away, Cassidy sprawled out on the floor at my feet looking satisfied with herself, and I flipped through the day's *New York Times*.

"Holy shit!"

"Miss, is the dog all right?" the Poison Control operator asked.

"What? Oh, yes, she's fine, sorry. The cosmic microwave background data are out."

"The what?"

"The microwave background radiation. It's the best picture they've ever taken of the early universe."

"Huh," she said. "You mean like outer space?"

"Sort of," I said. "It's a picture of the universe from, like, 14 billion years ago."

"Well," she said, "how about that."

I scanned the article as she continued searching. *The most detailed and precise map yet produced of the universe just after its birth confirms the Big Bang theory in triumphant detail and opens new chapters in the early history of the cosmos . . . gives a first tantalizing hint at the physics of the "dynamite" behind the Big Bang . . .*

"Your dog should be fine, miss. All the ingredients are toxic to insects but not mammals. The only thing I'm worried about is the plastic. It's cheap and it can splinter in their stomachs. Your best bet is to feed her a loaf of white bread. It will stick to the plastic and prevent it from puncturing her."

"A whole loaf?" I asked.

"Yes."

I thanked her, grabbed my gloves and a scarf, and headed to the corner store to pick up some bread. As I waited in line for the register, I texted my father: *WMAP!*

Back in the apartment, Cassidy greeted me with her tail wagging. "This is your lucky day," I told her, dumping the bread into her bowl. As punishment for eating roach poison, I was, apparently, giving her the biggest treat of her life. While she blissfully gobbled it down, I brought up NASA's website.

"NASA today released the best 'baby picture' of the Universe ever taken; the image contains such stunning detail that it may be one of the most important scientific results of recent years," their press release announced.

A decade earlier, when the COBE satellite had first snapped a picture of the nascent universe, Nobel laureate-to-be George Smoot said that it was like seeing the face of God. With thirty-five times the sensitivity, the Wilkinson Microwave Anisotropy Probe could now make out his faintest freckles, the subtlest laugh lines. The press release explained that the data confirmed the big-bang/inflation package that constituted cosmology's standard model. If you listened carefully, you could hear the champagne corks popping.

The standard model began with the big bang. With Edwin Hubble on Mount Wilson, watching the galaxies recede into the vastness of space and time. With a horrified Einstein realizing his greatest blunder, having let his philosophical prejudices eclipse his equations and missing the opportunity to make what would have been the most extraordinary prediction ever ventured: that the universe was expanding. That spacetime was stretching, growing bigger all around us, and that we were growing smaller by comparison, an ever dimming light in an increasingly magnificent void.

It didn't take long for physicists to run the film backward in their minds, to watch the galaxies hurtling toward one another, the universe growing smaller, denser, hotter, squeezing toward a single, infinite point.

If the universe had begun in fire, they figured, it should still be smoking. Radiation from the nascent universe should still permeate the cosmos, stretched to microwave wavelengths by 14 billion years of expansion, keeping the temperature of empty interstellar space just a nudge above absolute zero. Arno Penzias and Robert Wilson, two radio astronomers working at Bell Labs in 1965, discovered the radiation by accident while hunting the source of a persistent static in their antennae. They thought it was pigeon shit. It turned out to be a relic from the origin of time. They won a Nobel Prize.

But something about the cosmic microwave background, the CMB, didn't add up. It was the same temperature everywhere across the sky. Measure some patch of sky 12 billion light-years away in one direction and it's 2.7 Kelvin. Measure another patch 12 billion light-years in the opposite direction and, again, 2.7 Kelvin. Separated by 24 billion light-years, those two regions could never once have met in the mere 14-billion-year history of the universe. Yet they seem to be in equilibrium. It was too precise to be a coincidence. Something was missing.

The solution came late one December night in 1979 when Alan Guth, an unknown postdoc at Stanford University, had a spectacular realization. He had been thinking about monopoles. At the time, physicists believed that at extremely high temperatures, such as those near the big bang, the forces governing the interactions of particles would merge into a single superforce, one that would splinter into its constituent pieces as the universe expanded and cooled. The idea had one flaw. As temperatures plummeted, the superforce wouldn't be the only thing splitting apart—spacetime itself would suffer topological damage. Like water freezing into chunks of ice, fragments of spacetime would freeze out and form exotic particles such as the hypothetical monopole—a magnet with a single pole, north without south. Physicists searched the universe for monopoles but never caught sight of one, which was a serious problem for a theory that predicted a universe overrun with the things, monopoles more numerous than atoms.

Guth sat down, resolved to tweak the big bang in such a way that spacetime would no longer manufacture monopoles. When the solution dawned on him, he took out his notebook and wrote, "Spectacular realization."

Guth had realized that if the universe underwent a sudden burst of expansion in the first fraction of a second, inflating at speeds faster than light, any monopoles produced in the big bang would quickly be pushed out to the far reaches of space, far beyond any region we could feasibly measure. That would explain why no one had ever seen one. As a bonus, it would also explain why the temperature of the CMB was so uniform.

The problem with far-off regions of space having the same temperature was that they didn't have time to equilibrate—but by having spacetime expand faster than the speed of light, the inflationary theory buys the universe extra time. Superluminal expansion sounds like a blatant contradiction of relativity, but it falls into a sort of cosmic loophole—while nothing *in* spacetime can travel faster than light, there's no law against spacetime *itself* exceeding the limit. That distant regions are in equilibrium seems impossible because photons wouldn't have had enough time since the big bang to travel between them, but by expanding faster than light, spacetime gives those photons a boost, carrying them out to the far corners of the cosmos, much farther than they ever could have gotten on their own.

Of course, for the theory to work, you need some physical mechanism that will cause the universe to suddenly balloon outward like a blowfish. Guth came up with one. The newborn universe, he suggested, might have been filled with a hypothetical field called the inflaton, which found itself in a false vacuum—a momentarily stable state, but not the state of lowest energy, like a ball precariously perched on a plateau midway down a mountain. The smallest push will send that ball rolling downward until it reaches minimum energy, otherwise known as the ground. As for the inflaton, quantum fluctuations could provide the push, sending it tumbling into a lower energy state. As it plummets, spacetime takes on a negative pressure, creating a kind of antigravitational force that pushes outward and causes the universe to expand exponentially, growing a million trillion trillion times bigger in just a fraction of a second. As Guth put it, inflation explains what went bang.

When the universe's energy hits rock bottom in the true vacuum, all the kinetic energy of the inflaton field pours out into the nascent universe, heating it to a thousand trillion trillion degrees and flooding it with radiation. What were once tiny quantum fluctuations in the density of matter and energy on the order of 10^{-33} centimeter are now stretched out to astronomical proportions, creating subtle ripples and valleys throughout space, laying a gravitational blueprint for what will eventually become a network of stars and galaxies.

The hot universe continues to expand, propelled by inflation's inertia. For the first 380,000 years, a dense, hot plasma permeates space, so dense that even light can't shine through it. Any photon that tries to muddle through is quickly thwarted by a rogue proton or electron. But as the universe expands, the scalding temperature cools, and particles slow down long enough to bond together. As matter organizes itself into nuclei and then atoms, the photons are set free from the opaque plasma and stream out into the universe on their own. This first generation of emancipated photons are the same ones that constitute the CMB, a snapshot of the universe from the time they were first released.

While the photons travel unimpeded, the matter particles begin to collect in the overdense regions carved out by the quantum fluctuations, setting off a chain reaction of gravitational collapse. Matter piles upon matter, its temperature steadily rising until, after 200 million years, it triggers nuclear fusion. Suddenly the landscape changes dramatically: stars burst forth from the darkness and burn scattered across the sky. The chain reaction continues its cascade—stars coalesce into galaxies, galaxies into clusters, clusters into superclusters.

All the while the universe continues to expand at a leisurely rate. Eventually one particular star is born, and around this star some rocky debris assembles into a planetary system. On one particular rock, the third from the star, elements such as oxygen, hydrogen, and carbon come together, elements forged in the furnaces of other stars gone supernova, spewed remnants that journeyed through empty space to one day land on a lucky planet where they would combine in just the right way within just the right environment for life to emerge from some primordial goo, to grow and replicate and evolve until—voilà—we are born.

Only the story doesn't end there. For if inflation really happened, it didn't happen just once. While the history of our humble universe was unfolding, something much, much bigger was going on. Thanks to quantum randomness, the false vacuum from which our universe was born couldn't decay everywhere at exactly the same rate. While one region of the false vacuum rolled downhill to form our universe, other regions were left behind. Eventually they, too, would decay, forming other universes, permanently disconnected from our own. While these regions decayed, still others were left behind, and when in turn they plummeted toward universehood, still more lingered in their wake, and so on and so forth, ad infinitum. No matter how many universes it creates, there's always more false vacuum left over, and the creation process never ends. Inflation is eternal.

If we allow inflation to occur even once in our cosmic history, we are suddenly stuck with an infinite number of universes beyond our own, an ever-growing multiverse, a meta-Universe with a capital U composed of causally disconnected small-u universes, one sprouting from the next in a ceaseless process of birth and reproduction. While they are all governed by the same fundamental laws of physics, each universe is born with its own local laws: its own geometry, its own array of physical constants, its own set of particles, its own spectrum of force strengths, and its own unique history. Reality as a whole begins to resemble a vast cosmic patchwork quilt, wildly diverse and fast approaching infinity.

With the WMAP data, cosmologists now had in their hands a detailed map of the microwave sky, one that revealed subtle fluctuations from the uniform temperature, hot and cold spots that differed by a mere one part in a hundred thousand. The spots had been formed when the dense plasma still permeated the infant universe, imprints of a struggle between gravity and electromagnetism. While gravity had tried to squeeze the plasma tighter, electromagnetic radiation had tried to expand it, resulting in a tug-of-war that compressed and expanded the plasma like an accordion. When it compressed, it heated up ever so slightly, and when it expanded, it cooled, leaving subtle hot and cold spots for WMAP to find some 14 billion years later.

The fluctuations were fingerprints, forensic evidence of the universe's beginning. It was hard to believe that what looked like a bunch of random blotches on the map actually contained detailed information about the universe's origin, its makeup, and its evolution. With hard data like WMAP, cosmology had gone from a speculative science to a rigorous field on par with astronomy and astrophysics. The golden age had dawned, and the cosmologists were ready to party.

In fact, they already had one planned. Searching around online, I found that they were preparing for a big four-day conference the following month at the University of California, Davis. I had to be there.

I called my father. "Four days of sun and cosmology!" I said. "If we want to understand the origin of the universe, these are the people who can explain it. You've got to come with me!"

He sighed. "I wish I could. There's no way I can get out of work on such short notice. You should go, though! Be a journalist. You'll take lots of notes for me."

I hung up, unsure whether I really wanted to go by myself. This had always been intended as a joint mission; going solo didn't feel quite right. I mean, crashing physics conferences was *our* thing—if doing something once could count as a "thing." Of course, I had interviewed Fotini Markopoulou on my own, but that didn't seem quite the same at the time. After all, the idea had been to get some journalism credentials just so we could worm our way into more conferences. Thinking back, I wasn't entirely sure how those credentials were going to get my father anywhere, but it had seemed like a minor technicality, a bridge we'd cross when we came to it. Now I began to wonder if the five seconds I had given to devising that plan hadn't been enough. It probably should have occurred to me the moment I decided to spontaneously will myself into a science journalism career that my father's trajectory and mine would soon diverge, two blissfully unaware parallel world lines that unknowingly find themselves moving through a curved space, looking over to see the other receding into the distance despite all our best efforts to move in a straight line.

But what choice did I have? I was a twenty-two-year old kid who could afford the time to pursue a quixotic dream. It was like taking time to backpack through Europe, except that you couldn't pay me enough to wear a backpack, let alone fill it with months' worth of cute outfits and shoes. So I chose ultimate reality. And what did I have to give up? A job working for an ontologically questionable magazine in some guy's apartment. For my father, it was totally different. He had already chosen his path. He had a family and a house and a career. He was in the hospital every day, morning to night, helping to save people's lives. He wasn't going to just up and ditch his job. Even if he wanted to, there was no way in hell my mother was going to let him.

I was just going to have to go it alone, I thought, and know that I was doing it for the both of us. I sent an email to the people in charge of press registration. *I'm a freelance physics writer,* I told them. *I write for* Scientific American. It was vaguely true, I figured, if you ignored my use of the present tense. My profile of Markopoulou had been published in the previous month's issue. The conference organizers granted me a press pass right away, and I booked my flight to California.

*   *   *

A few weeks later I arrived in Davis and thawed my Brooklyn skin in the warm California air. Each morning I walked several blocks from my hotel to the UC campus for eight hours of physics lectures punctuated by coffee breaks and lunches. As each speaker took to the stage I hunkered down in my seat with my notebook, furiously scrawling, struggling to keep pace while trying to cut through layers of jargon and figure out what the hell everyone was talking about. I couldn't get enough. In a world in which I didn't belong and didn't speak the language, I had never felt more at home.

That's not to say I didn't stand out. My gender, age, and questionable occupation were all liabilities. Still, I did my best to blend in. I covered my tattoos with slacks and button-down shirts. I wore loafers. I tried to lie low.

Most of the talks focused on what the WMAP data meant for our understanding of the universe. To start, it pinned down the universe's age to 13.7 billion years. Even better, it decided the universe's geometry.

Because gravity can confer a net curvature to the overall shape of space, there were three possible cosmic geometries: the universe could be positively curved, like the surface of a sphere; negatively curved, like the inverted swoop of a saddle; or flat, like ordinary, Euclidean space where parallel lines never diverge nor meet.

The best way to find the geometry of a space is to draw a big triangle on it and add up its angles. If they sum to more than 180 degrees, you'll know the space is positively curved; if they sum to less, the curvature is negative.

Listening to the talks, I learned that the microwave background radiation provided the perfect cosmic triangle, with the WMAP satellite sitting at its pointed tip. The paths of two incoming photons from opposite sides of a hot or cold spot could be used to form two equal sides of a long, thin triangle, their length given by the light's travel time since the photons were freed simultaneously from the plasma. The length of the triangle's third side was determined by the distance a sound wave can travel in 380,000 years—that is, the stretch of space that the accordion compressed or expanded to form the hot or cold spot in the first place.

Knowing the lengths of all three sides, physicists used some basic trigonometry to calculate the angles at the triangle's base: 89.5 degrees apiece, summing to 179. Now they just needed the third angle—the one at the near tip. If the photons had traveled in straight lines to get there, the angle would be 1 degree, bringing the total to a flat 180. If their paths had bent outward as they journeyed through a positively curved universe, the angle would be larger, and if their paths had bent inward due to negative curvature, it would be smaller. According to WMAP, the third angle was precisely 1 degree. Way to call it, Euclid.

There was just one problem. The universe's geometry is determined by the amount of mass—or, using $E = mc^2$, energy—it contains. As Wheeler would say, mass tells space how to curve. A flat universe requires a critical density of mass to flatten it, the equivalent of an average of six hydrogen atoms per cubic meter. It didn't sound like much. You'd think there'd be more than enough stuff out there, considering all the galaxies swirling around. But there's not. Not even close.

Ordinary matter—particles such as protons, electrons, and quarks—accounts for a pathetic 4 percent of the total you'd need. Our planet, the stars, ourselves, everything we see and know, is virtually negligible in the cosmic scheme of things, the sad, shining tip of a larger, darker iceberg.

So what else was out there? The physicists had a few ideas.

For one, they already knew that there's more to matter than meets the eye, thanks to the simple fact that galaxies aren't bursting at the seams, sending billions of rogue unshackled stars flying off in all directions. Somehow gravity is holding them together in tight spiral and elliptical formations, despite the fact that the total mass of all the stars in a given galaxy doesn't provide nearly enough gravity to do the trick. Something else had to be lurking there, hidden in the dark spaces between stars or encircling each galaxy like an invisible fence, preventing stars from wandering off. In order to provide the necessary gravity but also to have remained unseen all this time, it had to be something sturdy and solid, like matter, but indiscernible to electromagnetism. Dark.

Astronomers calculated how much of this dark matter was skulking out there, but when you add it to the ordinary luminous stuff, you're still only at 27 percent of the total mass and energy needed to flatten the universe. A disturbing 73 percent is still missing.

Enter dark energy. In the late 1990s, two teams of astrophysicists—one led by Saul Perlmutter, the other by Brian Schmidt and Adam Riess—went supernova hunting, hoping to measure the expansion rate of the universe. They knew it had begun in a burst of inflation, but figured it had been slowing down ever since, reined in by the grip of gravity.

Perlmutter, Schmidt, and Riess realized that the expansion history of the universe is encoded in light from exploding stars. Certain kinds of supernovae—so-called standard candles—always burn with the same intrinsic brightness, even though they appear dimmer when they're farther away. Just how dim a standard candle appears reveals exactly how far away it is. As its light travels through an expanding space, it gets stretched out, its wavelengths shifted toward the red end of the electromagnetic spectrum. This redshift measures how much the universe expanded during the time it took the light to reach us. By collecting light from many standard candles at varying distances, the teams mapped a history of the universe's expansion. Only it wasn't slowing down at all. It was speeding up.

What could accelerate the universe's expansion in spite of gravity's best efforts to slow it down? Some mysterious dark force had to be out there, permeating the emptiness of interstellar space, hiding in the depths of the vacuum, pushing it apart, exerting a kind of antigravity, causing spacetime to expand faster and faster. Exactly how much of this dark energy is out there according to the supernovae? The answer was a near miracle. It was exactly the amount you need to fill that 73 percent hole and flatten the universe.

All the data were in rather spectacular agreement. As for inflation, WMAP had confirmed its most generic predictions. The temperature fluctuations didn't exhibit any characteristic scale, and the hot and cold spots were distributed randomly. Plus, a flat universe was exactly what inflation had ordered, because even if it were curved, the radius of curvature would be blown up to such large proportions that it would look flat anyway, just like the Earth looks flat around my feet. The physicists were so pleased with WMAP's vindication of inflation, they were practically glowing. There were rumblings that Guth would win a Nobel Prize.

But beneath the celebratory mood, something wasn't quite right. One piece of the WMAP puzzle didn't add up. Inflation predicted that the temperature fluctuations would occur at all scales, but at scales larger than 60 degrees across the sky, they abruptly stopped. Whenever this problem, known as the "low quadrupole," was mentioned, every physicist's face seemed to darken with worry, and while I wasn't sure why, I had the sense that it might be a bigger problem than they were letting on.

If there is one larger-than-life superstar, a Michael Jackson of physics, it's Stephen Hawking. Seeing him in the flesh was surreal. Even the other physicists, many of whom had known him for years as a colleague and a close friend, seemed a bit slack-jawed in his presence.

During one of the lectures, I sat directly behind Hawking. I was trying my best to pay attention to the speaker, but I found myself mesmerized by the words flashing on the computer screen mounted to the arm of his wheelchair. Paralyzed to the brink by a motor neuron disease, Hawking had one last functioning muscle, in his cheek, and by twitching it he could control the cursor on his monitor. The cursor constantly scrolled through a catalog of Hawking's most commonly used words, and with a properly timed twitch he could select one from the list. Twitch by twitch, Hawking could slowly, arduously build sentences, which were then sent to a speech synthesizer that spoke the words for him in a robotic voice that lacked not only a sense of humanity but also, as Hawking lamented, a British accent.

Seeing him there in front of me, his body slumped over in his wheelchair like a deflating balloon, I found myself in even greater awe of all he'd been able to accomplish. And as I watched the words flash across his monitor, knowing full well that they were nothing more than random lists, I couldn't help thinking that if I watched them closely enough, I'd glimpse the answers to the universe.

* * *

When the meeting broke for lunch, everyone headed outside. Lunch wasn't provided, so we were free to go off on our own. I noticed Lisa Randall, the Harvard physicist, standing on her own, likely waiting for someone, so I approached and introduced myself. In her talk, Randall had pondered the mysterious origin of the inflaton field, which I was glad to hear, as I was sitting there pondering it myself. The inflaton, in its false vacuum condition, was responsible for triggering inflation and spawning the large, uniform, star-speckled universe we know and love—but what had spawned the inflaton? Some other mysterious field? And behind that? Was it turtles all the way down? I was about to ask her when a few other physicists approached. "We found a restaurant. Let's go get lunch."

Chatting with Lisa Randall at UC Davis   Dan Falk

They seemed to be speaking to me, too—or at least they hadn't specifically asked me *not* to come, which I figured was as good as an invitation. So I tagged along and soon found myself at a long table in a casual Italian restaurant with Sir Martin Rees, Britain's Astronomer Royal; David Spergel, a Princeton physicist who played a key role in analyzing the WMAP results; Randall; and a few legit journalists.

After everyone placed their orders, the conversation turned to the dreaded *A*-word—dreaded but looking less avoidable by the day, given its ability to explain the inexplicable.

Like dark energy. Physicists knew from the supernova data, and now from WMAP, that dark energy is extremely sparse, a meager 10^{-23} gram in every cubic meter of space, barely a whisper in the dark void of the vacuum, but a whisper that builds with distance and at large enough scales crescendos to an audible howl.

That's because the most likely identity of the dark energy is the energy inherent to the vacuum of space itself, christened by Einstein as the "cosmological constant." Its power lies in its constancy—as space grows, everything in it gets diluted out, *except* for the dark energy, whose density remains constant. More space, more dark energy: the kind of feedback loop that takes off running.

You'd think that physicists could have predicted the observed strength of the dark energy, given everything they already know about the quantum vacuum. Quantum field theory provides all the tools you need to calculate the vacuum's energy. Unfortunately, the calculation comes out wrong. *Really* wrong. According to the theory, the vacuum energy should be infinite. Clearly it's not infinite, otherwise we'd all have been ripped to shreds by the blazing expansion of space. Since objects aren't spontaneously combusting all around us, the vacuum must be a reasonably calm place, at least at atomic scales and bigger. So if it's not infinite, physicists had figured, it ought to be zero.

That sounded like a weirdly big leap, but zero and infinity are more similar than you'd think. They are the simplest and most elegant quantities to calculate. It's far tidier to come up with a theory that suggests that some number should be zero or infinity as opposed to, I don't know, 3,746. Finite numbers can seem pretty random. So if infinity was off the table, zero seemed like the next best choice. Physicists figured that there could be some feature of the vacuum with positive and negative contributions in equal numbers, canceling out to a perfect zilch.

But that was before astrophysicists traded pencils for telescopes and actually measured the value of the dark energy, finding that it was almost zero, but not quite. It was the worst kind of number: tiny but finite. Getting the right value would require some mechanism that could take quantum field theory's infinity, cancel it to zero out to 120 decimal places, and then miraculously stop, leaving some minuscule crumbs behind. Crumbs that could hijack the universe.

A number that fine-tuned is rare, to say the least, and physicists hadn't been able to dredge up a single good explanation. In desperation, they turned to the A-word. As it happens, dark energy's bizarrely fine-tuned value fits squarely in the narrow range that would allow atoms, stars, carbon, and eventually life to exist. A little larger or a little smaller and our Goldilocks existence would be shot. In itself, that observation makes the whole situation worse—now not only is it an incredibly unlikely value, it's also, coincidentally, exactly the kind of unlikely value that life requires. Lucky us. It was the kind of coincidence that carried the unpleasant whiff of fate and teleology. But there was a catch. Dark energy's value is only a coincidence so long as our universe is the only universe around, and according to inflation, getting a single isolated universe is virtually impossible. Once you inflate one, you're stuck with an infinite number of them, a vast and varied multiverse. If every one of those infinite universes has a different amount of dark energy, then the tiny amount in ours is not only more likely, it's inevitable.

It was an answer, but not the kind physicists were hoping for, an explanation that placed disturbing limits on the very nature of explanation. Physicists want the laws of physics to be beautiful, basking in unity and inevitability. They want to perform elegant calculations, derive singular solutions, and know that the world had to be exactly as it is because it reflects the harmony and order that permeates a cosmos built of Platonic perfection. No one wanted to think that it was all a fluke, a petty accident of location. It was depressing.

Rees, who was extraordinarily polite and appeared to be carved of wax, explained that he takes the multiverse idea seriously and believes that anthropic reasoning is not only justified but necessary. Still, he said, physicists should go about their work as if it isn't, otherwise they risk getting lazy. They should still keep on trying to calculate physical laws from first principles, even if it's not going to happen. Spergel wasn't so enthused. Anthropic reasoning, he said, was nothing but scientific surrender.

Sitting there quietly, I couldn't help thinking back to something Wheeler had once written: "If an anthropic principle, *why* an anthropic principle?" For Wheeler, the *A*-word wasn't an explanation, but a *clue*—a clue to the role of observers in the origin of the universe, a clue to the nature of ultimate reality.

I was building up the courage to bring this up when Rees suddenly steered the conversation to politics, bioterrorism, and nuclear war. Over paninis and espressos, he explained that humanity has a fifty-fifty shot of destroying itself by the end of the twenty-first century. For a knight, he was a serious buzzkill.

Of all the brilliant people gathered at the conference, I was most intimidated by the prospect of talking to Timothy Ferris. Maybe it was because Ferris was a writer, not a physicist. His book *Coming of Age in the Milky Way* was one of my all-time favorite physics reads. When it came to the physicists, I was in awe. With Ferris, I was a fan.

So the next day, as everyone filed into the auditorium for a lecture and I spotted Ferris taking a seat in the front row, I quickly slid into the seat behind him, hoping that I would eventually think of something brilliant to say to him. I didn't. But when the lecture ended, Ferris turned around and asked, "How are you getting to the banquet tonight?"

The conference organizers had planned a banquet at the California Railroad Museum, about a half hour away in Sacramento's historic district. "I think they're busing us out there," I said.

Ferris gave me a look as if to say, *Do I seem like the kind of guy who rides a bus?* "I have my car here," he said. "I'll have to get directions. I don't want to get stuck there waiting for a bus. These conferences are great for the physics, but for social events . . ." He gave me a knowing smile. "If you decide you want to cut out early, just look for me. I'll give you a ride back to Davis."

I was itching to find out more about the worrisome low quadrupole, and during a coffee break I found my chance. As everyone milled around outside enjoying the California sun, I introduced myself to Lyman Page, a Princeton physicist and one of the lead investigators on the WMAP team.

"What's so problematic about the quadrupole?" I asked him.

Page explained that the lack of temperature fluctuations at scales larger than 60 degrees seemed to imply some kind of cutoff on the size of space itself.

That made sense. The temperature fluctuations had been formed when the hot plasma of the early universe was compressing and expanding, and that cosmic accordion was playing throughout all of space. If there were no fluctuations at scales larger than 60 degrees on the sky, it was as if there was no *space* at scales larger than 60 degrees on the sky. As if the universe were finite. Of course, those 60 degrees corresponded to the size of the universe at the time the CMB photons were first released. That region of universe has since undergone 13.7 billion years' worth of expansion. So the question was, if the size of space back then was capped at 60 degrees across today's sky, where does space end *now*?

The answer was shocking. Not only did the low quadrupole imply that the universe is finite, it implied that it's *small*—claustrophobic by cosmological standards. Stranger still, it implied that it was almost *exactly* the size of our observable universe. That if we could somehow peer just beyond the edge of our light cone, there would be nothing there to see.

"Could it be a glitch in the data?" I asked.

"No," Page said. "It's there. It's there to stay. It was there in the COBE data, too, but the signal-to-noise ratio wasn't high enough. Seeing it in the WMAP data is a wake-up call that there's really some potentially new stuff going on."

I had to wonder about inflation. The whole idea behind the theory was that spacetime gets stretched out far beyond our cosmic horizon, blowing up to such a huge size that monopoles disappear and curvature becomes negligible. "If that's true, and the universe really is small," I asked, "what happens to inflation?"

"Ninety percent—well, I'm leaving out Linde here—but ninety percent of inflationary guys would say, now we need a different model because a finite universe is just too weird," Page said. "It would mean the whole mechanism is off. I think it bothers all of us."

I was curious to know why Page had singled out cosmologist Andrei Linde as the one guy who wouldn't give up on inflation even in the face of a finite universe, so when I spotted him standing in the courtyard, I headed his way. I thought perhaps he had an idea of how inflation could explain such a phenomenon—I had no clue that it was simply because Linde was some kind of inflationary fundamentalist.

After introducing myself, I asked him whether physicists would abandon inflation should the low quadrupole turn out to be real. Apparently, asking Andrei Linde to give up inflation was like asking the Pope to spit on a Bible.

"No one should abandon inflation!" he yelled in a thick Russian accent. I cowered and frantically looked around, expecting everyone to have stopped what they were doing or bolted in fear, only no one seemed fazed. "If you have a model that shows why the universe is isotropic and why you get these density fluctuations, then you don't abandon that theory unless you have another theory that can account for those things. Inflation can suppress power on large angular scales; it just requires fine-tuning and it's ugly. But the universe is ugly—the standard model is ugly, the cosmological constant is ugly, dark matter, dark energy, ninety percent of the universe, what the hell is it? It's ugly. But that doesn't mean you abandon inflation."

Leonard Susskind, Alan Guth, and Andrei Linde enjoying the California sunshine   A. Gefter

* * *

Alan Guth, the man of the hour, seemed oddly approachable for a guy who had been tapped to win a Nobel Prize. He was in his fifties but exuded a cartoon-like youthfulness with his mop of brown hair and giant yellow backpack. He was famous for sleeping through every lecture, then waking up just in time to ask a bizarrely insightful question—a phenomenon I had already witnessed more than once. I asked him if he had any spare time to talk with me, and he graciously agreed. So in between lectures, wide awake, we sat outside in the sunshine.

"Inflation tells us what might have happened in the first fraction of a second after the universe was born," I said, "but what do we know about the actual birth?"

"We clearly don't have a definitive theory of how the universe originated," Guth said, "but the kinds of speculations people have, which I think are vague enough to be true even though we don't really know what we're talking about, is that the universe began as some sort of quantum event."

Understanding that event, he explained, would require a theory of quantum gravity.

"The general idea would be to have complete quantum descriptions of spacetime geometry. Then we would want to have some notion of what it means to have nothing, and nothing would have to be one of the quantum states. A state that describes no space, no time, no matter, no energy, nothing. But it would still be a state, a possible state of existence. That's the key feature. I'm assuming, without necessarily any right to, but I'm assuming that the laws of physics are somehow in place even before the universe. If we don't assume that, we can't get anywhere."

"Making that assumption means the origin could be knowable?"

"That's right. The origin could be knowable within the framework of the laws of physics. Right now I have no idea where we should look to uncover the origin of the laws, but we should worry about that later. So in this system that describes the ultimate laws of physics, hopefully there will be some quantum state of existence which will correspond to nothing. We know that quantum systems can undergo spontaneous transitions from one quantum state to another—atoms do this all the time when they decay. In a quantum system, any state can make random transitions into any other state, so you could start there, at nothing, and transition to a small universe, then inflation could take over and turn that small universe into a big universe. In vague terms, I think that's a plausible picture of how the universe could have begun."

"In that sense, it's possible to get something from nothing?" I asked.

"Our thinking about that question has changed rather dramatically since I was in graduate school," Guth said. "At that time everyone believed that the universe had many conserved quantities that had large values, and that the only way to produce the universe was to start out with that much stuff. But those conservation laws have all more or less disappeared. Today we think that the universe has zero values for all conserved quantities."

Conserved quantities are the features of nature that can never change, enshrined as they are in inviolable laws—laws such as the conservation of energy, which says that whatever happens, the amount of energy that comes out of an interaction had better equal the amount that went in. Energy can neither be created nor destroyed, only redistributed. Conservation laws are what keep the universe running smoothly from moment to moment. Without them, atomic bombs could appear in your bathtub, or your dog could suddenly blink out of existence. Physics would be impossible. Its equations would fall apart before you reached the other side of the equals sign.

But now Guth was saying that all conserved quantities are zero. That was rather shocking. You'd think the laws of physics are here to conserve *something*—like the "something" that came into existence 13.7 billion years ago. But if all conserved quantities are zero, it's as if the laws of physics are here to conserve *nothing*.

"Quantities like energy?" I asked.

"Energy was the most problematic because if you count up the mass in the universe and use $E = mc^2$, there seems to be a huge amount of energy. But the key realization was that gravity has a negative contribution to the total energy. It's not hard to prove that, but a crude way of thinking about it is to compare gravity to Coulomb's law of electrostatics. In electrostatics, if you want to push two positive charges together, they'll repel each other, so to build up a large charge you have to put a lot of work in to push together a large number of charges. It costs energy. For gravity it's exactly the opposite. Mass only has one kind of charge: positive. It always attracts. You can build up a large mass by pushing lots of mass together. It costs energy to pull them apart. So gravity's contribution to the total energy of the universe cancels out the positive energy of all the mass.

"The other important quantity in the history of this was baryon number," Guth continued, referring to the number of protons and neutrons that constitute every atom. "When I was in graduate school everyone thought that baryon number was conserved, and that the observed universe had a very large baryon number—that is, a large number of protons and neutrons and, as far as we could tell, very few antiprotons or antineutrons. Some people thought that maybe there was some large amount of antimatter out there somewhere that we hadn't found yet, but that idea never worked. With the development of grand unified theories in the 1970s, physicists realized that we didn't really know that baryon number was conserved. Later it was discovered that even in the so-called Standard Model of particle physics, where everybody thought baryon number was exactly conserved, it actually wasn't because of peculiar quantum effects. The evidence today seems overwhelming that baryon number is not a conserved quantity."

So energy was conserved but it didn't matter, because gravity always canceled it out, and the number of baryons wasn't conserved. If it was, then the total number of protons and neutrons in the universe today would have to be the same number that you started with at the beginning of the universe, and there would be no way to explain how all those protons and neutrons got there in the first place.

"Does that mean matter can spontaneously appear from nothing?"

"Yes." Guth nodded. "In the early days of inflation I made the statement that the universe could be the ultimate free lunch. Since then the idea of inflation in our visible universe has been elevated into a whole multiverse that just keeps growing and growing. If that picture is right, it's abundantly clear that you're getting something for nothing, and you just keep getting it. And it's all based on the idea that the universe does not have any nonzero conserved quantities."

"Gravity cancels out positive energy throughout the whole multi-verse?"

"That's right," he said.

"What about things that *are* supposedly conserved, like, say, angular momentum?"

"We believe that angular momentum is conserved, but as far as we can tell the total angular momentum of the universe is zero. If you add up the spins of all the galaxies spinning in different directions, as far as the astronomers can tell it really is zero. Electric charge is another quantity that we believe is absolutely conserved, but the universe, as far as we can tell, is electrically neutral."

"So if we observationally discovered that there was some conserved quantity with a nonzero value, that would mean it's impossible to get something from nothing?"

"That's right. That would change everything. The idea of eternal inflation would not be conceivable anymore. If our universe really needed a nonzero conserved value in order to make it something we would call a universe, then you could not make more and more of them without violating the conservation law."

"But as long as the only conserved quantities have zero values, you can get something from nothing."

"Maybe a better way of saying it is that something *is* nothing," Guth said. "Everything we see is in some sense nothing."

When it came time for Hawking to deliver his talk, I could barely contain my excitement. Hawking was notoriously stubborn, mischievous, and iconoclastic. He was a world-class troublemaker, and I couldn't wait to see what kind of trouble he was going to make today.

He was wheeled out to the center of the stage. "Can you hear me?" his computer politely inquired.

"Yes," the audience replied.

"In this talk, I want to put forward a different approach to cosmology that can address its central question: why is the universe the way it is?"

A different approach to cosmology? This was going to be good.

How can we ever figure out how the universe began? Hawking asked. "Some, generally those brought up in the particle physics tradition, just ignore the problem. They feel the task of physics is to predict what happens in the lab. . . . It amazes me that people can have such blinkered vision, that they can concentrate just on the final state of the universe and not ask how and why it got there."

Those who do attempt to explain the origin, he said, use a bottom-up approach, starting from some initial state and then evolving it forward to see if it develops into something that remotely resembles our universe. Inflation is just such an approach, he said, but even for a bottom-up theory, it doesn't make any sense.

This was just getting better. Here everyone was, celebrating the great successes of inflation, and now Hawking gets up and says it never made any sense to begin with.

Inflation, Hawking explained, lacked general covariance, the key ingredient of Einstein's theory, which ensured that every reference frame contains an equally valid description of the universe. Rather than working with the fully unified four-dimensional spacetime, inflation required spacetime to be broken apart into three dimensions of space and one of time. But whose space? Whose time? Breaking spacetime apart amounted to choosing a preferred reference frame—the ultimate crime against relativity. Worse, he said, if you choose certain coordinates to play the role of time, the inflaton field no longer expands. In other words, the theory only works in certain reference frames to begin with.

This was fascinating, but the talk was slow going. Minutes passed between sentences, eternal minutes during which the audience did their best to stay respectfully silent, the shifting of weight in seats and the clearing of throats resounding against the painful silence.

Suddenly his right leg began shaking violently, causing the computer mounted to his wheelchair to vibrate. His aide rushed over and knelt on the floor, holding Hawking's foot down as he continued with his talk.

Beyond the problems with inflation, Hawking said, there's a fundamental problem with the bottom-up approach as a whole. "The bottom-up approach to cosmology is basically classical, because it assumes that the universe began in a way that was well defined and unique. But one of the first acts of my research career was to show with Roger Penrose that any reasonable classical cosmological solution has a singularity in the past. This implies that the origin of the universe was a quantum event," he said.

Quantum events are described not by unique states but by superpositions of every possible state. It's not simply that we can't know which of those states the universe was actually in—it's that the universe wasn't actually in any of them. For that reason, Hawking said, we need to work from the top down, from the present to the past. By looking at the features our universe has today we can figure out all the possible histories that could have led to such a universe. Somehow, in doing so, we *create* the history of the universe. "This means that the histories of the universe depend on what is being measured, contrary to the usual idea that the universe has an objective, observer-independent history," he said.

No observer-independent history, I scrawled in my notebook, and on further contemplation underlined it. I wasn't sure exactly what it meant, but I had a hunch that it was going to be important.

That evening I boarded the bus headed for the banquet. I took a seat next to a man who was also wearing a press badge.

"Michael Brooks, editor at *New Scientist,*" he introduced himself with a charming British accent.

I recognized the name immediately. I was an avid *New Scientist* reader and a recent article, "Life's a Sim and Then You're Deleted," had made such an impression on me that I had torn it out of the magazine and pinned it to the wall above my computer. Written by one Michael Brooks, the article had discussed a paper by the philosopher Nick Bostrom, who argued that in all likelihood we are living inside a Matrix-style computer simulation. Bostrom's idea was that eventually our computers will be powerful enough to simulate conscious creatures, like humans. When that day comes, future programmers will be able to simulate entire societies, even entire universes, and watch how various scenarios play out, either for research purposes or as some kind of hyperreality TV. Once one simulated reality is created, hundreds, thousands, millions will follow. So given the inevitable existence of millions of simulated worlds, the odds that we are living in the one true original reality are pretty close to zero.

In the article, Brooks had wondered if there would be any way to tell whether this was in fact a simulated world. Programmers, he reasoned, wouldn't bother wasting resources designing every last microscopic feature of the fake reality. If the simulated observers start poking around, the programmers could always fill in the gaps on the fly. Thus, he argued, the microscopic realm of a simulated world might look a little nonsensical. "If you've ever wrestled with the weird nature of quantum mechanics, alarm bells may just be starting to ring," Brooks wrote.

I told Brooks that I was a freelance writer; we chatted about cosmology and the lectures we had seen so far.

"Pitch me some articles," he said as the bus pulled up to the Railroad Museum. "I reject ninety percent of the pitches I get, so don't be discouraged, just keep pitching."

"I will!" I promised.

As I stepped off the bus into the warm California evening, I couldn't help thinking it was all too good to be true. That this probably *was* a simulation. Then I remembered how Brooks's article had ended. The best chance for this to be the true reality, he had said, is if humanity should destroy itself before our computers grow sufficiently powerful to simulate complex societies and conscious minds. I thought back to yesterday's lunch, and Rees's laundry list of doomsday scenarios. Maybe Sir Buzzkill had the answer to reality after all.

Nervous at the thought of having to socialize with the world's most eminent physicists, I quickly downed two glasses of wine. Big mistake. My tolerance for alcohol was embarrassingly low. Two glasses of wine—I might as well have been doing shots of tequila.

Eventually everyone began sitting down for dinner, quickly filling up the round, linen-clad tables that had been set for the occasion. I grabbed the first empty seat I could find. I smiled politely, but the physicists talked amongst themselves as the waiters topped off our wineglasses and then left to fetch the salads.

Emboldened by the wine, I decided to strike up some conversation. "Did any of you read the profile of João Magueijo in *Discover* magazine?" I asked. I had read the piece on the plane ride over. It was the first thing I could think of. The article had discussed Magueijo's theory that the speed of light had been much faster in the very early universe. He had proposed the idea as an alternative to inflation, though I couldn't figure out the difference between the two. Speed up light but keep spacetime expanding at subluminal speeds, or keep the speed of light the same and speed up the expansion of spacetime— they seemed to be two ways of looking at the same thing, so why muck with Einstein? "Is his variable speed of light theory just bogus?"

The physicist directly across from me gave me a stern look. "I hope not, since I was his collaborator on that theory."

No one said a word.

Oh, dear God. Where were people's name tags when you needed them? The physicist, I now realized, was Andy Albrecht, the second man behind the variable speed of light. Had I really just suggested that his theory was bogus? I frantically searched my mind for some way to recover. Why did the magazine have to showcase a giant photo of Magueijo's face without ever showing Albrecht? I wanted to apologize. I wanted to explain that I was just trying to make conversation, that I had some brute allegiance to Einstein, that I didn't really think his theory was bogus, that perhaps I was having some kind of stroke. Instead I said this: "Wow, he really stole your spotlight."

Had that seriously just come out of my mouth? What the hell was I doing? *Just shut up,* I urged myself. *Just stop saying words.*

"Like I really care," Albrecht said, annoyed.

I nodded and smiled. I wanted to crawl underneath the table and hide. I scanned the room, pathetically searching for an escape route.

And that's when a telepathic miracle occurred. From across the expansive room, through a sea of physicists I had yet to accidentally insult, I locked eyes with Timothy Ferris.

Ferris stood up, looked directly at me, and gave a little nod toward the back door. Without saying a word, I stood up from the table, walked quickly to the back of the room, and quietly slipped through the glass door. He was outside waiting. "My car is around the corner," he said.

Okay, I thought, this has to be a simulation.

We walked together down the empty street. Ferris asked me whom I was writing for. "Well, I wrote my last article for *Scientific American*," I said. I didn't mention that it had also been my first. "How about you?"

"I'm doing a piece for *The New Yorker*," he said.

I felt unworthy just sharing the sidewalk.

We rounded the corner and parked there on the otherwise deserted cobblestone street was a small, shiny Porsche. I looked around, searching for some other car, the kind that would belong to a writer. But Ferris pushed a button on his keychain and the Porsche beeped a friendly greeting in reply. Seriously? I thought. A writer? This pretend career was looking better by the minute.

I climbed into the passenger seat and buckled my seatbelt. Ferris revved the engine and turned on the stereo, cranking the volume. The car filled with the pounding of a snare drum, the wailing verve of an electric guitar.

"Is this Bowie?"

Ferris smiled, grabbed hold of the stick shift, and peeled out onto the street. The force of acceleration pinned me to my seat, and as he navigated the tiny streets of old Sacramento like a racetrack, Ferris showed no signs of slowing down. Soon we were speeding down the California freeway, weaving in and out of traffic, blazing through the warm night air, a blur of palm trees passing by my window.

Five minutes later, we arrived back in Davis. Ferris dropped me off at my hotel and told me to keep in touch. I stepped out onto the sidewalk, unsteady on my feet, bummed that the party was over but happy to be alive and on solid ground.

I pulled my cell phone from my purse and hit the speed dial to call my father.

It was the last day of the conference. I didn't want it to end. I had learned so much I thought my brain might overflow, but I wanted to keep going, push further. I couldn't help thinking that something was missing in the way everyone was talking about the universe. Something . . . quantum.

"Any satisfactory theory of quantum cosmology has to refer to observations that can be made by observers inside the universe," Markopoulou had told me. But inflation referred to spacetime regions beyond our observable universe, and worse, eternal inflation conjured an entire multiverse that no one, not even in principle, could ever observe. The fact was, the standard model of cosmology was not a quantum cosmology. Sure, the inflaton's descent from the false vacuum was a quantum process, but otherwise the whole thing had a distinctly classical feel. That had been Hawking's point: "The bottom-up approach to cosmology is basically classical [but] the origin of the universe was a quantum event." I needed to find out more. What was his "top-down" cosmology all about? And how did it account for the fact that we're stuck inside the universe?

I also couldn't stop thinking about my conversation with Guth. All signs pointed to a universe that came from nothing, he had said. A universe that *is* nothing. And the most exciting part was that he presented it as a falsifiable claim: find one nonzero conserved quantity and nothing goes out the window. If the universe is nothing, I thought, then everyone's been walking around asking the wrong question. The question isn't how you get something from nothing. The question is, why does nothing *look* like something?

With so many legendary thinkers in one room, the monumental weight of the dawning of cosmology's golden age had inspired the conference organizers to hire a photographer to take a group photo, one sure to go down in the annals of science history.

"At the break, we're going to ask everyone to go outside and assemble on the steps for the photo," Albrecht announced from the stage.

While the physicists slowly gathered on the steps, I snuck away from the crowd to call my father.

"What's the scoop?" he asked.

"They're all worried about this low quadrupole thing," I said, speaking just above a whisper like a spy reporting back to Reality Headquarters. "This lack of fluctuation power at large angular scales in the CMB. It's almost as if the universe isn't big enough to hold them."

"How big would it have to be?" he asked.

"That's just the thing. It's looking like it's only the size of the observable universe."

"Well, that would be rather suspicious," he said.

"Exactly! It's crazy. . . . Shit, I gotta go. They're taking this group photo of all the physicists and I want to snap a few pictures of my own."

"Get in the picture!"

"Mom?"

"Get in it!" she repeated, half Jewish mother, half weirdly bossy cheerleader.

"Okay, okay," I said, humoring her as I rolled my eyes.

But as I stood off to the side with the other journalists watching the physicists assume their positions, I couldn't get my mother's voice out of my head. Why couldn't I be in the picture? There were no physics bodyguards milling around—who was going to stop me? Frankly, who was even going to notice?

I looked over at the photographer; he was still fiddling with his camera. As inconspicuously as possible, staring down at my feet so that no one's eyes would catch mine, I slunk along the side of the steps and quickly made my way to the back corner of the crowd. I was sure no one would notice, and that even though I'd barely be visible in the photo, I'd be able to point to a tiny bit of shoulder peeking out from behind someone important and say, "Look, that was me."

The photographer finally looked up and put his eye to the camera. Everyone held a collective breath and smiled. But then he paused, lowered the camera and scanned the crowd looking for . . . Jesus, *me*?

"You there!" he shouted, pointing toward me. If there had been a record playing, it would have come to a screeching halt. I felt my face growing hot. Was he going to out me as a fraud right here in front of everyone? Announce that not only was I not a physicist, I wasn't really even a journalist? That I was on a covert mission to figure out the nature of reality, and that I was willing to go to any lengths necessary to do it? How did he know? And where the hell was Timothy Ferris with the getaway car?

"You! You're too short," he shouted. "Get down in front."

Avoiding all eye contact once again, I shuffled down toward the front, where he promptly grabbed me by the shoulder and inserted me into the spot of his choosing—front row and nearly center. To my right was Guth, and one over to his right was Hawking. Why not just have me sit right on Hawking's lap? I thought. "You there," Guth muttered to me, imitating the photographer, "Your nose is crooked—get that fixed!" I laughed gratefully.

"Okay, people, here we go!" the photographer shouted. And with that, there was nothing left to do but smile.

The Davis Meeting on Cosmic Inflation
March 22-25, 2003

Not so sly, at the Davis Meeting on Cosmic Inflation, 2003   Debbie Aldridge, UC Davis

4

Delayed Choices

Back in my apartment, I couldn't get Nick Bostrom's simulation argument out of my mind. If the world around us really was a virtual reality simulation on a computer living in some higher reality, could we ever find out? Would it even matter?

Descartes had struggled with the same question. Of course, there weren't computers back then, but there were evil demons, and Descartes wondered if one might be tricking his senses into perceiving a false reality. He worried that everything around him, including his own body, might be a fraud. But in a sea of demonic doubt, he could say one thing for certain: He was perceiving. He was thinking. He was real. Even if everything that presented itself to his consciousness were an illusion, the very fact of his consciousness would remain true. He was thinking, therefore he existed. *Cogito ergo sum.*

That was it? The one thing I could say for sure? I exist. Game over.

It was a depressing thought. Descartes never really clawed his way out of the cogito, not using reason anyway. He had to take a leap of faith and invoke a benevolent God who wouldn't be so cruel as to fool us with a fake world. But if you're willing to take a leap of faith, I thought, why add a middleman? Why not just believe in reality and call it a day?

I wasn't particularly worried about evil demons, but Bostrom's computer simulations seemed like a more viable threat. Flipping through an issue of *New Scientist*, I came across an article by cosmologist John Barrow arguing that if we are in a simulation, we ought to see glitches in reality. "If we live in a simulated reality, we should expect to come across scientific phenomena, such as occasional glitches in experimental results that we cannot repeat, or very small drifts in the supposed constants and laws of nature that we cannot explain," Barrow wrote. "Tantalisingly, we do have a few such results: the apparent astronomical variation in the fine structure constant by a few parts in a million, for example. Clearly, finding explanations for these phenomena is something of a priority. If we can't, then the flaws of nature may turn out to be at least as significant as the laws of nature for our understanding of true reality."

It was tantalizing, but even if we observed glitches, how would we know they were evidence of a simulation? Couldn't they just be flaws in reality itself? Barrow seemed to be assuming that the true reality must be flawless, a pristine specimen of logical consistency. And if that's the case, perhaps there's only one possible reality anyway, one unique ideal of logical perfection. After all, physicists have yet to stumble upon a single complete and logically consistent model of a physical universe, and they've really been trying. If we can't even come up with one, what are the odds that the simulation's programmers can come up with several, even infinite variations? If there's just one possible world, it might be knowable—demons and programmers be damned.

Then again, maybe human brains aren't up for the task. Maybe the programmers have no trouble inventing universes. And if this is a simulation, who's to say it's not a simulation simulated by simulated beings, and who's to say that their reality isn't just a simulation from within another simulation, which in turn is just . . . When you start questioning the reality of reality, it's easy to spin out. My mind was racing. Was reality unknowable? Was this whole mission flawed from the start? *Cogito ergo panic*.

I was spiraling when I landed on a strange thought: if reality is not a simulation, what is it? Simulation is an unnerving word as an antonym for something else—but what? Simulation is all we know. Our brains are our sole portals to so-called reality. There is nothing in the universe that we can access without first filtering through the labyrinthine lumps of gray matter in our heads. We are literally, hopelessly, eternally trapped inside our minds. All the things we see, hear, touch, smell, and taste are nothing other than perceptions generated by our brains. Cats, dogs, trees, other people . . . all astonishingly realistic neural simulacra. Then again, who's to say they're astonishingly realistic? Compared to what?

Our eyes are not transparent windows to the outside world, despite evolution's brilliant illusion. When we think we're walking the streets of a city, we're really strolling the neural paths of our brains. Everything that appears to be outside is really inside. For all intents and purposes, there is no outside. The brain is a universe unto itself: billions of twinkling neurons, dendrites splayed like fingers reaching for the beginning of time, chemical messengers leaping across the mindless darkness of deep intracranial space. As the cosmologist James Jeans said, "The universe begins to look more like a great thought than a great machine."

Of course, it's tempting to think that our brains' simulations are simulating *something*, some external reality that impinges on our senses, nudging our neural cogs and gears to churn up a trusty illusion. But who knows? We hallucinate, we dream. Chuang-tzu dreamed he was a butterfly, then awoke to find he was a butterfly dreaming he was a man. I suddenly understood the moral of that story: we're all screwed.

Bishop Berkeley embraced the dilemma and ran with it, claiming that the world was mind-dependent—a physical reality built of abstract thought. To Descartes's *cogito ergo sum*, Berkeley replied, *esse est percipi*: to be is to be perceived. The world stops at perceptions—beyond them, there's nothing. Perceptions, he said, are the be-all and end-all of existence, not representations of external, physical things. This didn't go over well. Outraged by Berkeley's idealist philosophy, Samuel Johnson famously kicked a rock, declaring, "I refute it thus." How can we refute Bostrom? I wondered. Who was going to kick him?

Berkeleyan idealism had one fatal—and, frankly, kind of obvious—flaw: the bishop's mind-dependent world was dependent upon minds. Minds that were somehow separate from the world they perceived.

There was a categorical dualism: observer and observed. Two fundamentally different kinds of things. But what are our brains if not physical objects conceived of the same stuff they simulate? After all, we are just pieces of the universe looking at itself, and if we are a simulation, then we are a simulation simulating itself. So was it all just a cosmic hall of mirrors? Mirrors reflecting mirrors, infinite regress of images of . . . nothing? Was that what Wheeler meant when he talked about a self-excited circuit? Or my dad, imparting his lotus-style wisdom? *You think there's you, and then the rest of the world outside you. But it's all just one thing.*

I was ready to resign myself to life in Plato's cave, mistaking shadows for reality, when it dawned on me: *The brain is a universe unto itself. For all intents and purposes, there is no outside. One-sided coin, the side of things . . .*

Smolin had said that the first principle of cosmology must be that there is nothing outside the universe. Maybe we were in need of an analogous slogan here: there is nothing outside reality. Suddenly the simulation problem looked an awful lot like quantum cosmology's observer problem in a different guise. You can't step outside the universe, you can't step outside your brain, and you can't step outside reality. If I'm a simulation, there's no way I can step outside the simulation and look down on it from a higher level of reality, nor could I then step outside that level to the next level up. And if I'm not a simulation, I likewise can't step outside reality and look back to confirm that it's real. There is simply no vantage point from which we can assess the reality of the reality we're inside. The simulation argument begs for an impossible God's-eye view. Does that mean we'll never know the truth? Or is the truth that reality is a one-sided coin?

Leibniz once said, "Although the whole of life were said to be nothing but a dream and the physical world nothing but a phantasm, I should call this dream or phantasm real enough if, using reason well, we were never deceived by it." Well, sorry, Leibniz, but I was looking for something a little better than "real enough." I wanted ultimate reality and I wasn't going to settle for anything less.

* * *

Several months later, I got a call from *New Scientist*. They wanted me to write an article about a group of physicists in Long Island who had created some kind of fireball. I had already written one article for them—a story about loop quantum gravity, which I had pitched to Michael Brooks, the editor whom I had met on the bus. Despite his warning of nearly inevitable rejection, he had not only accepted it, he had put it on the cover. Now they were calling me for a story? It seemed too good to be true.

"It's got a lot of complicated particle physics," explained the editor, one whom I had never met. "We all agreed that you were one of the few writers who could handle the difficulty. Are you up for it?"

We all agreed?

I cleared my throat to stifle my excitement. "Of course."

"They suspect they've created the quark-gluon plasma," she continued.

"Ah, yes, the quark-gluon plasma," I said. "Fascinating."

When I hung up the phone, I immediately set to work on the story. I needed to call the physicists in Long Island and ask them about the details of their experiment. And I had to call other physicists in the field to discuss the implications of the discovery for our understanding of the universe. But first things first—I needed to find out what the hell a quark-gluon plasma was.

"I just had the most surreal night."

I had been curled up in bed with a book about quarks when the phone rang. It was my father calling from a hotel room in Chicago, where he was attending a radiological society conference.

I dog-eared my page and closed the book. "What happened?"

"I was invited to a reception at the Field Museum tonight," he said. "Everyone was at the cocktail party in the atrium, but I just sort of wandered off into the museum. It was after hours so it was completely empty. But it turns out the current exhibit is the Einstein exhibit! So there I found myself alone in a room surrounded by all of Einstein's things—his handwritten manuscripts, photographs, and letters. It was so strange. It was completely silent and I was alone with all of his stuff.

And for some reason I just kept staring at his compass. I wanted to grab it and run."

"You should have!" I said.

When we hung up the phone I giggled as I pictured my father busting open the glass case, snatching the compass, and sprinting through the crowd of confused radiologists as a growing swarm of museum guards followed in close pursuit, shouting, "Stop that man!" I pictured him clutching it on the plane as he flew back to the East Coast. And then, since it was my imagination, I pictured him placing it in a small box, wrapping it with a bow, and giving it to me.

Einstein had been only four or five years old when his father gave him that compass. It was one of those small tokens that somehow change the world. Watching the way an invisible force always guided the compass needle north had convinced Einstein that "something deeply hidden had to be behind things." He spent the rest of his life trying to find it.

My father, too, had offered me my first clue that reality is not what it seems. Only in my case the clue wasn't an object but an idea, and instead of turning out to be Einstein I grew up to be a counterfeit journalist with more questions than answers. Still, it occurred to me now that the best gift a parent can give a child is a mystery.

Quantum chromodynamics, or QCD, was the theory that described how gluons bind quarks together in groups of three to form the protons and neutrons deep in the core of every atom. Quarks, I learned, come with three possible charges, known metaphorically as red, blue, and green. If you combine all three, the colors cancel to neutral. In fact, the quarks *have* to remain color neutral, which means that they are stuck living in groups, bound together by the gluons. Never can a lone quark venture out on its own. Unless, that is, you turn up the heat. In extreme temperatures, such as those following the big bang, the gluons' grip loosens, the quarks wander freely, and matter dissolves into a primordial plasma.

To achieve such extreme temperatures, physicists at the Relativistic Heavy Ion Collider, or RHIC, at Brookhaven National Lab had taken gold nuclei, steered them around a 2.4-mile track at nearly the speed of light, and then crashed them together, releasing 100 billion electron volts' worth of energy and creating said fireball, 300 million times hotter than the surface of the Sun. A good 10^{-23} second later, it was gone. But in that fraction of a fraction of a second, the quarks roamed free.

It was an exciting discovery, but the plasma didn't look quite like what physicists had expected. Contrary to their calculations, the quarks and gluons seemed to be moving around in a coherent way. It wasn't the chaotic free-for-all motion of a gas-like plasma, but the synchronized swim characteristic of a liquid. In fact, its viscosity made it the most ideal liquid ever observed—nearly twenty times more liquid than water.

That was pretty weird, but what really grabbed my attention was something Johann Rafelski said. Rafelski was a quark-gluon plasma expert; I had phoned him to discuss the implications of the discovery. "The structure of the vacuum is the origin of quark confinement," he told me. "So the idea was to melt the vacuum and dissolve the binding, allowing the quarks to move freely."

Melt the vacuum? I couldn't get that phrase out of my head. It was so awesomely bizarre—you can *melt nothing*? Okay, I knew that the vacuum wasn't really "nothing." Nothing, presumably, would be a state of zero energy, and zero was way too precise a number for quantum mechanics. Quantum nothing seethes with activity, thanks to the uncertainty relation between energy and time—the shorter the time period, the larger the energy that can spontaneously spring from the depths of the vacuum only to disappear again in far less than the blink of an eye. This energy can take the form of fleeting pairs of virtual particles and antiparticles that boil up from the vacuum, then meet and annihilate. But how did those virtual vacuum fluctuations bind quarks together? I had to do a lot more research—quickly.

As I gorged on all things QCD I learned that, as Rafelski had said, it's the vacuum that forces the quarks into confined living. From quantum uncertainty in the gluon field, virtual gluons emerge. But the thing about gluons—even virtual ones—is that they carry a charge. A gluon's job is to transmit a sticky force—the so-called strong force—to quarks.

The gluon recognizes a quark by its color charge. Photons act in much the same way, transmitting the electromagnetic force between electrons, which they spot by their electric charge. But whereas photons carry no electric charge of their own, gluons have color charge—so in addition to interacting with quarks, they also interact with themselves and with other gluons. As they boil up from the vacuum, the virtual gluons stick to one another, twisting and contorting into complex structures—structures that wall in the quarks, making it impossible for them to move freely through the vacuum. Stuck in a gnarled sea of virtual gluons, the quarks huddle together—red, blue, and green— their neutral charge protecting them from the dangerously sticky gluons. Imprisoned in triplicate, they form protons and neutrons—the massive cores of atoms. If it weren't for the structure of the vacuum, atoms would fall apart.

The force of the virtual gluon field restrains the quarks' movement; if you tried to grab one and move it, it wouldn't budge. As if it were *heavy*. In that way, the vacuum's virtual gluon field endows quarks with 95 percent of their mass, which in turn gives protons and neutrons their mass, which in turn gives atoms 99 percent of *their* mass . . . all of which means that the mass of everything around us, including our own bodies, is little more than the weight of the vacuum. The material world is nothing incarnate. Lucretius had said that "nothing can be made out of nothing." Quantum chromodynamics begged to differ.

To set the quarks free, you have to dissolve the vacuum's virtual gluon structures. Look to higher temperatures and energies, inching ever closer to the conditions of the big bang, and the vacuum's structure melts away. As its convoluted forms dissolve, it begins to look more and more like nothing. Smooth and simple. Undifferentiated. Symmetric.

If there was one thing to know about symmetry, I learned, it's that it tends to break. As all the books explained, a pencil balanced on its tip has perfect rotational symmetry—it looks the same from every angle, 360 degrees around. It's also about to fall. Even though the pencil is in a kind of equilibrium state, it's not going to last, because there exists a state of even lower energy: the state in which it's horizontal.

The slightest breeze is going to knock that pencil over. And even though it has an equal chance of falling at any angle around the circle, it chooses only one. When it lands, rotational symmetry is broken.

One way to get symmetries to break is to turn down the temperature. A puddle of water is highly symmetric—look at it from any angle and it looks the same. But cool it down enough and it freezes, forming ice crystals with more structure and less symmetry.

Physicists, I learned, think about the universe the same way. In the heat of the big bang, the vacuum is symmetric. As the universe expands and cools, structure freezes in, like the twisted forms of virtual gluons. With structure comes mass. With mass comes everything else. The world we see around us and the people we see in mirrors are nothing more than broken shards of symmetry. Shards of *nothing*.

I picked up the book *Longing for the Harmonies* by Frank Wilczek, who had won a Nobel Prize for helping to formulate QCD. He explained that spontaneous symmetry breaking occurs whenever there are an infinite number of equally valid vacuum states for a single, unique higher-energy state—like the continuum of possible positions in which the single pencil can land.

"The most symmetric f phase of the universe generally turns out to be unstable," he wrote. "One can speculate that the universe began in the most symmetric state possible and that in such a state no matter existed: the universe was a very empty vacuum, devoid both of particles and of background fields. A second state of lower energy is available, however, in which background fields permeate space. Eventually, a patch of the less symmetric phase will appear—arising, if for no other reason, as a quantum fluctuation—and, driven by the favorable energetics, start to grow. The energy released by the transition finds form in the creation of particles. This event might be identified with the big bang. . . . Our answer to Leibniz's great question 'Why is there something rather than nothing?' then becomes 'Nothing is unstable.'"

But the symmetry isn't really broken, Wilczek said. It's just hidden. You can always find it again if you look hard enough—in the fundamental equations, say, or perhaps inside a fireball.

The quark-gluon plasma glimpsed at RHIC was evidence that the vacuum really did start out more symmetric. Still, the vacuum was

more resilient than anyone had expected, the coherent, liquid movement of the quarks displaying some kind of residual asymmetry, rather
than the symmetric free-for-all of particles in a gas. To get to the nothing, physicists were going to have to melt the vacuum even more.

As I interviewed various physicists I found that no one seemed to
know what to make of the unexpected result. But when I searched
around online I stumbled across an obscure clue. Apparently, something known as the "AdS/CFT correspondence" could explain the ul-
traliquid plasma. There wasn't enough time to figure out what that
meant, nor enough space in the article to mention it, but I jotted it
down in my notebook so I wouldn't forget. *Look into AdS/CFT correspondence . . . something to do with string theory . . . explains liquid fireball?*

I wrote my article and sent it in just before the deadline. But I
couldn't stop thinking about Wilczek's idea that nothing was unstable.
It was kind of awesome, and smelled an awful lot like an explanation.
My father and I had spent so much time wondering why nothing—that
state of infinite, unbounded homogeneity—would ever change. If it
were so perfectly homogeneous, so perfectly *symmetric*, why would it
ever break? Why would it ever become a universe? Wilczek seemed to
have the answer. Nothing was unstable. Universe solved.

Almost. The problem with invoking spontaneous symmetry breaking to explain that primordial alchemy, the transformation of nothing
into something, of symmetry into structure, is that it requires some
external force, the breeze that knocks over the universe. But there is
nothing outside the universe. Wilczek had suggested that quantum
fluctuations could provide the breeze, but that wasn't any better. If you
use the laws of quantum mechanics to tip the universe into existence,
you leave the existence of the laws themselves unexplained. Guth had
acknowledged that. "I'm assuming, without necessarily any right to . . .
that the laws of physics are somehow in place even before the universe," he had admitted. "If we don't assume that, we can't get anywhere."

That was pretty discouraging. A real answer to existence would
have to start with nothing, and then, somehow, explain why the laws of
physics pop right out. We can't just assume the existence of quantum

mechanics and then use it to explain other things, like universes. We need to explain quantum mechanics. *Why the quantum?*

The story in which the universe starts out in a perfectly symmetric state that promptly breaks, creating our elaborate, frozen world, can't be the real story, because there's no one to tell it. It's a story that requires an omniscient narrator, a narrator with a God's-eye view, the kind that Smolin's slogan had strictly forbidden. Wheeler and DeWitt's damned equation didn't work because you ended up with a universe trapped in an eternal instant, a universe where nothing can ever happen—no big bang, no quark-gluon plasma, no computer simulation. It occurred to me now that perhaps my father's H-state was stuck in the same trap. The nothing could never change, because what would it be changing in reference to? You'd need some reference frame outside the nothing, which you can't have, at least not according to my father's definition of nothing as infinite and unbounded. Nothing was a one-sided coin.

What we desperately needed, I realized, was a story told from here inside the universe. Here inside the nothing, if Guth was right. *Something* is *nothing*. And if the universe *is* nothing, maybe the nothing never changes at all. Maybe the universe was never really born. Maybe nothing just *looks* like something when you're inside it.

If nothing was by definition unbounded, I thought, then all you'd need to make it look like something was a boundary. Markopoulou had said that when you're stuck inside the universe, you can't see the whole thing, only the region within your light cone. Could a light cone provide the boundary you need to turn nothing into something? I wasn't sure. After all, light cones grow with time. At best they can provide a temporary boundary. I wasn't sure if it would be enough. Besides, a light cone wasn't a *thing;* it was simply the delineation of a reference frame. How could it ever do any physical work, let alone the heavy lifting you'd need to drag a universe—even the *appearance* of a universe—out of nothing?

After a whirlwind tour of symmetry breaking and quantum chromodynamics, I finally had a chance to relax. Instead I masochistically surfed the Internet for more on Nick Bostrom and the simulation nightmare. In the midst of my existential flagellation, I came across a website called Edge.org.

How had I not seen this before?

The site was an intellectual salon, a kind of virtual Algonquin Round Table where the most brilliant scientists, writers, and thinkers discussed and debated everything from consciousness and the origin of life to game theory and parallel universes. The site showcased the very latest in scientific thinking as it was unfolding in real time in a way that any nonscientist could understand but without dumbing it down or packaging it into sound bites.

Poking around, I discovered that the man behind Edge was one John Brockman, literary agent and self-made cultural impresario. Brockman had started off in the avant-garde art and film scene in 1960s Manhattan, where, at twenty-five, he was hanging out with the likes of Andy Warhol, John Cage, Robert Rauschenberg, and Bob Dylan, organizing events for multimedia artists and running the independent leg of the New York Film Festival.

After Cage lent him a copy of *Cybernetics* and Rauschenberg recommended books by James Jeans and George Gamow, Brockman became interested in science and developed a hunch that scientists would emerge as the leading public intellectuals who, like the avant-garde artists, would shape public discourse by forcing people to question their most basic assumptions about the world. This couldn't happen, though, until the scientists had a direct route by which to engage the public. So in 1973 Brockman founded Brockman, Inc., a literary agency that specialized in encouraging scientists to write books for a lay audience.

Five years later, with physicist Heinz Pagels, Brockman created the Reality Club, an intellectual salon that met in restaurants, museums, and living rooms across Manhattan. The club met for fifteen years before Brockman moved the group online to Edge.org. Meanwhile, he completely transformed the world of science books; his agency's clients included huge names, such as Richard Dawkins, Steven Pinker, Sir Martin Rees, Daniel Dennett, Jared Diamond, Craig Venter, and Brian Greene. Though the workings of the Reality Club had gone virtual, Brockman still hosted a few live salons. Once a year he'd bring a handful of scientists and writers out to his sprawling farm in western Connecticut.

The Reality Club? There was an actual Reality Club? How did one become a member of such a club? I wondered. I wasn't a scientist or a public intellectual. I wasn't anything at all, really, unless you counted budding fraudulent science journalist. But I didn't care. All I knew was that I wanted in. I wanted to engage in intellectual debate on Edge.org. I wanted to hang out at Brockman's farmhouse. And most of all, I wanted John Brockman to be our agent for the book that my dad and I would someday write about the nature of ultimate reality. Unfortunately, Brockman's world didn't seem like the kind of place you could weasel your way into by pretending to be something or someone else.

I clicked on Brockman's picture. There he was, gruff and imposing, sporting a linen suit and a Panama hat, looking like a cross between a mob boss and a member of the Buena Vista Social Club.

So Nick Bostrom was part of Brockman's crew. It made sense, given his penchant for taking reality and twisting it like a balloon animal. I clicked on his bio, curious to see what path had landed him at Brockman's virtual door. Apparently he had earned his Ph.D. in the philosophy of science at the London School of Economics, where he studied philosophy, logic, artificial intelligence, and computational neuroscience. But before all that, Edge explained, Bostrom had been a stand-up comedian.

You made my head explode, I thought, staring at his stern headshot. Very funny.

A few weeks later, I was back in the suburbs of Philly to spend a few days with my parents.

"Now that you're writing more articles, do you think you can make a real career out of this?" my mother asked at the dinner table.

I put down my fork. "A journalism career? I don't know. Maybe. That's not really the point."

"What is the point?" she asked.

"The point is to figure out the nature of ultimate reality. How to get something from nothing. The journalism thing is just a front. It's a means to an end."

I looked over at my father for some backup. He offered an agreeable nod.

"Well, I don't know about ultimate reality," my mother said, "but in this reality you're an unemployed coat-check girl."

"That's not really my fault," I said. "It's August. People stopped wearing coats."

"Even so," she said, "I think it's time you consider a more sustainable plan."

She was right, of course. I couldn't learn physics in the coat room forever. Luckily, I *did* have a plan. "I'm thinking about going back to school," I announced. "There's a program in the philosophy of science at the London School of Economics. Nick Bostrom went there. He says we're probably living in a computer simulation, and he hangs out on John Brockman's website. Not that Brockman's website is the simulation. *This* is." I waved my arms in the air to indicate our kitchen. "At first I was thinking, if everything is a simulation, what's the point of grad school? But I don't see how simulated learning can be anything other than actual learning, right? Anyway, I suspect the whole thing is an invalid argument because it presupposes a view from outside reality."

"You're moving to London?" my mother asked.

"*Simulated* London," my father corrected her.

"But we'll miss you," she said. "And our phone bill is going to be outrageous."

Ever since I had moved to New York my father and I had replaced our nightly kitchen-table cosmology talks with phone calls that ran on for hours.

"We'll switch to email," my father said.

"What about Cassidy?" she asked.

I gave her a sheepish grin.

"Oh, no," my mother said. "I told you when you got that dog that we were not going to end up taking care of it. I am *not* going to have dog hair all over my furniture. I am *not* going to be picking up poop."

"*Simulated* poop," my father said.

"There must be philosophy of science programs in the U.S.," said my mother.

"Sure," I said. "But there's no evidence that they lead to Brock-man."

"And you need him because . . . ?"

"So he can be our agent."

"For?"

"For the book we're going to write when we find the answer to the universe."

"Can't you just call him when the time comes?"

I laughed at her adorable naïveté. "Um, no. You can't just *call* John Brockman. Do you know what you get if you go to the website for his literary agency? A blank page that says 'Brockman, Inc.' That's it. There's nothing to click on! It's so badass."

"So you're going to move to London so you can go to school in the hopes that for some inexplicable reason it will lead you to an agent for a book you haven't written about an answer you haven't found."

I nodded. "Exactly!"

I looked over at my father. He grinned.

My mother threw up her hands. "Well, at least you have a plan."

Later that night, having trouble sleeping, I wandered into our physics library. It felt good to be back in that room, warm and homey with its broken-in leather couch and the spines of countless books painting brightly colored stripes across the walls. It was comforting, being sur-rounded by all that wisdom. I noticed that my father had added a new bookcase, and I wondered, as I always did, when he found time to read so many books. I knew his job didn't afford him many spare moments, and it was dawning on me that he was using all of them to pursue our strange mission. That this wasn't just a hobby for him. Despite his mel-low, Zen-like demeanor, the appearance of that bookcase betrayed an urgency. A hunger. This thing *meant* something to him. Of course I had always known that, but there was something about seeing it rendered in rich mahogany that gave it weight—not only the weight of the wood or the tomes on its shelves, but the weight of his ambition, an ambition that was now my inheritance. I wanted it to add up to something. I wanted to earn his words, the ones he had leaned in to tell me all those years ago in the Chinese restaurant, to prove to him that he had chosen well when he chose me to be the heiress of his secrets, the beneficiary of everything and of nothing.

Scanning the shelves, a single book caught my eye: *At Home in the Universe,* Wheeler's essays on physics. I hadn't looked at it since our mystifying conversation with Wheeler in Princeton, so I curled up on the couch with a blanket and began to read.

Wheeler was searching for reality's building blocks, the most basic units from which life, the universe, and everything emerge. "No search has ever disclosed any ultimate underpinning, either of physics or of mathematics, that shows the slightest prospect of providing the rationale for the many-storied tower of physical law," he wrote. "One therefore suspects it is wrong to think that as one penetrates deeper and deeper into the structure of physics he will find it terminating at some *n*th level. One fears it is also wrong to think of the structure going on and on, layer after layer, ad infinitum. One finds himself in desperation asking if the structure, rather than terminating in some smallest object or in some most basic field, or going on and on, does not lead back in the end to the observer himself, in some kind of closed circuit of interdependences. . . . Is the universe a 'self-excited circuit'? Does the universe bring into being the observership, and the observership give useful meaning (substance, reality) to the universe?"

I adored Wheeler's writing: poetic, prophetic, and raw. A superposition of science and art, fact and dream. In his quest for ultimate reality, he viewed every inexplicable mystery as a clue. Wheeler was never going to shut up and calculate. He wanted answers, and he wasn't going to rest until he found them.

In the book, Wheeler had drawn a diagram: a capital letter *U* for "universe." The top of the right-hand side was the big bang, the swooping curve the history of the universe, time marching right to left and culminating at the top left of the letter, where a giant eye, a present-day observer, was perched, the product of billions of years of cosmic

evolution. In turn, the eye looks back across the abyss to the far tip of the letter, present to past, its gaze, presumably, giving meaning (substance, reality) to the universe. A self-excited *U*.

The universe creates us so that we can create it? Reality, for Wheeler, was a kind of Möbius strip, like Escher's hands drawing themselves. Was it just circular logic, or was it the only explanation that stood a chance of satisfying? The alternatives sure weren't. Either you've got some infinite regression of turtles atop turtles, and you're left wondering where the hell all the turtles came from, or reality comes to a screeching halt at some particle or field, and again you're stuck wondering, why *that*? Where did *that* come from? There was something distinctly more palatable about a causal loop, but I couldn't help thinking that the most satisfying thing of all, the thing that would stop all the but-how-did-that-get-there's in their tracks, would be a loop made of nothing at all.

I read on as Wheeler trekked through several thorny and seemingly disconnected issues in physics before I realized that he was carefully piecing them together into a grand, if unfinished, vision of reality—one so mind-boggling that it would have seemed insane had it come from anyone but him.

At its heart was the central mystery: the quantum. By freely choosing what to measure—particle or wave, position or momentum—an observer brings a bit of information into existence, transforming a smoky haze of possibility into a single unit of reality. Such bits, Wheeler said, were the building blocks of the universe. Physical reality, at bottom, was made not of electrons or quarks or strings, not of space or time, but of information—and information, at bottom, was made of observation.

But what exactly did Wheeler mean by an observer? Without careful clarification, *observer* was a dirty word. Fotini Markopoulou had made it clear that by observers she meant reference frames, possible points of view. That was their meaning in relativity, too—reference frames centered on rods and clocks. But in quantum mechanics things always got murkier, especially in interpretations that sought to give observers a privileged role, like the ability to create reality. Wheeler himself acknowledged the problem. "Any exploration of the concept of 'observer' and the closely associated notion of 'consciousness' is destined to come to a bad end in an infinite mystical morass," he wrote. And yet at times he teetered dangerously on the banks of the morass, his view of observers skewed far more toward minds than toward rods or clocks.

"Unless the blind dice of mutation and natural selection lead to life and consciousness and observership at some point down the road," he wrote, "the universe could not have come into being in the first place . . . there would be nothing rather than something." And later, "There are many to whom the idea of a world without any purpose— except what we and our fellow men agree upon—comes at first as a dreadful shock. Later comes the feeling of challenge; and then at last an inspiration: a feeling that we who felt ourselves so small amidst it all are, in the end, the carriers of the central jewel, the flashing purpose that lights up the whole dark universe."

I smiled at the poetry but cringed at the thought. As much as I would have liked to imagine myself carrying around some purposeful, radioactive jewel, I couldn't see how bringing consciousness into the mix could possibly help—not least of all because scientists don't know what consciousness is. Whatever it is, it's governed by the same laws of physics and composed of the same particles, fields, or information-theoretic bits as everything else. Of course, Wheeler agreed with that—in the first arc of his loop, the universe gives rise to observers through the blind dice of mutation and natural selection. Nothing mystical or supernatural happening there. But if that was the case, what could privilege certain physical objects (brains) and not others (rocks) as the "observers" capable of turning around and, with a long gaze to the past, creating the universe? I was confused, but I decided to accept Wheeler's premise and see where it led. I read on.

The O-word notwithstanding, Wheeler's view of a universe built bit by bit through observation and measurement suffered an obvious flaw: how could one observer possibly make enough measurements to create everything we see around us? Barring the hallucinatory deceptions of evil demons, the universe appears to be composed of far more bits than could possibly have been born through the handful of observations that a single observer—or even a planetful of observers—could ever muster. "Mice and men and all on Earth who may ever come to rank as inter-communicating meaning-establishing observer-participants will never mount a bit count sufficient to bear so great a burden," Wheeler wrote.

He offered a two-pronged solution. First, the total number of bits in the universe must be finite. I knew that general relativity ruled out this possibility—its spacetime manifold was continuous, with an infinite number of points between any two points, the kind of continuum that made my teenage brain reel with Zenoesque rebellion. You'd need an infinite number of bits just to specify the gravitational field, let alone the rest of the universe. But I also knew that general relativity didn't have the last word on spacetime; my foray into loop quantum gravity had taught me that. Zoom in on spacetime to a millionth of a billionth of a billionth of a billionth of a centimeter, and thanks to quantum mechanics the continuum dissolves. Zoom further and the notion of a point loses meaning as the fabric of reality is torn to shreds, threadbare as the dark center of a black hole or the singular breakdown at the beginning of time.

"Spacetime," Wheeler wrote, "often considered to be the ultimate continuum of physics, evidences nowhere more strikingly than at big bang and collapse that it cannot be a continuum." What's more, he said, "quantum fluctuations of geometry and quantum jumps of topology are estimated and calculated to pervade all space at the Planck scale of distances to give it a foam-like structure."

Second, count the contributions of all observers, not merely those living today but all who ever have been and ever will be. That was a bold move, considering its flagrant disregard for the usual rules of time, like that one that says the future comes *after* the past. But quantum mechanics already violated that rule, and no one knew that better than Wheeler.

In the late seventies, he had proposed a thought experiment known as the delayed choice, a twisted take on the classic double-slit experiment that was even more of a mindfuck than the original. In the classic version, the observer has a choice: either he can measure the interference pattern produced on a photographic screen after the photon passes simultaneously through both slits or he can place detectors at each slit to find out which path the photon travels, destroying any interference in the process. In Wheeler's updated version, the observer makes that choice *after* the photon has already passed through the slits. At the very last second, he can remove the photographic screen, revealing two small telescopes: one pointed at the left slit, the other at the right. The telescopes can tell which slit the photon has passed through, and they will always show that the photon took only one path. But if the observer decides to leave the screen in place, interference bands will form, showing that the photon has traveled through both slits. The observer's delayed choice determines whether the photon took one route or two . . . *after* it has presumably done one or the other.

In case that wasn't creepy enough, Wheeler proposed an even more extreme version. Imagine light traveling toward Earth from a quasar a billion light-years away, he said. A massive galaxy sits smack between the quasar and Earth, diverting the light's path with its gravitational field like a lens. The light bends around the galaxy, skirting either left or right with equal probability. Imagine, too, Wheeler said, that the arrival rate of the light is low enough that we can measure a single photon at a time. So we have our usual choice: we can keep a photographic plate centered at the arrival spot, where an interference pattern will inevitably emerge, or we can point a telescope to the left or right of the galaxy to see which path the photon took. Our choice determines which of two mutually exclusive histories the photon lived. It determines whether the photon traveled both paths or just one. We determine its route (or routes) from start to finish, right now—despite the fact that the photon began its journey *a billion years ago*. There's no sense in asking which path(s) the photon "really" took; there's simply no "really" until we choose which measurement to make. When we do, we create a past that stretches back billions of years.

"We used to think that the world exists 'out there' independent of us," he said, "we the observer safely hidden behind a one-foot thick slab of plate glass, not getting involved, only observing. However, we've concluded in the meantime that that isn't the way the world works. In fact we have to smash the glass, reach in."

Delayed-choice experiments have been carried out in laboratories, and each time they've worked just as Wheeler suggested. It's an established scientific fact: measurements in the present can rewrite history. No, not rewrite. Just write. Prior to observation, there is no history, just a haze of possibility, a past waiting to be born. "There is no more remarkable feature of this quantum world than the strange coupling it brings about between future and past," Wheeler wrote. If observations we make today can create a billion-year-old past, so, too, can observations made in the future help build the universe we see today.

If the total number of bits that constitute the universe is finite, and if we can count the bit-building contributions of all observers who will ever live, including those in the future, then it's at least plausible to suspect that observers create reality. That was Wheeler's vision, anyway. "Except via those time-leaping quantum phenomena that we rate as elementary acts of observer-participancy, no way has ever offered itself to construct what we call 'reality,'" he wrote.

All in all it was a pretty incredible story—way more interesting than the usual bottom-up one in which the universe starts out in some hot, dense state, inflates, evolves, and 13.7 billion years later accidentally gives rise to moderately clever chunks of gray matter in a boring, linear march from past to future, cause to effect. But Wheeler's story left dangling a host of unanswered questions. Like, what counts as an observer? What gives the observer its special reality-building status? What physical mechanism allows observers to create bits of information through measurements? What does the "boundary of a boundary" have to do with a self-excited universe? And if observership is required to give meaning (substance, reality) to existence, who observes the observer?

Wheeler's privileging of the observer followed Bohr's view of quantum mechanics, with observers standing outside the systems they're observing. At the same time, his self-excited circuit was a closed loop, with internal observers looking back to observe the same past that bore them, an ouroboros swallowing its tail. So which one was it? Inside or out?

Finally, I was left wondering about the very aim of the thing. If, as Markopoulou had told me, we each have our own light cones, rendering ordinary binary Boolean logic unfit for cosmic use, how could the grand total of all observers that will ever live join together to create a single object called "the universe"?

Wheeler's book didn't have all the answers, but I had the sense that it was raising the right questions. "Can we ever expect to understand existence?" he asked. "Clues we have, and work to do, to make headway on that issue. Surely someday, we can believe, we will grasp the central idea of it all as so simple, so beautiful, so compelling that we will all say to each other, 'Oh, how could it have been otherwise! How could we all have been so blind so long!'"

A soft knock on the door jolted me out of Wheeler's world.

"You're up?" my father asked, peering in.

"I couldn't sleep."

"Come outside," he said. "There's supposed to be a meteor shower."

I grabbed a sweater and some sneakers and we tiptoed downstairs so as not to wake my mother. Outside, we walked up the driveway toward the street, stopping once we had cleared the canopy of the maple tree and had a perfect view of the cloudless, starry sky. It was 3:00 A.M. The houses were dark and the street was drenched in a uniquely suburban quiet, the click of cicadas and the hum of air conditioners saturating the thick summer air.

We stood side by side, looking up, waiting for the stray dust of a comet's tail.

"I think it's great you're going back to school," my father said, his eyes on the sky.

"*New Scientist* is based in London, too," I said. "I'm hoping that by being there I'll get to write more articles so I'll have a steady stream of excuses to talk to physicists."

I blurred my gaze, trying to expand my peripheral vision.

"There's one!" we shouted simultaneously as a flash of light streaked across the sky.

"Jinx," I said with a laugh.

"Do you ever think you want to be another kind of writer?" my father asked.

The question caught me off guard. "What do you mean?"

"You went to New York with the intention of becoming a novelist or a poet," he said. "And it's so great what you're doing with journalism. I'm just worried we're taking it too far, that I'm steering you too far off course. Are you sure this is what *you* want to be doing?"

I thought for a moment in silence. He wasn't wrong. I *did* want to be another kind of writer. It had never been my dream to report on physics, an ill-fitting journalist hat balanced precariously on my head. My dream had always been to *write*—to take amorphous thoughts and morph them, to ink them on the page, to give them form and heft and permanence, like turning nothing into something, however small. Writing, for me, was about muddling through ideas, turning them over, viewing them from every angle to see where they led, even if they only led back to themselves. My favorite stories and poems shined a spotlight on the writer's thought process, exposing all of its cracks and contradictions. But the writing I did as a journalist was just the opposite. Its light revealed only the end products of thought, the conclusions. Science journalism's express goal was to hang over the writer's mind a veil so opaque that the reader would mistake the writer's thoughts about the world for the world itself—the world as seen from an impossible God's-eye view, a paradigm of objectivity and at the same time a lie. For me, hiding the writer's thoughts strips writing of its greatest gift: its ability to grant us access to other minds. Writing has the potential to be magical because it lets us see the one thing we can never see; it opens our eyes to that feature of the world that is most profoundly invisible. Writing is the rock we can kick to refute our loneliness, to cure the claustrophobia that comes from being trapped inside a one-sided mind.

But it was okay with me that my journalism wasn't exactly the kind of writing I had dreamed about. Journalism wasn't my goal, it was my disguise. It was my laminated invitation to the ultimate reality party, and I wanted to see how far it could take us.

"This thing is important to me," I said as another streak of light caught my eye. "The writing will have to come later. When it does, I'll have something to say."

He smiled.

"What about you?" I asked.

"What about me?"

"You read every new physics book the minute it's published, every science magazine, every journal. You added a new bookcase. Is this taking you away from *your* work?"

"I guess I feel it's equally important," he said. "Maybe more. Fungal infection of the lung or nature of reality?" He paused. "Sometimes I wish I were an astrophysicist. If I were a little younger, I might consider switching."

"You still could," I said.

He didn't respond.

We stood in the street in silence. Comet dust blazed across the sky.

Thanks to yet another not-particularly-well-thought-out plan, I was moving to London. If Bostrom had gone to the London School of Economics to study the philosophy of science, decided that reality was probably a simulation, and befriended John Brockman, I thought, it had to lead to something. I had no intention of giving up the journalism gig, but I wanted to step back and see the big picture. I didn't want to lose sight of the goal of this thing.

"Never get too comfortable," my father had once told me. "As soon as you feel settled, it's probably time to take things to the next level."

Wheeler had said that philosophy was too important to be left to the philosophers. But I figured it couldn't hurt.

So I rented a flat on the first floor of a charming white townhouse in a posh little cul-de-sac in Notting Hill. My parents and brother came with me to check it out.

"It's a bit small," the real estate agent warned us as she jiggled the key in the lock, "but very modern."

She opened the door and I peered in. It was indeed a bit small.

I turned to my family. "Maybe we should go in two at a time?"

My mother and I walked into the studio apartment. It really was pretty modern. Everything was shiny and new, like a luxury apartment that someone had accidentally stuck in the dryer and shrunk.

"The wood floor is just lovely," the agent pointed out. I nodded. It was, although for that square footage they could have encrusted it in diamonds without having to jack up the rent.

I looked around: there was a one-seater couch, which someone more pessimistic might have called a chair, and a round desk/coffee table/dining table just big enough for a laptop. Or a plate.

"Where's the bed?" I asked.

The realtor pointed up over my head. I followed her gaze to a small, steep ladder leading up to a loft bed above the couch. "Well, that works," I said, noting the two-foot clearance between the bare mattress and the ceiling. "I'll just have to remember not to sit up."

"And this is the kitchen?" my mother asked, as if perhaps there were a real kitchen hidden in a closet somewhere.

The agent nodded. "And everything is new!"

"Everything" consisted of a miniature sink, miniature stove, and miniature refrigerator.

"Maybe there's a miniature grocery store nearby that sells miniature food?" I offered up helpfully.

"There's no freezer?" my mother asked, knowing full well the answer.

I shrugged. "But how convenient that you can reach everything in the kitchen from the couch!"

"It's brilliant how they mounted the telly on the wall here," the agent said. "It doesn't take up any space, and you can see it from every spot in the flat!"

"Yes." I smiled. "A real feat of engineering."

"I'll step outside so Dad can come in," my mother said, sounding defeated.

My father came in and gazed around, unsure of what to say.

"Aren't the wood floors lovely?" I said, nudging him.

He nodded, then whispered, "Do you think this place is subject to quantum effects?"

"How much is the rent?" I asked the real estate agent, then winced at her reply. The place was more expensive than any apartment I had rented in New York, and a fraction of the size. But it was in a beautiful neighborhood and a minute's walk to the Tube station that would take me directly to campus. Besides, I wasn't planning on bringing much stuff. Looking around, I couldn't imagine what kind of place I'd get for less.

"Okay," I said, "I'll take it."

5

Schrödinger's Rats

If I had come to London to ponder the nature of reality, I had clearly come to the right place. In my philosophy of science class, we discussed it endlessly. Is there a reality? Is it sitting "out there," independent of us? If so, what is it made of? How can we distinguish it from mere appearances? Is there any hope we'll ever know it at all?

In class, we debated the merits of realism and antirealism. Realism is the commonsense belief that scientific theories describe true things about the world—a real world that exists whether or not we're looking—and that electrons, quarks, dark matter, and whatever other objects appear in our best theories, whether or not they can be observed directly, are *real* objects, the true ontological furniture of a singular, mind-independent world.

Antirealism is an umbrella category for all sorts of ideas that reject realism in one way or another. There's Kantean antirealism, which says that while there is a real world out there independent of us, there's no way for us to know it. There's Berkeley's *esse est percipi,* the more radical claim that behind appearances lurk more appearances, that objects are made not of atoms but of thoughts. There's social constructionism, which says that reality and truth are whatever we agree to call reality and truth, a theory that reminded me of something my New School postmodernist friends would say and then argue that it had to be true because they believed it, even after I had pointed out that by not agreeing with them I had, by their own definition, proven them wrong. On the saner side there's instrumentalism, which simply states that science is a tool for predicting the outcomes of experiments; whether or not there is a reality, and whether or not we can access it, is entirely beside the point.

I had already discovered that instrumentalism was a common position among physicists, who always seemed to squirm at any mention of the R-word. *It's the philosophers' job to worry about reality,* they'd say. *We just calculate and predict and test.*

No matter how many times I heard that, it always struck me as total bullshit. Okay, maybe if you were an electrical engineer or a surgeon or a meteorologist you'd just be concerned with predictions and the outcomes of experiments, but the people I was hearing this from were physicists. *Theoretical* physicists. People who were dealing with black holes and multiple universes and glitches in the simulation. Maybe when you work in theoretical physics, you feel the need to overcompensate by pretending to be as no-nonsense as a refrigerator repairman, but at the end of the day, who are you kidding? You stay awake nights worrying about how matter behaves at length scales a millionth of a billionth of a billionth of a billionth of a centimeter in six extra dimensions undetectable by any foreseeable experiment, but you don't care at all what reality is? Please.

Given my propensity for worrying about simulations and shadows and butterfly dreams, I wouldn't have guessed that I would find myself advocating a strict realist view. Then again, I was a self-proclaimed reality hunter, so entertaining any antirealist ideas would be like shooting myself in the foot. Besides, at times the arguments for antirealism struck me as utterly absurd. The pinnacle of absurdity came one afternoon when a girl in my class argued her antirealist position from a feminist standpoint.

"Wait, did she just say 'feminist'?" I asked the guy next to me. "Feminist physics?" I couldn't imagine where this was headed.

"Not only is science a socially constructed enterprise, it is also explicitly male-centric," she explained to the class. "Think about the terminology. Particles are represented as *balls,* and they interact with each other through *forces.*"

Seriously? Balls? I coughed to cover my snickering. Judging by her expression, this was a very serious matter.

"So if physics is socially constructed," one guy began, "regardless of whether it's constructed by men or women, you don't think it corresponds to reality at all?"

"No, I don't," she replied.

I couldn't stop myself from joining in. "So how exactly do, say, airplanes fly?"

"Because we all agree that they do," she responded.

I blinked. "Are you serious?"

Somehow, seemingly instantaneously, the classroom had divided itself into teams—realists versus antirealists. We even shuffled our desks around to make it known exactly which side of this fight we were on.

Antirealism had seemed a rather insane position until I felt the sting of its best right hook: every previous scientific theory ever devised in the history of science has, until now, turned out to be wrong. So what kind of morons would we have to be to believe that our current theories are the exception, the one time mankind—or womankind— has ever gotten it right? And if theories are always turning out to be wrong, how can they possibly be telling us anything about the true nature of reality? I learned that this rather fatal blow is known in philosopher-speak as the "pessimistic meta-induction," which just means that with some solid inductive reasoning it becomes obvious that science is a hopeless enterprise.

It was a depressing thought, but luckily realism had its own uppercut ready, the argument that I had unknowingly made against the girl who was mad about balls: if scientific theories don't describe at least part of the true reality out there, then the success of technology— not to mention the success of a theory's bold, novel predictions that go way beyond whatever observations were fed into it in the first place— has to be chalked up to a miracle.

Okay, so all theories turn out to be wrong, but the technologies we build based on those theories miraculously work. The pessimistic meta-induction and the no-miracles argument formed a kind of stalemate, and philosophers had been bickering about it ever since. One philosopher, however, had found a middle ground. He happened to be sitting in an office down the hall.

I had barely unpacked my things when I started hearing the noises. Scurrying noises. A rustling. On a few occasions I swore I saw a blur of motion out of the corner of my eye. Then one night, lying in the loft, half asleep, I heard a guttural sound, the kind of sound a cat makes before it jumps, a sort of revving of the motor. It startled me and I sat up without thinking, bashing my head into the ceiling. By the time I managed to turn on the light, whatever had made the sound was gone.

It wasn't hard to guess what was happening. This was London, after all. I had read somewhere that wherever you stand in the entire city, you're never more than twenty yards from a rat. There were 50 million of them. That was like seven rats per human. Could seven rats even fit in my flat? Not if this one was big enough to make guttural sounds, I thought. I tried to go back to sleep, assuring myself, unconvincingly, that rats can't climb ladders.

In the morning I went to the hardware store, where I found a disconcertingly large selection of rodent control devices, a whole wall of them. I was gazing at it in awe and confusion when the sales guy asked if he could help.

"I don't want to be cruel," I said. "I mean, I want them out. If I could reason with them, I would. I just want something that doesn't make me a terrible person."

He nodded. "Then I'd avoid the glue traps."

He showed me a trap that consisted of a box that you rig with bait and when the rat goes in to get it, it triggers this sort of garage door that falls shut, locking the rat inside, where it waits for you to take it outside and set it free. Not into the wild per se, but at least headed for someone else's flat. I bought two.

I heard them rustling around down there as I drifted off to sleep that night. *Esse est percipi. Esse est percipi.* I chanted the phrase like a spell, hoping it might transform any ontologically valid rats into vaporous thoughts I could sleep off by morning. Perhaps the real estate agent had meant to say that this flat was modern *and* mind-dependent. Reassuringly, I had yet to actually perceive any living creatures; their existence was nothing more than a pessimistic induction. *Cogito ergo rats.* Maybe the programmers were messing with me. Maybe the strange sounds were just glitches in the simulation. Or maybe my dad had been right and this place was subject to quantum fluctuations, the sudden but fleeting appearance of rodents from an ever-churning vacuum. Maybe as long as I didn't observe them, they'd be stuck suspended in a kind of quantum mousetrap, half real, half illusion. Schrödinger's rats.

But in the morning, when I observed them, the traps were empty.

John Worrall had a sweet look about him, like the kind of guy who could broker peace among feuding academics, or the kind who would one day become the leader of a philosopher-of-science-made rock band called Critique of Pure Rhythm. He had started out in statistics but been lured to philosophy by Karl Popper, who had founded the philosophy of science department here. In 1989 Worrall published an article in the journal *Dialectica* arguing for a middle ground between realism and antirealism. He called his view structural realism, and claimed that it held the best of both worlds: it could explain science's success without invoking miracles and account for its pessimistic progression from one wrong theory to the next.

The problem, Worrall explained, was that realists were being realists about the wrong things. In fact, "things" were precisely the problem. Realists talked about a real mind-independent world, out there, composed of real things such as atoms and tables and rats. But when you look closely, scientific theories aren't about "things" at all. They're about mathematical structure.

A mathematical structure is a set of isomorphic elements, each of which can be perfectly mapped onto the next. The notations 25, 5^2, and $(27 - 2)$ all share the same mathematical structure. The structure isn't any particular number—it's the whole set of equivalent representations of a number, the steady, singular truth behind a multitude of mere appearances. Sets are more fundamental than the numbers themselves.

All of mathematics—all of structure—comes down to sets? I wrote in my notebook. I remembered reading somewhere that the entire number line could be built from the empty set: the set containing nothing. Inside the empty set is nothing. Zero. But the set that contains the empty set is not empty. It contains one element: the empty set. It's the number 1. Not merely equal to 1, but the very definition of the number 1. The set that contains both the empty set and the set that contains the empty set is 2. Ad infinitum. Or ad nothing.

The number line was nothing more than a series of nested sets, and in its hidden center was nothing. Worrall said that physics was about mathematical structure. Set theory said that mathematical structure was about nothing.

The idea that you could build the number line from the empty set—was it a clever trick or was it telling us something profound about the universe? Was it telling us how to turn nothing into something? *Put brackets around it. A boundary.* Somethingness emerges from a change in point of view. Inside to outside.

I wasn't sure how you'd apply that lesson to something like a universe, something that doesn't have an outside. *One-sided coin, the side of things.* How do you make something out of that? Even if you could, you'd still be stuck with Russell's paradox. The barber shaves every man who doesn't shave himself—so who shaves the barber? If he shaves himself, he doesn't shave himself, and if he doesn't, he does. The issue wasn't about facial hair. It was about the paradoxes that arise if sets can contain themselves. When you take the view from outside the brackets and try to shove it back inside.

Worrall attributed structural realism to Henri Poincaré, who in 1905 wrote, "Equations express relations, and if the equations remain true, it is because the relations preserve their reality. . . . The true relations between these real objects are the only reality we can attain." Theories are just sets of mathematical relations—equations related by isomorphisms. By equals signs. Quantum field theory doesn't talk about hard little (*cough*) balls called particles; it talks about "irreducible representations of the Poincaré symmetry group." If it makes it easier to picture those irreducible representations as little spheres, that's your right. But if that picture doesn't hold up in light of new evidence, don't get mad at the theory. Quantum field theory is a group of mathematical structures. Electrons are little stories we tell ourselves.

Of course, we need stories. There's a reason "42" is not a satisfying answer to life, the universe, and everything. Structure alone doesn't quench our existential thirst. We want meaning. And for our brains, meaning comes in the form of stories.

Still, it's important to separate what theories mean to us from what they actually say. That was Worrall's point. Theories never talk about objects—only our *interpretations* of theories do. Theories themselves only talk about mathematical structure. And if we're realists about structure, the pessimistic meta-induction no longer applies.

When theories turn out to be wrong, Worrall said, it's usually our interpretative story that's wrong—not the structure. Take gravity. According to Newton, gravity is a force that masses exert on one another from a distance. According to Einstein, it's the local curvature of spacetime. The two ideas are contradictory. Both couldn't be right, so clearly, the antirealists said, Newton's theory wasn't describing reality at all, a fact that made it pretty hard to explain how he was able to predict the motions of the planets. Worrall disagreed. If you take away the interpretations and just look at the math, it's a whole different game. When gravity is weak and velocities are low, Einstein's equations give way to Newton's. Newtonian gravity is the low-energy limit of general relativity. Newton had the wrong story but the right structure—only it turned out to be a tiny corner of something much, much bigger. We don't need miracles to understand why Newtonian gravity worked; it was successful because it had homed in on a small piece of reality's structure. Einstein discovered a bigger piece, and there's still more to be found.

The same went for quantum mechanics. Although its *description* of the world is drastically different from that of classical mechanics— where particles have simultaneously defined positions and momenta, cat obituaries are far more straightforward, and demons can predict the future to infinite accuracy—its mathematical structure reduces to that of classical mechanics when physical systems are large compared to the size of Planck's constant. As one theory surrenders to the next, physical interpretations are left behind in ruin, but mathematical structure persists. Scientific progress isn't a parade of miraculously wrong theories—it's an optimistic snowball, gathering the structure of reality as it rolls.

The Kantian Structures

Several more rustling nights were followed by several more ratless mornings.

I hunted around the flat, looking for any rat-sized entryways. I taped up the tiniest cracks in the walls and stuffed the openings around pipes with steel wool. To be extra-cautious, I stacked books all along the perimeter. Just in case they could jump the books, I created various obstacles for them to encounter on the other side. The whole setup got quite elaborate, with makeshift forts and moats and the garage-door traps in the center. The rats might be clever and resilient, I thought, but I had physics books and duct tape and opposable thumbs.

Still, the rustling continued, and one night I was awoken by the thump of a book falling from its fortress. In the morning I saw that it was Julian Barbour's *The End of Time*. I wondered if the rats were trying to tell me something.

According to Worrall, I didn't have to lend any ontological credence to the individual rats—all I had to worry about were the structural relations among them. That made me feel slightly better, but I still wished I had it in me to be a social constructionist. Then I could get rid of the damn things just by refusing to believe they were there—philosophical extermination. Unfortunately, I believed that physics was what made planes fly and rats scurry. Given the data points of shuffling noises, motion in my peripheral vision, falling books, apocalyptic messages, and London, I had to face facts: the existence of rats, quantum or otherwise, was the simplest explanation.

If I couldn't excise them with Occam's razor, I was going to have to resort to more conventional methods. "Okay," I told the hardware guy, "give me the traps that will kill them. But kill them quick, so they don't suffer."

He helped me load my basket with rat traps—standard spring-loaded mousetraps, only bigger. I bought seven.

I went home and rigged the traps. It wasn't as easy as it looked. They're supposed to be this perfect, impossible-to-improve-upon invention, but I nearly lost a finger. Eventually I got them all set up and baited them with peanut butter. I had read somewhere that rats love peanut butter. Then I grabbed my suitcase and got the hell out of there.

I was sitting in a Japanese restaurant in Holborn, in central London, waiting for Michael Brooks.

After laying the traps, I had checked in to a hotel a few blocks down the road on Notting Hill Gate. I didn't want to be around when the rats discovered the peanut butter and I figured I'd enjoy a few extra square feet. After settling in, I had emailed Brooks about a *New Scientist* article and mentioned that I was now living on his side of the pond. "Since you are here in London," he had replied, "why don't we meet for lunch?"

Brooks arrived at the restaurant along with Valerie Jamieson, another physics editor from *New Scientist,* who introduced herself with a melodic Scottish brogue. We ordered drinks and sushi, which arrived at our table on a large wooden boat. As we plucked fish from the deck with our chopsticks, we chatted about life in London and in the universe at large.

"What's your view of inflation?" Brooks asked me.

Having just shoved a piece of salmon into my mouth, I had a moment to think. Inflation. On one hand, I understood the appeal. As Guth liked to say, it was the ultimate free lunch: a universe that blossoms from some primordial seed and just keeps on growing, gravity's negative energy offsetting the limitless creation of infinite space, which stretches quantum ripples into astronomical veins, the gravitational lifeblood of stars and galaxies.

On the other, inflation couldn't explain why the universe exists at all. From whence the primordial seed? It assumed the existence of the inflaton from the start, not to mention the very laws of physics, and in its heart it wasn't quantum. It didn't account for what internal observers could see or explain why nothing looks like something. Its logic was Boolean, its view God's-eye, its approach bottom-up; it was helpless in the face of quantum dragons. Besides, there was that disturbing low quadrupole. WMAP hadn't found any large-scale temperature fluctuations—not what you'd expect from an inflated universe.

I swallowed the salmon. "I think it has more problems than people admit."

I felt weird offering my opinion, as if maybe I wasn't supposed to have one, and as the conversation continued, I couldn't help feeling a little guilty. Brooks and Jamieson had PhDs in physics, and they were real journalists to boot. I was nothing but a poseur trying my best to fit in. But the strange part was, I felt like I *did* fit in. As we compared views of inflation and its discontents and swapped stories about our run-ins with quirky cosmologists, it dawned on me that there was a whole community of people out there—*writers*—who actually wanted to talk about physics over sushi. Science journalism was supposed to be my disguise, but the mask fit a little too perfectly today.

As I snagged a piece of port-side tuna, I couldn't help but wonder what my father was doing now, on the other side of the ocean. It was morning there. He was probably getting ready for work.

One . . . two . . . three. Turn the key. Take a deep breath. Open the door.

After a week living it up in a hotel, it was time to return to my miniature flat and dive back into ultimate reality. But as I stood frozen outside my door it occurred to me that when I had set up the traps, I hadn't fully considered the end result. I had wanted the rats gone, but they weren't gone. They were right there on the other side of the door, possibly seven of them, with snapped necks and shocked faces, the traps sprung and sated like bottomed-out guillotines, the morbid remains of a rodent revolution, a noble troop brought down by Sainsbury's peanut butter. And what exactly was I supposed to do with them? Hold a mass funeral? Fire twenty-one shots from a tiny, tiny cannon? Run?

One . . . two . . . three . . .

Fuck.

Is there anything in there I can't live without?

After several more failed attempts, I finally turned the key, cringing as I pushed open the door. Inside, I surveyed the gruesome scene. It was even more horrifying than I had imagined. Every last lick of peanut butter was gone, and the traps, still rigged, were empty.

Worrall's structural realist philosophy had struck a chord in me. If I wanted to find the truth about ultimate reality and the nature of the something that allegedly came from nothing, it was going to be crucial to separate our descriptions of the world from the world itself, what physics really says from the meanings we ascribe to it. But I was confused. Worrall had said that theories talk about mathematical structure, and not about objects. Did that mean that objects don't exist at all, or merely that our scientific theories can never tell us which objects are the real ones? Was it a claim about what we can know or about what actually exists? Was it epistemological or ontological?

"Epistemological," Worrall answered definitively when I asked him. "I have a lot of trouble with the idea of relations without relata. And anyway, I feel that we should generally be silent about metaphysical issues. We think about what reality is probably made of via physics. All structural realism does is insist that we should not think that we have any grasp of reality over and above what our current theories tell us."

At first, Worrall's objections to an ontological structural realism seemed fair enough. After all, what *could* it mean to talk about relations without relata? If the world is made of mathematical relationships, mathematical relationships among *what*?

Maybe they're not among anything. Maybe the relationships are all that exist. Maybe the world is *made* of math. At first that sounded nuts, but when I thought about it I had to wonder, what exactly is the other option? That the world is made of "things"? What the hell is a "thing"? It was one of those concepts that fold under the slightest interrogation. Look closely at any object and you find it's an amalgamation of particles. But look closely at the particles and you find that they are irreducible representations of the Poincaré symmetry group—whatever that meant. The point is, particles, at bottom, look a lot like math.

If structure is all that our theories can ever tell us about the world, forever veiling some unknowable ontology beneath, then our pursuit of ultimate reality was completely hopeless. Accepting Worrall's epistemological structural realism was like retreating right back into Bostrom's computer waving a simulated white flag.

On the other hand, if structure is all that exists—if the world really is made of math rather than things—then physics *can* tell us everything there is to know about ultimate reality. Ontic structural realism was our only hope. My father's and my mission hung in the balance.

"Does *anyone* think that structural realism is ontological?" I asked my philosophy professor after class one day.

He thought a moment, then nodded. "You ought to talk to James Ladyman."

The fact that all the peanut butter was gone was pretty damning evidence that the rats were ontologically valid, but I knew that I couldn't logically defend my inference to the best explanation. Sure, it seemed the most likely conclusion, but blunting Occam's razor was the undeniable fact that there were an infinite number of possible unobservables that could explain the peanut butter's disappearance—though I was having trouble imagining what the hell they might be. Was British peanut butter especially prone to rapid evaporation? Had seven dollops of anti–peanut butter spontaneously sprung from the vacuum, annihilating the store-bought stuff in a sudden burst of light? This underdetermination of theory by the data was bolstered by the null results of the traps, which just sat there, empty, rigged, full of potential energy, itching to go kinetic. I had learned in philosophy class that inductive reasoning was totally indefensible; all the clues in the world just wouldn't amount to much. The only way to claim that the rats were categorically real was to logically deduce their existence from some set of self-evident axioms, to render them necessary and not merely contingent. Of course, by those standards, a rat could be sitting in front of me waving its paw and my existence claims still wouldn't hold water. I could hear the contingent bastards scratching at the walls, scurrying in the ceiling two feet above my head.

"Okay," I told the hardware guy, "I'll take the glue traps."

"I'll tell you what reality is not. It's not made of little things."

James Ladyman was sitting on the floor of his hotel room. "We can't help but think that way, but it's not what reality is like."

I was swaying in a creaky swivel chair. We had met in the bar of the Holiday Inn, where Ladyman was staying while he was in town for a conference on metaphysics. Despite Worrall's warning that we should be silent on metaphysical issues, it seemed a whole slew of philosophers weren't ready to keep their mouths shut. The hotel bar had proved too noisy for a discussion of the nature of reality, so we had retreated to his room, where he was now sitting on the floor, stretching his legs. With a headful of dreadlocks spilling halfway down his back, it would be easy to mistake Ladyman for the bongo drummer in a reggae band, though his British accent carried the distinct melody of academia.

"But how do you go from saying 'structure is all we can know' to 'structure is all that exists'?" I asked.

"The motivation for me was looking at contemporary physics and realizing that it doesn't support an intuitive picture of unobservable objects. You could say particle physics is about mesons and quarks and baryons and electrons and neutrinos and so on, but when you get beyond the pictures they draw and just look at the theories, it's very difficult to interpret those theories as being about particles, right?" Ladyman said. "So the thing about particles is that they're not particles. . . . If you want to know what the ontology is, look at what the theory is saying. Don't try to overlay the mathematical structure with some kind of folky, homely imagery."

Like *balls*?

"So physics itself steered you to interpret structural realism ontologically?" I said, grinning.

Worrall had developed structural realism to respond to a philosopher's squabble. If Ladyman's version was driven by physics rather than pure philosophy, it stood a better chance of being true.

"Both quantum mechanics and relativity profoundly challenge our intuitive idea that the world is made of objects," he said. "Quantum particles have all sorts of problems about their individuality: entangled states, quantum statistics. Then in general relativity, spacetime points don't seem to be the ultimate reality; the reality is something more like a metric field. In both cases we are pushed away from an ontology according to which you drill down and find little things that everything is built of."

It was a good point. Not only were quantum statistics weird, but they made it pretty much impossible to think of particles as "things." If you have two electrons, there's no way to distinguish between them. Electrons have no known substructure; they're defined solely by their rest mass, spin, and charge, which are the same for every electron. Electrons, by definition, are identical. Of course, you'd think you could distinguish them just by their locations in space and time—an electron *here* is not the same particle as an electron *there,* by virtue of their being in different places. That trick might have worked in classical physics, but not quantum. Quantum particles don't have well-defined positions in spacetime, only probabilities for appearing in various locations, the locations themselves smeared out by uncertainty. The result is that quantum physics renders elementary particles literally indistinguishable, a fact that becomes pretty important when you're calculating probabilities. If each of the seven rats in my flat inevitably ended up stuck to a glue trap, then I'd say there was a one in seven chance of finding a given rat on a given trap. But if the rats really were quantum, there would be a 100 percent chance of finding any rat on any given trap. If you're placing bets, knowing whether you're dealing with classical or quantum statistics makes a pretty big difference. And what would it even mean to call a rat a "thing" if it has no individuality on which to pin its "thingness"?

General relativity only exacerbated the situation. My father had taught me that to keep accelerated and inertial reference frames on equal footing—to turn a curve into a line—you have to bend the paper. The problem is that you can bend it in endlessly different ways and produce the same results, a fact made possible by Einstein's central principle, general covariance. Different configurations of the paper can all correspond to the exact same physics, a kind of underdetermination that led not only Ladyman but also Einstein himself to believe that the paper itself—the "thingness" of spacetime—wasn't ultimately real. The only reality lay in the spatiotemporal relationships traced by the paper's curves. The metric. The structure.

The more I thought about it, I realized that such underdetermination in ontology runs rampant in physics. It reminded me of Dirac's holes. In the early days of quantum mechanics, Paul Dirac had come up with an equation that made the Schrödinger equation compatible with special relativity. The only problem was that the equation allowed particles such as electrons to have negative energy, something that clearly didn't happen in the real world. To save his equation, Dirac imagined that the quantum vacuum was a sea in which every possible negative energy state was already filled, leaving only positive energy states accessible to electrons. But a new problem arose when Dirac realized that, if excited, the negative energy states could transform into positive energy states, leaving an empty hole in the negative energy sea. The hole would have all the properties of an electron, but with a positive charge.

With his holes, Dirac had predicted the existence of antiparticles. What Dirac had considered a positively charged hole physicists nowadays think of as a positron—an object in its own right, not merely a hole. But the point is, the math never changed. Only the interpretation did. Physicists could just as well stick with the hole picture and they'd still come up with all the predictions for anything they might test in a lab. You can think of a positron as a thing or as an absence, two ontologies about as opposite as you can find, but from the point of view of mathematical structure, they're exactly the same. I wanted to run to philosophy class to tell my classmate the good news: *You don't have to talk about particles as little balls! You can talk about them as holes!*

"How do you define structure?" I asked Ladyman.

"I'd say it's a system of relations. But then people say, 'Well, a system of relations is among objects so related,'" he said, echoing Worrall's critique. "But quantum mechanics and general relativity don't seem to be based on an ontology of objects first and then relations between them sort of sitting on top. It's really more the other way round. The objects are just nodes in the relational structure or something."

Balls and holes are merely *descriptions;* they're instantiations of structure, not the structure itself. The real thing is a mathematical relationship. If you're a realist about structure, the underdetermination crisis is averted.

"Does that mean the physical world is made of math?"

"It might be that at a certain level of description it becomes impossible to adequately represent the world other than mathematically. If you read popularizations of, say, quantum field theory, at a certain point the writer has to say, 'We can't explain this but it turns out that such and such . . .' The resources they've got to communicate are not adequate because they make people think that we're talking about little particles, and we're not. So the more fundamental a description of reality becomes, the more mathematical it becomes, and the distinction between the abstract and the concrete becomes sort of unstable. On the other hand, I don't want to say that the concrete universe is made of maths. But its nature might be so far removed from our common-sense notion of a concrete physical object that maybe it is less misleading to say it's made of maths than to say it's made of matter. These are very difficult issues. I really don't know."

"The way I picture it is like reality is the bottom layer, and then you have a layer of mathematics on top, and there's a one-to-one mapping between the two," I said. "And on top of that you have language, but there's not a one-to-one mapping between the mathematics and language, so something gets lost in translation, like you said. But then my question is, if there's really a one-to-one mapping between math and reality, doesn't that by definition mean that they are the same thing?"

"I suppose the problem at the moment is that we don't have a one-to-one mapping, because even our best theories aren't completely accurate," Ladyman said. "So yeah, you might think, if we eventually did have a one-to-one mapping, what would be the grounds for denying that reality was mathematical? I'm not really sure. I suppose I'm very skeptical of anything in philosophy that purports to explain the difference between abstract maths and maths that's substantiated. Because in the end, what could we possibly explain that difference in terms of? Like, I reject the question, 'What breathes fire into the equations?' Because anything you say is just gonna be figurative, right? Because you'd say, 'Well, there's the abstract maths and then the actual universe is a sort of substructure of all the possible structure there could be. So what's the difference between the uninstantiated structure and the instantiated structure?' Well, the philosopher will say there's a primitive instantiation relation or something—you could invent some metaphysical language to talk about it, but to me that's no different from saying that some of the maths has pixie dust in it. It's not going to do any work. Because what could it possibly connect to that would have any meaning? If you ask questions in science like 'What causes an earthquake?' you appeal to conceptual resources and those are nonempty because they're tied to observation. But maths—pure maths isn't tied to observation. If the theory of everything is a mathematical theory, how would you test it? It would have to have some content that has to do with something other than mathematics."

"I've heard some people say that if you really had a theory of everything, it wouldn't be testable," I offered up.

"Right, hmmm," Ladyman said, thoughtful. "That's interesting."

I could hardly believe I was defending the notion that the world was made of math, given my teenage years as a strict nonbeliever. I was glad my mother wasn't there to get the satisfaction.

But like Ladyman, I didn't see what the other option could be, not if we followed Worrall's advice and listened to "what our current theories tell us." As far as I could see, our current theories really were telling us that reality is made of math. That objects give way to equations, that thingness melts to abstraction. Given the drastic underdetermination of ontology in general relativity and quantum mechanics, Ladyman's version of structural realism seemed to be the only lifeboat capable of keeping us afloat in a sea of existential crisis and contradiction. As I thought about it, I realized how surprising that was. I mean, you'd think it would be the other way around—that as our theories of physics got better, snowballing ever closer to ultimate reality, they'd offer us increasingly clearer pictures of the objects that ultimately constitute reality. Instead it seemed that the only clear-cut message they offered was that "objects" aren't the right ontology at all. Not only was physics undermining every intuition we have about the world, it was also weeding out philosophies. From where I was sitting in a nondescript room in a nondescript hotel, ontic structural realism seemed to be the only one left standing.

As I walked the London streets, the sky a dull gray overhead, the pavement slick with rainwater, I looked around at the so-called world. It was crazy to think that everything—the majestic townhouses and double-decker buses, the sprawling green of Hyde Park and the white stone at Marble Arch—was made not of physical things, but of math. Then again, wasn't that exactly what Wheeler had been saying all along?

It from bit: the world is made of information. Not *described* by information, but *made* of information. *A house is made of bricks but the bricks are made of information*. And what was information if not mathematical structure?

Being a realist about objects was kind of like believing that *love* and *amor* are two totally different things just because they look and sound different. You have to know the rules of translation between English and Spanish to discover that the two words are equivalent—there's a one-to-one isomorphic mapping from one word to the other, a mapping that preserves some underlying structure, not *love* or *amor* but the concept to which they both refer. *Love* and *amor* are words. Descriptions. What's real is what survives the translation, the structural relationship between them. We can't give it a name. Giving it a name would trade structure back for description. Giving it a name would require choosing a single language, a preferred coordinate system, violating general covariance, breaking the symmetry of a linguistic spacetime.

Science is about structure. The stories we tell and the images we create to describe the structure are up to us. The key is to not mistake description for reality. But how do we sort them out? We have to look at all the varied descriptions and find their common denominators, the structure they share, the thing that remains unchanged when you go from one description to the next. And then it hit me.

* * *

I nearly ran from the cab to the door, hurriedly dragging my suitcase behind me, and rang the bell.

On the other side of the door, Cassidy launched into her best rendition of a ferocious bark. "You're a good girl," I heard my mother reassuring her as she made her way toward the door.

"Oh my God!" my mother shouted when she discovered me standing on the other side, suitcase in hand. "What are you doing here?"

She tried to hug me, but Cassidy pushed past her, hopping and whimpering, her butt wiggling so fast that for a second she lost her footing. She jumped up, put her paws on my chest, and licked my chin. "Cassideeeeeeeee!" I squealed, grabbing her floppy ears and planting a kiss firmly on her snout. She wiggled with delight, then bolted into the yard to pee.

As I gave my mother a big hug, I saw my father emerge from the doorway behind her, trying to figure out what the commotion was about.

"Surprise!" I said.

He hugged me, looking happy and shocked. "What are you doing here?"

I grinned. "I know what we're looking for."

6

Fictitious Forces

"Are you hungry?" my mother asked as my father grabbed my suitcase from me and carried it into the house.

I followed them inside. Cassidy trotted alongside me, her tail happily whacking my legs as I walked.

"You must be hungry after your flight," my mother continued. "I can't believe you flew all the way here without telling us."

I could see the anger dawning on her face.

"In this family," she said with a stern voice, staring me down, "we do not fly across the ocean without telling each other."

"Sorry," I said. "It was a last-minute decision."

"Too last-minute to make a phone call?"

"I wanted to surprise Dad. I had an epiphany."

"Epiphanies can be shared over the phone."

"I guess," I said, pouting. "It would have been so much less dramatic."

I followed her into the kitchen, where my father joined me at the table. Cassidy flopped down on the floor at my feet.

"So are you hungry?"

"I've been in England," I said. "I'm *starving*."

"What's the epiphany?" my father asked.

"I can make chicken," my mother said, peering into the refrigerator. "And I have those spicy noodles you like. Let's see what else. There's fruit salad. There's peanut butter. . . ."

Cassidy's ears perked up, but I shuddered at the thought. "No peanut butter. Never peanut butter."

"What's the epiphany?" repeated my father.

"I can make a salad with feta and walnuts."

"That sounds great."

"What about dressing? I have a raspberry vinaigrette—"

"For the love of God, *what* is the epiphany?"

"Okay," I said, turning to my father. "Are you ready for this?"

He offered up a look of cartoon-like suspense.

"Something is only real if it's invariant," I said.

He stared off into space, mumbling the words back to himself. "Something is only real if it's invariant. . . ."

"Think about it. Invariant means it's the same in every reference frame. It's a feature of the world that all observers would agree on. It's how we intuitively define what's 'objective.' Here's the reality test. If you can find one frame of reference in which the thing disappears, then it's not invariant, it's observer-dependent. It's not real."

He sat quietly for a moment, thinking. "So if something is invariant, it's real. And if it's observer-dependent, it's what? An illusion?"

"Yeah. I mean, it's not like a hallucination, it's not subjective. But it's not *ultimately* real."

"Like a rainbow."

"Exactly! It's caused by physics, it's not subjective, but it's not real, either. Right? Wait. How do rainbows work?"

"The Sun shines from behind you and the light gets refracted by water in the atmosphere."

"Right, okay. So you need the Sun and you need the water, so it's objective, but it's dependent on your reference frame. If you move to another spot you might not see it anymore. It's a legit physical phenomenon, but it's a product of your viewpoint. There's no tangible, physical rainbow-colored object hanging in the sky. You can't grab hold of it. It's like a mirage. It's not real."

"It's like the color of a galaxy," my father said. "Color isn't a real feature of the galaxy; it's a feature of how the galaxy is moving relative to an observer. The relative motion changes the wave's frequency, and frequency is what we see as color. So if a galaxy is redshifted, we know it's moving away from us. If it's blueshifted, it's moving toward us. It's a Doppler effect. It's observer-dependent."

I nodded. "If we want to find ultimate reality, we have to eliminate all the features of the universe that are observer-dependent until we're left with the ones that are truly invariant."

My mother plunked the salad bowl on the table, along with plates and forks.

Cassidy whined. I looked down to find her staring up at me, her tongue hanging out of her mouth in what looked uncannily like a smile. She offered me her paw.

"Really?" I asked her. "Salad?"

I tossed a piece of lettuce into the air; her jaws snapped shut around it. My mother flashed me a disapproving look.

That evening, I dug some books and papers out of my suitcase, then headed toward our physics library. In the hallway, my mother was lying on the floor with the dog, nuzzling her face into Cassidy's and whispering in motherese. *Yes, I love you. Yes, I do.*

"Still hate dogs?" I asked.

"Yes, I do," she cooed as Cassidy licked her nose.

In the library, my father was lounging in his leather chair, flipping through a book. I made myself comfortable on the couch.

"Check out this paper," I said. It was written by Max Born, one of the founders of quantum mechanics, published in the *Philosophical Quarterly* in 1953,

My mother rendered powerless by Cassidy's charm W. Gefter

and entitled "Physical Reality." I read the first line aloud. "'The notion of reality in the physical world has become, during the last century, somewhat problematic.'"

My dad laughed. "You think?"

I continued to read aloud, and my father listened intently.

Cut a circle from a piece of cardboard, Born wrote, hold it in the light of a distant lamp, and observe its shadow on the wall.

"'The shadow of the circle will appear in general as an ellipse, and by turning your cardboard figure you can give to the length of an axis of the elliptical shadow any value between almost zero and a maximum. That is the exact analogue of the behavior of length in relativity which in different states of motion may have any value between zero and a maximum. . . . It is evident that the simultaneous observation of the shadows on several different planes suffices to ascertain the fact that the original cardboard figure is a circle and to determine uniquely its radius. This radius is what mathematicians call an invariant.'"

"That's basically how a CT scan works," my father mused.

I continued to read. "'The projection (the shadow in our example) is defined in relation to a system of reference (the walls, on which the shadow may be thrown). There are in general many equivalent systems of reference. . . . Invariants are quantities having the same value for any system of reference.'"

"They're observer-independent."

"Exactly. And here's the kicker," I said, continuing. "'The main advances in the conceptual structure of physics consist in the discovery that some quantity which was regarded as the property of a thing is in fact only the property of a projection.'"

"That's a really interesting point," my father said. "Progress in physics comes from realizing that things once believed to be invariant are really just observer-dependent. Shadows."

"Yup. Born goes on to say, 'I think the idea of invariant is the clue to a rational concept of reality.' Then he talks about quantum mechanics, arguing that a measurement is a projection onto some reference system, the measuring apparatus. And he ends by saying, 'Invariants are the concepts of which science speaks in the same way as ordinary language speaks of "things." . . . The feature which suggests reality is always some kind of invariance of a structure independent of the aspect, the projection.'"

"What's real is what's invariant."

I nodded. "What's real is what's invariant. It almost sounds too obvious, like it's trivial, but it's incredibly profound."

"I can see that," my dad said, flipping through a book of Einstein's collected papers. "It's the whole idea behind relativity. Listen to this. Einstein was thinking about electricity and magnetism. If you move a magnet, you generate an electric field, and if you move an electron, you generate a magnetic field. But how can you say what's really moving? Motion is relative—are you stationary to the frame of the electron or to the frame of the magnet? He wrote, 'That these were two, in principle different cases was unbearable for me. The difference between the two, I was convinced, could only be a difference in choice of viewpoint and not a real difference. Judged from the [moving] magnet, there was certainly no electric field present. Judged from the [ether state of rest], there certainly was one present. Thus the existence of the electric field was a relative one, according to the state of motion of the coordinate system used, and only the electric and magnetic field together could be ascribed a kind of objective reality, apart from the state of motion of the observer or the coordinate system. The phenomenon of magneto-electric induction compelled me to postulate the (special) principle of relativity.'"

As my father read Einstein's words, I realized that of all the things for which physicists had Einstein to thank, they probably owed him most for demonstrating the deep connection between invariance and reality.

Because motion is relative and yet the laws of electromagnetism required light waves to move at 186,000 miles per second, space and time themselves had to vary from one reference frame to the next. That is, space and time were observer-dependent. They weren't real.

By weeding out what was observer-dependent, however, Einstein had discovered what *was* real: the unified, four-dimensional spacetime. Different observers could slice spacetime in different ways, calling some bits "space" and others "time," but they were just different ways of looking at the same invariant thing. If you had a world line that spanned, say, ten units of spacetime, I might call five of those units space and the other five time. From another reference frame, my father might call seven units space and only three time, which means two of the units that he sees as space, I see as time. Light calls all ten units space and has zero left for time. That's why you can't go faster than light. You can't allocate less than zero units of spacetime for time. If you did, you'd have a negative number—you'd be traveling backward in time.

The point was, no matter how you slice it, spacetime is spacetime. It's invariant.

That's why Hermann Minkowski said, "Henceforth space by itself and time by itself are doomed to fade away into mere shadows and only a kind of union of the two will preserve an independent reality." Space and time were shadows on the wall; spacetime was the cardboard.

Einstein thought the latter point was more important than the former—he was not so concerned with what was relative as with what was invariant, because he knew that what's invariant is what's *real*. In fact, he regretted having called his theory a theory of relativity, wishing instead that he had named it *invariantentheorie*: the theory of invariance.

The thing is, we can never see spacetime. Like prisoners in Plato's cave, we are compelled to experience the world through its shadows, a universe broken into pieces of three-dimensional space and one-dimensional time. But by spotting the invariant in Einstein's equations—the spacetime interval that holds steady through Lorentz transformations from one inertial frame to the next—we can glimpse the true reality behind appearances. Spacetime is a symmetry, but the universe of our perception is a symmetry broken. We are living among the shards.

Things got even more observer-dependent when Einstein upgraded from special to general relativity. People say that his inspiration—what Einstein described as his "happiest thought"—came when he saw a workman fall off the roof of a building near the patent office. That makes Einstein sound like kind of an asshole. But it's probably not true. Either way, it occurred to him that a man falling off a roof would, while in free fall, experience weightlessness, as if gravity had somehow disappeared. It was his happiest thought because it contained an unbelievable epiphany: if gravity could disappear in one observer's reference frame, then it couldn't be a fundamental ingredient of reality. It had to be an artifact of perspective.

From the ill-fated roofer's perspective, he's in an ordinary inertial frame, a frame without gravity. It's not that he's delusional—from his point of view, he *really is* in a gravity-free inertial frame, and if he bothered to do some quick science experiments on the way down, they would all confirm it. If, for instance, he took his keys from his pocket and dropped them, they would not fall toward his feet, as gravity would have it, but simply hover alongside him, since they're falling at the same rate. The only thing out of the ordinary would be the massive planet that happened to be accelerating toward his face.

An inertial observer traces a straight trajectory through spacetime. But from the perspective of the onlookers pointing and laughing on the ground, the falling man is crossing more and more space in less and less time as he plummets toward the Earth. According to them, he's accelerating, his world line tracing a curve. So which is it? Line or curve?

Einstein knew the answer had to be *both,* since the line and the curve are merely different descriptions of the same falling man. But how can they both be right? How can a curve also be a straight line? *To turn a curve into a line, you have to bend the paper.* Equating the roofer's perspective with the onlookers' requires a diffeomorphism transformation. It requires spacetime to be curved. It requires gravity.

Einstein's principle of general covariance demanded that all observers see the same laws of physics. Gravity upholds general covariance at the edges of mismatched reference frames; it turns curves into lines. "We are able to 'produce' a gravitational field merely by changing the system of coordinates," Einstein wrote. "The requirement of general covariance . . . takes away from space and time the last remnant of physical objectivity."

Newton had believed in the reality of absolute space because without it acceleration didn't mean anything—acceleration relative to what? But Einstein's general theory of relativity had shown that what looks like an accelerated frame from one point of view looks from another like an inertial frame with gravity. There's no ontological difference between accelerated and inertial frames, which in turn meant that you didn't need absolute space. That is, you didn't need space to be *real.*

It also explained what had always been a curious fact, one that I assumed would have the girl from my philosophy class foaming at the mouth: if you dropped two balls off the Leaning Tower of Pisa, say a bowling ball and a Ping-Pong ball, they'd hit the ground at the exact same time, assuming they were falling in an airless vacuum. You'd think the heavier one would fall faster, but it doesn't. Because if heavier things fell faster than lighter things, you'd be able to tell whether you were in an accelerated frame or an inertial frame with gravity.

How? If you found yourself in a windowless elevator and felt your weight being pulled toward the floor, you might wonder whether the elevator was accelerating upward, causing the floor to push up against your feet, or the elevator was at rest on a planet with a strong gravitational pull. To find out, you could drop a really heavy vagina and a really light vagina at the same time. If the heavy one hit the floor first, you'd know you were in a gravitational field. If they hit the floor simultaneously, you'd know that the elevator was accelerating upward, so the floor rose up to meet the hovering vaginas at the same time.

It's only because vaginas of different weights fall at the same speed that Einstein's equivalence principle holds: you can never tell the difference between acceleration and gravitation. If you could, "space" would mean something. It would be real. But it's not.

"Special relativity shows that space and time aren't real—they're observer-dependent," I said to my father. "And general relativity shows that gravity isn't real, because it disappears in certain frames. But here's the crazy thing—it doesn't stop with Einstein. It applies to all the forces. None of the so-called 'fundamental' forces is real!"

There are three forces in addition to gravity. Electromagnetism is the most familiar, since it operates on the scales of our everyday lives. The other two are more remote, governing subatomic matter. The strong nuclear force binds quarks into the protons and neutrons at the heart of every atom. The weak nuclear force swaps protons for neutrons, and vice versa, by changing the flavors of their constituent quarks, mediating radioactive decay and allowing the Sun to shine.

Despite all the talk of gravity being the odd man out in a world governed by quantum mechanics, all the forces arise in essentially the same way—specifically, to account for the fact that things appear differently in different reference frames.

When it comes to the quantum forces, the "things" are no longer space and time but quantum wavefunctions. And the key thing about wavefunctions is that, like all waves, they have a phase.

"Let's say you have some kind of matter particle, like an electron," I said. "It's described by a wavefunction, which has a phase. But the phase isn't a physical thing. It's just a measure of how far along the wave is in its cycle—whether it's halfway up its peak or heading down its trough, or whatever—*relative* to some measuring apparatus. To some observer. If you're watching a wave go by and you take a step to the left, you've now changed its phase, so obviously phase can't be an intrinsic feature of the wave; it's observer-dependent." Of course, phase *differences* mattered—that was the source of the interference pattern in the double-slit experiment. But phase in and of itself had no intrinsic meaning.

"The phase defines a reference frame," my father said.

"Exactly! So imagine that this electron's wavefunction is spread throughout all of space. I mean, its amplitude is probably peaked around a specific location, but technically it extends forever because of the uncertainty principle—it can't have zero probability of being in any location. So you're looking at this electron and you take two steps to the left. You've changed its phase. But you haven't changed its phase throughout all space, because your perspective only spans your own light cone. Shifting the phase of the entire wavefunction everywhere in the universe at once would require moving faster than light. If you could, that would be like making a Lorentz transformation or something. But you can't. You're only shifting a local portion of it. So now you have a portion that's shifted and a portion that isn't—the phases don't line up properly. They're mismatched, like a curve and a line. So you need to introduce a force to account for the mismatch. You have to bend things in just the right way so that the phases match up smoothly—like a diffeomorphism transformation."

"You need the equivalent of gravity."

"Exactly. And in the case of the electron, the equivalent of gravity is electromagnetism."

Electromagnetism is a gauge force. *Gauge* is just another word for phase. It's a point of view, a reference frame. Like Einstein's principle of general covariance, the principle of gauge symmetry demands that all gauges are created equal; no reference frame is truer than the next. But local gauge shifts—shifts in points of view—leave a wavefunction with misaligned phases. To account for the mismatch and keep all reference frames on equal footing, you need a gauge force.

Many of the books and articles I had read claimed that forces affect particles by changing the phase of their wavefunction, but really it was the other way around: a change in reference frames creates misaligned phases, which *produces* a force. In other words, the mismatched reference frames *are* the force. In the case of the electron, the force that arises from its mismatched phase is electromagnetism; its particle excitations are photons.

The electromagnetic force ensures that we don't confuse two different *descriptions* of one electron for two different electrons, just as gravity ensures that we don't confuse two different descriptions of spacetime for two different universes. The strong and weak nuclear forces are gauge forces, too, existing solely to account for the way that the phase of a quark's wavefunction can appear misaligned when seen from another point of view. And the similarity of gauge transformations to the diffeomorphisms of general relativity was no coincidence: gravity is a gauge force.

I had learned about the nuclear forces back when I was writing my quark-gluon plasma article, but I hadn't appreciated the deeper significance of gauge theory until my epiphany about the link between invariance and reality. Because the point is, gauge forces aren't invariant. Just like the falling roofer, you can find reference frames in which they disappear. In fact, in any single reference frame, they don't even exist. They only *appear* to exist when you compare one frame with another. They're observer-dependent. *They're not real*.

"They're fictitious," my father said, excited.

"Right! They're not real."

"No, I mean they're fictitious forces," he said, leaning forward in his chair.

"That's a thing?"

"Let's say you're stuck at a red light. It turns green, you press your foot down on the gas, and as the car lurches forward you feel like there's a force pushing you back into your seat. Physicists call that a fictitious force—like the centrifugal force that seems like it's pushing you toward the side of the car when you career around a corner. The forces aren't real—they're artifacts of being in an accelerated reference frame and not knowing it. Take the first example, when you step on the gas, and look at it from the perspective of a guy standing on the sidewalk watching. He's in an inertial frame, right? He sees your car jolt forward and sees you collide with your seat, but according to him the explanation is simply that the car is accelerating and it's trying to carry you along with it. From his point of view you're not being pushed back into your seat, your seat is pushing forward on you. But from inside the car, you have no way of knowing you're really accelerating."

"Well, I can see that I'm moving faster and faster away from the stuff outside my window," I said.

"But for all you know you could be at rest and everything outside could be rushing faster and faster away from you. And if you covered up all the windows, you could assume that you weren't moving at all, because relative to you, everything inside the car, including your seat, isn't moving. You would have every right to assume you're at rest, in which case it would be really strange to suddenly be thrown back into your seat. The only way to explain it would be to assume that a force is pushing you."

"But it's not a real force. . . ."

"Right, it's fictitious, because it disappears when you switch to the point of view of the inertial guy on the sidewalk. For him there's no force; there's just an accelerating car. Physicists call it fictitious because you can switch frames and make it go away. But really, based on what you're saying, *all* the forces, even the ones we thought were real, are fictitious."

"Yes, exactly! Gravity, electromagnetism, the nuclear forces . . . they're all fictitious. They're gauge-dependent, which is just another way of saying observer-dependent. They're not invariant. But you said the fictitious force arises because you're 'really' accelerating even though you don't know it. Isn't the point of relativity that you can't say that you're 'really' accelerating? You might be in an inertial frame with a force or you might be in an accelerating frame with no force, but they're equivalent. You can't privilege the sidewalk guy's frame as the 'real' one—every observer's view is equally valid."

"That's absolutely right." My father nodded. "The idea of fictitious forces comes from Newtonian physics, where the inertial guy on the sidewalk is considered to be in the absolute space relative to which the car is accelerating. Einstein made the sidewalk guy's frame and the driver's frame equivalent."

"By making space and time observer-dependent!"

We discussed the issue for a few hours until the jet lag was tugging on my eyelids.

"Come sleep with me, girly!" I said to Cassidy as I headed toward my old bedroom. She started to follow me but then turned around, headed back down the hallway, and lay down at the entryway to my parents' bedroom, keeping guard. "I see how it is," I told her, shaking my head. "Traitor."

Lying in bed that night, happy to be in a room large enough to be governed by the laws of classical physics, I thought about ultimate reality. Einstein had said, "Physics is an attempt conceptually to grasp reality as it is thought independently of its being observed. In this sense one speaks of physical reality." "Real," to Einstein, meant observer-independent, and the only way to figure out what was observer-independent was to compare all possible viewpoints and hope to find those rare keystones that don't change from one to the next. *What's real is what's invariant.*

It is a philosophical truth that everybody already knows, at least instinctively. If we see something so strange that we can't believe our eyes and we want to make sure that we haven't lost our minds or been served a dosed cocktail, what do we do? We turn to the guy next to us and ask, "Do you see that, too?" If he says no, then we know that the thing's not invariant across reference frames, and that it's probably time to panic.

As a newly enlisted structural realist, I knew I had to be careful to disentangle our stories about physics from its underlying mathematical structure, to not mistake different *descriptions* for different *things*. And now with invariance as my sole criterion for ultimate reality, I understood that the descriptions are what vary from one reference frame to the next—only structure has the potential to be invariant.

Ladyman had been right to turn structural realism into an ontological claim—structure, stripped of the baggage of our individual perceptions, was the only candidate for reality that stood a chance. Because the point was, there are infinitely many ways of looking at the same thing, of describing the same structure. That was obvious from general relativity alone. You could describe a curved world line in a flat spacetime or a straight world line in a curved spacetime. You could talk about a warped manifold with a simple grid-like metric or a distorted metric and no manifold at all. You could describe the cosmos with non-Euclidean geometry or you could stick with Euclid and toss in some extra forces. You could label and relabel spacetime points in infinitely varied ways. And none of it made any difference. The underlying structure was always the same. Our creativity for describing reality is probably boundless. The trick is to know what's mere description and what's the real thing underneath.

Luckily, I had discovered a simple rule of thumb: anything that serves to uphold gauge symmetry is mere description. And yet, mere *descriptions* give rise to what seem like very real and often dramatic physics. Simply shifting from one point of view to another can transform space into time, make gravity disappear, or produce an electromagnetic force. It can spark a nuclear reaction. It can cause the Sun to shine.

In addition to the four fundamentally fictitious forces, there's one more thing you need in order to preserve gauge symmetry: a Higgs field.

All particles have a property called spin, a kind of intrinsic rotation, which accounts for how the particles look from different reference frames. I liked to picture it as a beach ball flying past me, with different partial views consecutively coming into sight as it approaches and then recedes, making it appear to be rotating, even though in its own reference frame it's not rotating at all. Of course, it doesn't make any sense to ask whether the ball is "really" rotating, because motion is relative. An observer walking 360 degrees around an object that's standing still and an observer who is standing still while the object rotates by 360 degrees are two equivalent descriptions of the same thing.

A particle's spin is described as either right- or left-handed—right if it is spinning in the same direction in which it's moving through space and left if it is spinning opposite its direction of motion. But handedness is relative, too: if you have a particle that's spinning to the right, you can always run faster than it, turn around, and you'll see it spinning to the left. Handedness depends on the reference frame from which it's viewed.

That's a problem. Handedness is an observer-dependent property, which means it's not real. There's no true difference between left- and right-handed particles. And yet experiments in the late 1950s had proven that the weak nuclear force, which acts on quarks and electrons, acts differently on left-handed particles than on right-handed ones, in direct defiance of Einstein's governing principle and its incarnation in the mandate of gauge symmetry. Swap spacetime for its mirror image, interchanging left and right, and you'll see a different world. As if left and right actually mean something. As if they're invariant. Why would the weak force see handedness as invariant when it's really observer-dependent?

There was only one possibility: if the particles move at the speed of light, no one can ever outrun them, which means there's no possible reference frame in which their handedness appears reversed. Even though handedness is still fundamentally observer-dependent, it will always *appear* to be invariant. A left-handed particle traveling at light speed will be left-handed in *every* reference frame.

It seemed like an easy enough fix: just have all quarks and electrons travel at the speed of light. But there was a major snag in that plan, namely, that quarks and electrons have mass. You can't have mass and travel at light speed—the tiniest bit of heft will slow you down. If the particles move slower than light, however, there's no way to explain the weak force's preference for left-handed particles without violating gauge symmetry.

Unless you have a Higgs field. Physicists hypothesized* that there exists a space-filling field constructed in such a way that when a left-handed particle interacts with it, the particle comes out right-handed, and vice versa. So while the weak force *thinks* it's only acting on left-handed particles, the Higgs field is there in the background swapping left and right, so that the weak force is really acting on left- and right-handed particles equally. Now you can swap spacetime for its mirror image and the world will appear unchanged. Thanks to the Higgs field, particles like quarks and electrons can have mass without violating gauge symmetry.

If you look carefully at what the Higgs field is doing, you'll notice that something strange is happening to time. When a left-handed electron interacts with the Higgs field, it comes out a right-handed antipositron. And an antipositron is nothing other than an electron viewed in a different frame—namely, a frame in which time's arrow has been reversed.

Two observers will always agree on the time ordering of events, so long as they occur in regions where their light cones overlap. They might not agree on *when* the events occur, but they'll always agree on the order. For overlapping observers, "before" and "after" are invariant. But relative to some observer outside my light cone, those words lose all meaning. My before could be her after, her cause my effect. You wouldn't think you'd have to worry about that, given the fact that the two of us could never compare notes. But that's not quite true once quantum mechanics comes into play. According to the uncertainty principle, a particle that should be outside my light cone always has some nonzero probability of showing up inside it, skirting the laws of relativity. In doing so, it would appear to be traveling faster than light—which is to say, it would appear to be traveling backward in time.

It was Wheeler who first realized that an antiparticle is just an ordinary particle for which time's arrow has been reversed. Antiparticles have to exist to account for the fact that, for some observers, a

* In 2012, a particle matching the description of a Higgs boson was observed at the Large Hadron Collider at CERN.

particle might look like it hitched a ride in a DeLorean. Particles and antiparticles—they're not two different things. They're two different points of view.

It's no coincidence that the Higgs field has just the right properties to patch up the differences created by our changing reference frames, because the Higgs field, it now dawned on me, isn't something that exists out there in ultimate reality. Like gravity, electromagnetism, and the nuclear forces, the Higgs is fictitious—it's something we're forced to add to our *description* of reality to ensure that we can treat all reference frames equally and don't confuse different views for different things.

That's what physics does, I realized. Every time we break the world apart with our reference frames, physics offers a way to piece it back together. Reverse the direction of each spatial coordinate, turning the universe into its mirror image, and physics changes. Reverse charges, swapping every particle for its antiparticle, and physics changes. Reverse the arrow of time, trading future for past, and again, physics changes. But do all three at once and physics remains the same. CPT symmetry, as the triplicate invariance is known, is a direct consequence of the Lorentz symmetry of spacetime. Charge, parity (or handedness), and the arrow of time work together to maintain the structural equivalence of reference frames, to ensure that we can't mistake different descriptions for different realities.

CPT symmetry revealed a deep connection between the structure of spacetime and the structure of matter. Whenever I'd asked physicists to define a particle, they'd say it was an "irreducible representation of the Poincaré symmetry group"—which, I figured, sounded better than saying it was "a little ball." But now I finally understood what they meant. They meant that the symmetry of spacetime defines everything in it. Poincaré symmetry is the symmetry of the flat, gravity-free spacetime of special relativity, the symmetry that enforces an equivalence between inertial frames that are rotated relative to one another or that are moving at different uniform velocities or whose origins are at different locations. What we call "particles" are the most basic invariant structures that, in flat spacetime, won't disappear in any frame.

That reference frames matter so much—that they define what was beginning to seem like all of physics—was all Einstein's doing. The limits he imposed with a finite speed of light and the relativity of space and time meant that different observers have different views of the same ultimate reality. In Newtonian physics, where space was absolute and the speed of light was infinite, you didn't have to think about what different observers see, because they'd all see the same thing. Not in Einstein's world. In Einstein's world, you need rules for comparing different frames, for filtering out the artifacts of perspective. You need Lorentz and diffeomorphism transformations; you need gauge forces. In Einstein's world, our individual points of view shatter the unity of reality. Physics pieces it back together. It has to. Because reality is never really broken—it only appears that way.

Suddenly, the moral of the story about that falling pencil became clear to me. A paradigm of spontaneous symmetry breaking, I had read again and again, the pencil balanced on its tip can be knocked over by the slightest breeze, landing in one of an infinite number of equivalent ground states that encircle it, states of the lowest possible energy, where there's nowhere left to fall. The ground states, the story always goes, do not exhibit the original rotational symmetry of the vertical pencil—the symmetry has been broken.

But I now saw that the ground states are *gauges*. They're points of view. Which means *the pencil never really falls*—it only looks like it has fallen from the vantage point of an observer stuck in a gauge on the ground. From such a vantage point, the pencil casts a horizontal shadow, a shadow we mistake for the real thing, a shadow that doesn't enjoy the original rotational symmetry. To see the full symmetry, you'd need a God's-eye view from which you could see the pencil from every point around its circumference simultaneously. Since that's impossible, we're stuck having to infer the rotational symmetry from our limited vantage point. And we can do it by traveling 360 degrees around the pencil, transforming from one reference frame to the next as we go, gauge after gauge, remembering to account for the slight shifts in angle required to keep the pencil in view as we make our way around the circle. Gauge symmetry ensures that such transformations are possible. Gauge forces do the shifting.

Wilczek had suggested that perhaps spontaneous symmetry breaking created the universe by shattering the symmetry of nothing. That explanation had bugged me, though, because it wasn't really an explanation at all—by invoking some kind of preexisting quantum breeze, it violated Smolin's slogan: "The first principle of cosmology must be 'There is nothing outside the universe.'" But if the pencil never really falls, maybe the universe never really begins. Maybe it just looks like it did from here inside it.

In and of themselves, symmetries don't break—they just appear broken when our reference frames are finite and the full symmetries of ultimate reality can't fit within our view. If you could see the whole of spacetime from some Archimedean point outside the universe, every wavefunction's phase would be aligned and every angle of the pencil would be simultaneously visible. Symmetry would reign. Forces would disappear. And what would be left behind, invariant? That, I knew, was the ultimate question. The answer, whatever it was, was the ultimate reality.

Here—inside the universe, under the covers, stateside—I was left to view things through a funhouse mirror, hoping to reconstruct a single reality from a deceptive multiplicity. Still, I had to admit the distortions were pretty extraordinary. Spin, charge, handedness, speed, causality, mass . . . they all fit together in precisely the right way to keep reality intact despite our fractured viewpoints, and in doing so give rise to the world. From afar, physics looks so messy, so replete with ingredients, with so many arbitrary parameters. Only the truth is, none of it is arbitrary. It's all working toward the same goal: to account for how some singular reality appears from every possible point of view.

That was what I loved about physics—that moment of pure surprise when you suddenly realize that what you had thought was one thing is really something else, or that two things that seemed so different are really two ways of looking at the very same thing. It was the perennial comfort that comes from discovering that the world is not remotely what it seems.

I had grown pretty adept at faking the stuff of everyday living, but I had never been very good at it. At paying bills or doing dishes, at meeting for coffee or making small talk, at any of the daily proceedings that constitute life here on the surface of things. Sometimes I'd walk down the street feeling like everyone else was gliding past me, unfettered, like my feet were somehow heavier than theirs and I could feel the ground buckling beneath me, like I might plunge through at any moment, like I would give anything to plunge through, only you're not allowed to, because life is here on the surface and you just have to hang on and try not to slip through the cracks. It was a feeling that on some days left me wondering whether I was a trespasser not only at physics conferences and editorial meetings but here in the world, on the surface, too. Then, on nights like tonight, I'd catch a glimpse of the underlying structure, of the world beneath the world, the truth below the surface. I'd see the way everything was so perfectly connected to everything else, the way it was all governed by simple notions of singularity and symmetry—and it was just so fucking beautiful. "I believe that nature is a perfect structure," Einstein wrote. Lying there in bed in the dark, I was beginning to understand exactly what he meant.

"I've been thinking more about invariance and its relation to symmetry," my dad said, passing the syrup across the table. We had come to IHOP for a late breakfast. "Noether's theorem says that for every continuous symmetry there's some conserved quantity—an invariant. If we're hunting for invariants, the symmetries will help us find them."

"That makes sense," I said. "Symmetries tell you what remains the same when you transform between frames." I couldn't decide which to dig into first, the omelette or the pancakes. They looked symmetrically delicious. Wasn't this how some philosophical donkey starved to death? Buridan's ass?

"Right. A snowflake has 60-degree rotational symmetry because if you rotate it by any multiple of 60 degrees it looks the same as the original. But that's a discrete symmetry—there are still reference frames, say a rotation by 64 degrees, in which its appearance isn't invariant. That's why you need a continuous symmetry to find real invariance, so that there's no possible frame in which the thing changes."

"Okay," I said, "so let's look at continuous symmetries." I went for the pancakes. Symmetry broken. No asses were going to starve today.

"Well, the translational symmetry of space gives you conservation of momentum, and the rotational symmetry of space conserves angular momentum," my father said. "Time translation conserves energy. Rotational symmetry of four-dimensional spacetime conserves spacetime intervals. And gauge symmetry conserves charge."

"All right, that gives us some candidates for what's real. Let's make a list," I said, digging a pen out of my purse. On my napkin I wrote, *Ingredients of ultimate reality*. "Let's just list anything that could possibly be real, and then we'll look at them more closely. Let's see . . . space, time, spacetime, gravity, electromagnetism, the nuclear forces, mass, energy, momentum, angular momentum, charge . . . what else?"

"How about the number of dimensions?" my dad asked. I wrote it down. "Or particles? We should assume particles are real, right?"

"Unless they're strings," I said.

"Well, particles are excitations of fields, so particles and fields really should go together. And the fields are defined in terms of the vacuum."

I nodded, adding them to the list. *Particles/fields/vacuum. Strings.* "How about the universe? You'd hope that was real. Maybe that goes without saying?"

My dad shook his head. "Nothing in physics goes without saying."

I added *universe* to the list. I paused, then added *multiverse,* too.

"The speed of light," my dad said, pointing to the list as he took a sip of coffee. "That's the ultimate invariant."

I wrote it down. *Speed of light.*

"Bostrom would say that we need to question the reality of reality," I said. "But I think adding it to the list could send us into some kind of turtle-laden vortex."

"Skip it." He nodded. "It would be like listing cake as an ingredient in cake."

"Okay, so let's see," I said, turning the napkin so that we could both read the list. "Based on relativity, we can cross off space and time. They're both observer-dependent."

"You can cross off gravity," my dad said. "And the other forces.

They're all fictitious. What about mass? Mass is invariant, right? Or at least rest mass?"

I swallowed a gulp of coffee and shook my head. "It's not. Rest mass is invariant in special relativity, but in general relativity it's not well defined. In order to define it you have to break general covariance—you have to choose a time coordinate, and that sets a preferred frame. Mass is only defined within specified frames, and since $E = mc^2$ transforms mass into energy, the same goes for energy. They're observer-dependent." I crossed them off the list. "And momentum and angular momentum are defined in terms of mass, so they become observer-dependent under general relativity, too."

"Even in quantum field theory mass changes with scale," my dad said. "With the resolution at which it's measured."

I nodded. "The standard model says that all particles are ultimately massless—mass only arises as a consequence of symmetry breaking or the structure of the vacuum at low energies or interaction with the Higgs field. At high enough energies, mass disappears."

"Should we have put the Higgs on the list?"

"I think 'particles/fields/vacuum' covers it."

"Okay," my father said, scanning down to the next item on the napkin. "What about charge? Isn't charge conservation violated in some kinds of weak nuclear decay?"

"Yup," I said. "It's only conserved when you put it together with parity and time translation. But CPT invariance is really just Lorentz invariance, and Lorentz invariance conserves spacetime intervals. So we have to keep spacetime on the list."

"We can cross off spin," my father said. "Supersymmetry shows that what looks like a boson in one frame looks like a fermion in another."

That was a good point. Usually it's easy to tell the difference between a force-carrying boson, which has an integer spin, and a fermion, or matter particle, which carries a half-integer spin: just rotate the particle by 360 degrees, and if it looks exactly the same as when it started, it's a boson. If, instead, the amplitude of its wavefunction comes out inverted and you have to rotate it a second time—720 degrees in total—before it goes back to the way it started, it's a fermion.

To transform a fermion into a boson, and vice versa, you need some way to flip the amplitude of its wavefunction. And you can do that if you add a few extra dimensions. Not spatial dimensions, but mathematical ones. Rotate the particle through the extra dimensions and a positive amplitude will come out negative, and a negative amplitude positive, interchanging whole- and half-integer spins. In the higher-dimensional "superspace," bosons and fermions are identical. In ordinary space, they are different shadows of the same piece of cardboard, their distinction based only on the reference frame from which they're viewed.

"*Should* we assume supersymmetry?" I asked.

No experimental evidence for it had turned up yet. If reality were really supersymmetric, every boson would have a fermion partner and vice versa, a particle that is perfectly identical to its partner except that the amplitude of its wavefunction is inverted. Physicists were gearing up to start hunting for supersymmetric particles at the Large Hadron Collider in Geneva, but the accelerator hadn't been turned on yet. Supersymmetry remained theory.

My father shrugged. "There are good theoretical reasons to believe in it."

That was true. For one, in a supersymmetric vacuum, the forces can be unified. From our cool, low-energy perspective, the strong force appears 100 times stronger than electromagnetism, and the weak force 100 billion times weaker than that. But heat up the vacuum and their strengths begin to change. As the vacuum loosens its grip on quarks, the strong force weakens. At the same time, the electromagnetic and weak forces grow stronger. Keep cranking up the heat and soon all three forces are approaching the exact same strength. At around 10^{16} billion electron volts, the electromagnetic and weak forces merge into a single electroweak force, but the strong force is still a little too strong. But not when there's supersymmetry involved. In a supersymmetric world, the three forces are revealed to be broken facets of a single, unified, fictitious superforce.

It wasn't the only theoretical motivation. Supersymmetric particles don't interact with electromagnetism or the strong nuclear force, but they do interact with gravity. Just like dark matter.

"Besides, there's no reason to expect that experimentalists would have seen supersymmetric particles yet," he said. "They could well be out of reach, at higher energies than we can measure."

"Okay," I said. "Let's assume supersymmetry and cross off spin."

"So what's left?"

I grinned with excitement, picking up the napkin and reading from it as if it contained the Gettysburg Address and the overweight sweatpant-clad patrons of IHOP were brave soldiers in the Union Army. "The potential ingredients of ultimate reality are—"

"More coffee?"

My father laughed, and we both gave the waitress an appreciative nod. Our mugs full and steaming, I began again. "The potential ingredients of ultimate reality are: spacetime, dimensionality, particles/fields/vacuum, strings, the universe, the multiverse, and the speed of light."

"You know, my suspicion is that none of those is really invariant," my dad said, smiling mischievously.

"Then nothing would be real."

"Exactly. 'Nothing' would be real. If everything is ultimately nothing—and, really, it *has* to be—and we're defining ultimate reality as what's invariant, then the only invariant should be nothing. Which makes sense because it's the most symmetric thing there is."

"But we've got lots of invariants on the list here. Can it really all be nothing?"

"Well, look at how many things that physicists once thought were invariant have already been crossed off. Born said that's how physics progresses. I doubt it ends there."

"So if everything is ultimately nothing, then every remaining ingredient on this list should turn out to be observer-dependent."

"That's right. They'd have to."

I smiled, intrigued. "Well, I guess we'll see!"

I was rummaging through a drawer in my childhood bedroom, looking for a pen, eager to write down my racing thoughts about invariance and symmetry and reality, when I saw, sticking out from beneath a stack of papers, a blue folder. I slid it out, sat down on the bed, and opened it.

Your first years were so silent.
Waiting, waiting for the words.

I grinned. It was the Beat poem my father had written for me in honor of my high school graduation all those years ago. I had always thought it was sweet. But as I read it now, it finally dawned on me what he had done to write the thing. It wasn't just that he had paid attention to the books I loved and to the ideas I cared about.

And Kerouac, and On the Road
The rhythm, the rhythm of the words
And Ginsberg and "Howl" and "Kaddish"
Chanting the rhythm
And Kesey and Burroughs, Fitzgerald and Proust
The words, the words

It was that he had read the words. He had noted which books meant the most to me, and he had—in his exceedingly spare time between saving lives and marking nipples and reverse-engineering the fucking universe—read them, just so he could write me a poem in a voice that I would hear, a poem to send me off to New York City, to send me off into the world, only it wasn't *the* world, or even *his* world. It was *my* world. As if my world mattered. As if my words mattered.

The world's a big blank journal
Waiting for your words
Let all hear the rhythm, the rhythm of your words.

I closed the folder and placed it carefully back in the drawer. A nagging sadness pooled in my stomach. Like nostalgia, only counterfactual. Like the world was still big and still blank. Like I was still waiting, waiting.

A few days later I boarded a plane headed back for London. No matter how often I traveled, I couldn't seem to get through a flight without an anxiety attack. Takeoff was the worst. I forced myself to breathe deeply as the plane readied itself on the runway. *Physics works, physics works,* I repeated in my mind, my standard mantra. Suddenly the girl from philosophy class popped into my head. *Airplanes only fly because we all agree they do.* I rolled my eyes in disgust. We picked up speed, accelerating down the runway. A few rows behind me, a baby started to scream. The cabin shuddered violently. Suitcases banged about in the overhead compartments, the tray tables rattled on their hinges. Then the wheels were off the ground, the plane wobbling as it rose. *I agree that planes can fly, I agree that planes can fly,* I silently chanted. Anxiety trumps realism. Pascal's postmodern wager.

Soon we were cruising smoothly above the clouds. I unclenched my fists and reaffirmed my philosophy. Thirty thousand feet above the Atlantic Ocean, at rest relative to the obese man spilling over the seat next to mine, in 500-mile-per-hour motion relative to the slowly turning planet below, I thought about my mission to understand the universe. *Find the invariants, and you'll find reality.* I pulled the crumpled IHOP napkin from my pocket and stared at the handful of items that had survived the first round of cuts, the remaining candidates for the ingredients of ultimate reality. *Spacetime. Dimensionality. Particles/fields/vacuum. Strings. The universe. The multiverse. The speed of light.* They were good leads, and I felt a new surge of motivation now that we had a solid plan. A strategy.

Still, I couldn't help but think that it would be anticlimactic if any one of those ingredients turned out to be invariant. *Reality has ten dimensions and is made of tiny strings* would be a valid conclusion, but I was pretty sure I'd be left unsatisfied. The truth was, any ontology would seem awkward and arbitrary. *Reality is shaped like a trombone and is made of goldfish crackers.* I thought of Wheeler's words: "One therefore suspects it is wrong to think that as one penetrates deeper and deeper into the structure of physics he will find it terminating at some *n*th level . . . in some smallest object or in some most basic field." He seemed to think that the only ultimate reality was the observer. That when we look closely enough at the universe, we'll find ourselves staring back. But were observers any less arbitrary than goldfish crackers? I found myself asking the same old question: where do the observers come from? *The universe is a self-excited circuit*. We really needed to figure out what the hell that meant.

My father, meanwhile, seemed pretty convinced that nothing would be invariant. That is, that Nothing would be invariant. It did seem more satisfying. Questions stop at nothing. You don't have to ask, "But where did *that* come from?" Nothing doesn't come from anywhere. It's nothing. It needs no explanation. At the same time, I found it nearly impossible to imagine how this whole crazy universe, obese airline passengers and bottles of Xanax, press passes and Panama hats, oceans and rats and poems and pancakes . . . how it could all be just *nothing*.

Back on solid ground, in my shrunken flat, I grabbed a tiny soda from my tiny fridge and sat down to check my email. There in my inbox I found one with the subject *New Scientist*.

From: Michael Bond
To: Amanda Gefter
Subject: New Scientist

Dear Amanda,

I edit the comment and opinion sections of *New Scientist*. Michael Brooks gave me your name—he speaks highly of you! At the end of April, one of the people in my section is going off on maternity leave for six months and I'm looking for someone to take their place for that time. Would you be interested? The job is very versatile and involves editing, writing, and interviewing across all the different bits of the opinion section. It would be based in the London office.

Best wishes,
Michael

Seriously? A job on staff as an editor at *New Scientist*? We had just figured out our reality-hunting strategy, and now I was being handed the ultimate press pass? Hell, yes, I was interested! I began composing my reply. As I typed, I spotted something out of the corner of my eye. There on the lovely wooden floor, between the one-seater couch and the miniature sink, was a glue trap, and on the glue trap, a single, solitary, silvery tail.

7

Carving the World into Pieces

My work visa came through and I began my six-month stint as an editor at *New Scientist*. The timing couldn't have been better. By the end of April I had finished up my classes, leaving me the next several months to work exclusively on my thesis.

On my first day at work, I made the rounds, introducing myself to all the editors and reporters. I grew increasingly intimidated with each person I met. At twenty-five years old, I was noticeably the youngest editor there. All of them had graduate degrees in science or science journalism, often both, not to mention British accents that made everything they said sound smarter. They had held internships at major newspapers and scientific journals; they had worked alongside scientists in the field and in labs. They had paid their dues, worked their way up. I had faked my way into a conference or two and written a handful of articles. I was going to have to prove myself quick.

Everything seemed to be going great, but each morning when I'd stroll confidently into the office, the receptionist would invariably take one look at me and ask, "Are you all right?" He seemed friendly enough, but I couldn't pinpoint his cause for concern, so I'd respond with "Yes," then add, "Well, maybe I didn't get quite enough sleep," or "The commute was a little rough today," or "I'm being stalked by this troop of invisible rats." He'd give a polite but awkward smile, and I'd head over to my desk wondering if the color of my shirt made me look ill.

After *many* weeks of this, I was walking in one morning just a few steps behind another editor. "Are you all right?" I heard the receptionist ask her. She replied, "Are you all right?"

I quickened my pace until I was walking alongside her. "Sorry, what just happened there?" I asked.

"Where?"

"He asked if you were all right, and then you said, 'Are you all right?'"

She laughed. "It's an expression. A greeting. It's like . . ." She searched her mind for an American analogue. "It's like 'What's up?'"

"Ohhh."

"Is that not what it means in America?"

"No," I said. "'Are you all right?' is something you'd say to someone who just tripped on her high heels and spilled a purseful of tampons on the sidewalk. 'Are you all right?' means 'You absolutely do not look all right.'"

She laughed again, and we made our way to our desks. I sat down, relieved to know that presumably I did look all right, but also a bit concerned about some of my previous responses.

The next morning, I opened the large glass door to the reception area and took a deep breath. I was ready to nail it this time.

The receptionist looked up and smiled. "Are you all right?"

I opened my mouth to repeat the phrase back to him, but I just couldn't do it. Answering a question with the same question was too advanced for my small-talk skills. Instead, I gave him a little upward tilt of the chin and replied, "Whassup?" He smiled, but it still didn't feel right.

Meanwhile, I needed to choose a topic for my thesis. I knew that this was my opportunity to delve deeply into a particular question. It was what I had come to London to do. I needed to choose carefully.

As I sifted through ideas, I kept coming back to the arrow of time. We had discussed the mystery of time's arrow extensively in my philosophy of statistical mechanics class. Einstein's theories had supposedly put time and space on equal footing, sewing them together into one big block universe. Why, then, can we move backward in space but not in time? Relativity didn't have the answer, and particle physics wasn't any help, either. The laws of physics that govern particle interactions work the same way forward and backward. If particles don't see an arrow of time, why should we?

You need some kind of gross asymmetry on which to pin time's arrow. Luckily, there is one: entropy never decreases. And entropy, like the arrow of time, only exists here on the macro scale with us, not in the micro realm where particles live. I had always heard entropy described as a measure of disorder, but fundamentally, I learned, it's a measure of hidden information. If you want to describe a physical system, say a gas in a box, you have two options. You can track the constantly shuffling positions and momenta of every single molecule in the gas, or you can just take an average. Average the rate of shifting positions and call it temperature; average the changing momenta and call it pressure. Temperature and pressure are macroscopic shorthand—every last bit of information about the ever-changing micro states compressed into just two numbers.

There are tons of different possible microscopic arrangements that will all average out to the same macroscopic feature. The greater the number of microscopic possibilities, the harder it is for us to guess which arrangement is the real one, so the less precise our information about the micro state and the higher the entropy of the system. That's where the disorder comes in—there are way more microscopic configurations compatible with "disordered" states than with "ordered" ones. Countless arrangements of H_2O molecules correspond to a puddle of water; far fewer correspond to an intricately structured ice crystal. The puddle is messier—we have less information about its hidden inner workings, so it has more entropy. And more entropy means heat.

At first that didn't seem right. Why should our lack of information manifest itself as something as physical as heat? *We can literally be burned by our own ignorance?* I had written in my notebook. But it made sense once I thought about the scales involved. Temperature isn't part of bedrock reality—it's an emergent, collective macroscopic feature. An individual molecule doesn't have a temperature. So if you choose to study a system at the level of individual molecules, temperature vanishes. Average out the microscopic information to look at swarm behavior and now you've got heat. It's a matter of choosing a scale—choose a larger one and you trade information for temperature.

When I pour milk into my coffee, it swirls around for a second, then dissipates, settling into a monochromatic mocha color. Why? Why doesn't it ever spontaneously settle into the shape of the word *hello*? Because that cup of coffee has something like 10^{24} molecules sloshing around, and the number of arrangements that correspond to a uniform mocha color vastly outweighs the number of configurations that correspond to *hello*. "Vastly" doesn't even begin to capture those odds. I could sit here and wait for billions of years as the configurations shift every split second and it still wouldn't be long enough for my coffee to offer a warm greeting. What were the odds of all the air molecules in my flat stumbling into the fleeting configuration of a rat? I wondered. How about just the tail?

So that was that, the second law of thermodynamics: entropy never decreases. It's merely statistical, but it's enough that physicists treat it as sacrosanct. It's enough that it treats us to an arrow of time. Entropy always increases because, statistically, high-entropy states are overwhelmingly more likely than low-entropy states. If entropy decreased—if milk reswirled out of a mocha-colored coffee, if the air from my car's exhaust flew right back into the pipe whence it came, if cracked eggs healed—it would look like someone was rewinding the world. Like time was moving backward.

But saying that high entropy is more likely than low is not in itself enough to give you an arrow of time. After all, high entropy would be the most likely scenario in the past, too. Statistically, entropy should always be high—and once it's high enough it reaches a kind of mocha-colored equilibrium and there's nowhere else to go. In a universe in equilibrium, nothing would ever happen, save the rare statistical fluctuation every few billion years. But we don't live in equilibrium. We live in a universe where things happen all the time. Where entropy still has room to rise.

The only way to get an arrow of time is to assume that for some unknown reason the universe began in an extraordinarily unlikely low-entropy state. Milk mixes into my morning coffee because 13.7 billion years ago the universe began in a rare configuration. Breakfast has cosmic significance. The mystery is no longer why there's an arrow of time; it's why did the universe start out so improbably?

When I first heard this issue about the mysterious low-entropy origin, it didn't seem to add up. What about the cosmic microwave background? It was a snapshot from nearly the beginning of time and it showed a universe that's perfectly smooth to one part in a hundred thousand. That sure looked like equilibrium to me. I searched some books for an explanation, and eventually I found one. The entropy that's low at the beginning of time isn't thermodynamic entropy—it's gravitational entropy. Once you get to large enough scales, thermodynamic entropy doesn't run the show. Gravity does. And gravitational entropy has its own arrow that points in the opposite direction. From gravity's point of view, a smooth universe is ripe for clumping. Gravity always attracts, so a world where nothing is clumped together is hugely unlikely. If gravity had its way, the whole universe would be a giant black hole—gravitational equilibrium.

The cosmic arrow of time depended on gravitational entropy, but when I tried to look into the matter further, I found that physicists didn't really know what gravitational entropy was. If entropy is a measure of missing information about the micro scale, what kind of microscopic information does gravity encode? Of course, if physicists knew the answer to that—if they knew the microscopic structure of gravity—they wouldn't be thinking about the arrow of time. They'd have found quantum gravity.

There is, however, one place where gravitational entropy *is* well-defined: on the event horizon of a black hole. Suddenly I realized that there was something amazing, something profound lurking there. I wasn't sure what it was, but I knew I had found the topic for my thesis.

Einstein had discovered that mass and energy wrinkle space, but he never anticipated that there would be places where space becomes so twisted that it turns on itself, like a snake swallowing its tail. When a massive star burns through its fuel and buckles under its own weight, gravity ignites a runaway process of collapse. Growing denser and denser, the star can eventually collapse right through itself, burrowing into the very fabric of spacetime. In the aftermath of this implosive chain reaction, space and time are left warped beyond recognition, laying bare the bones of reality. It was Wheeler who gave this mangled bit of world a name: black hole.

Black holes bring together the three pillars of physics—general relativity, quantum theory, and thermodynamics—and force them to battle it out. If you're hunting ultimate reality, a black hole is the place to look. It is the breakdown of time and space, the beginning and end of the universe. It's where the shards are restored to symmetry. At the center lurks a singularity, a place where spacetime curvature approaches infinity and physics turns pathological. As the radius of spacetime curvature spirals down toward the Planck length, physics as we know it falls away, unveiling some terra incognita only quantum gravity can traverse.

Given the similarity of a black hole's singularity to the singularity at the origin of time, I had always assumed that singularities would be the most interesting aspects of black holes. I was wrong. I quickly learned that the real action was out on the edge: the event horizon. The horizon is the gravitational point of no return, a surface of spacetime where gravity's grip exactly counteracts the speed of light. It is a surface of light rays frozen in place by gravity. For an observer outside the black hole, the event horizon is a kind of cosmic wall. Since light can't escape it, an observer can never see anything on the other side. *For all intents and purposes, there is no other side.* It is fundamentally and eternally out of reach—nothing that takes place there can have any physical impact on the external world. It's what makes the black hole black. It carves the world into pieces.

The horizon is a one-way door: things go in, but they can never come out. For physicists, that was a huge problem. It meant entropy could go in and never come out, which would mean the entropy of the universe outside the black hole was decreasing. So much for the arrow of time.

The first step toward a solution came in 1970 when Stephen Hawking was getting ready for bed. Suddenly Hawking realized that in order for horizons to be stable, the light rays that constitute them must never travel toward one another, only parallel or apart. That meant that the area of an event horizon could never shrink. If matter or radiation fell into a black hole, the area of its horizon would inevitably grow, and if two black holes merged, the horizon of the resulting black hole would be equal to or greater than the sum of the original two.

The area of an event horizon can never decrease. When Jacob Bekenstein, one of Wheeler's students at Princeton, heard about Hawking's area theorem, he couldn't help but notice that it bore a striking resemblance to the second law of thermodynamics. Could they be related? It was just a hunch, but he knew it was heretical, and he might have swept it under the rug if not for Wheeler.

"I always feel like a criminal when I put a hot cup of tea next to a glass of iced tea and then let the two come to a common room temperature, conserving the world's energy but increasing the world's entropy," Wheeler told him. "My crime echoes down to the end of time, for there is no way to erase or undo it. But let a black hole swim by and let me drop the hot tea and the cold tea into it. Then is not all evidence of my crime erased forever?"

Jesus, I thought. Did Wheeler ever just speak in normal sentences?

Apparently Bekenstein understood what he meant. Months later he showed up in Wheeler's office with a bold claim: an event horizon's area, he said, is not only analogous to entropy, it *is* entropy. Wheeler replied, "Your idea is so crazy that it might just be right. Go ahead and publish it."

When Hawking read Bekenstein's paper, he was pissed. He felt like Bekenstein had misused his area theorem to push an intrinsically flawed idea. The problem was obvious. Entropy means heat. Anything with entropy has a temperature—it radiates. Only black holes can't radiate. They're *black*.

Annoyed, Hawking, along with physicists Brandon Carter and Jim Bardeen, wrote a paper explaining why Bekenstein couldn't possibly be right. But the idea kept nagging him, and after two years of calculations Hawking arrived at a shocking conclusion. In his now legendary 1975 paper "Particle Creation by Black Holes," he showed that when quantum mechanics meets gravity at the horizon, particles appear.

That means that black holes do radiate heat, with a temperature inversely proportionate to their mass. If black holes can radiate, they must have entropy. The arrow of time had been saved, black holes were no longer quite so black, and Bekenstein had been vindicated. Hawking produced an equation showing that the entropy of a black hole is proportional to one-quarter the area of its event horizon. He requested that it be inscribed on his tombstone.

Soon other similarities between black hole physics and thermodynamics began to emerge. The so called zeroth law of thermodynamics says that temperature is constant in a thermodynamic system in equilibrium. Likewise, gravity is constant across the surface of an event horizon. The first law of thermodynamics says that energy can change forms but is always conserved. Likewise in black hole physics. When an object is swallowed by a black hole, its mass and energy (which, via $E = mc^2$, are interchangeable) are transferred to the black hole itself, so that the total energy of the system is always conserved. For each law of thermodynamics, there seemed to be an equivalent law of black hole physics. A deep connection between thermodynamics and gravity was beginning to emerge. For physicists it was an intriguing connection, to say the least. After all, thermodynamics is about matter and energy. Gravity is about space and time. Find the connections between the two and you're headed fast down the road to quantum gravity.

In hindsight, the notion that an event horizon would be marked by entropy isn't that surprising—after all, entropy is a measure of hidden information, and an event horizon's entire job is to hide information. But why would the entropy of the black hole, which represents all the hidden stuff inside the three-dimensional volume, scale with the two-dimensional area of the horizon? And where the hell were those particles coming from?

From: Amanda Gefter
To: Warren Gefter
Subject: A hunch . . .

I think I found a topic for my thesis: Hawking radiation. I can't put my finger on it yet, but there's something seriously profound

there. The black hole's entropy is proportional to the area of the
horizon? That's weird, right? Why wouldn't it be proportional to
the volume? It's like you lose a dimension. And these particles . . .
where do they come from? Horizons just conjure them out of thin
air? There's something here, I'm sure of it.

From: Warren Gefter
To: Amanda Gefter
Subject: RE: A hunch . . .

Your thesis idea sounds radiant. The Hawking particles—aren't
they virtual particle pairs that get separated by the horizon?
Whatever is there, I have no doubt you'll find it. Keep me posted.
Mom sends her love, and a box of Balance bars. They are in the
mail and should arrive this week.

My dad was right about Hawking radiation. The usual story went
something like this. Thanks to quantum uncertainty, pairs of virtual
particles and antiparticles are constantly popping out of the vacuum.
Fleeting ghosts, they surface for an instant, then meet and annihilate,
disappearing back into the seething quantum sea. Should such a pair
happen to emerge near a black hole, they can be divided by the horizon. Unable to partake in their mutual annihilation, the particle outside the horizon escapes into space while its antiparticle partner falls
toward the singularity. Alone, separated from its partner, the escaped
virtual particle becomes real. To an observer outside the black hole, it
appears that the horizon is radiating. Meanwhile, the negative energy
of the antiparticle shrinks the black hole ever so slightly, so it loses
mass and appears to slowly evaporate.

Particles, however, are really excitations of fields—quantum fields
that, even in their lowest energy states, fluctuate around a mean value,
the zero-point energy. A positive frequency fluctuation corresponds to
the presence of a virtual particle, while a negative frequency fluctuation corresponds to a virtual antiparticle. But things get interesting
when there's an event horizon.

In an infinite, unbounded space, frequencies of every possible wavelength are equally represented, so they cancel one another out, leaving behind what appears to be calm, empty space. But when you stick an event horizon into that space, everything changes. The vacuum is totally different depending on which side of the horizon you're on. The space outside the horizon is now finite and bounded. Its energy changes and, with it, everything else. New vacuum, new fields, new particles.

Horizons create particles by restructuring the vacuum, I jotted in my notebook. And on further thought I added, *Like the Casimir effect?* The whole situation sounded familiar. In the Casimir effect, two parallel, uncharged metal plates hovering just microns apart get pushed together by a mysterious force. The force is really just the vacuum. Outside the plates, the vacuum's zero-point energy extends infinitely throughout space, so every possible wavelength of the field modes is present and accounted for. But inside the tiny gap between the plates, only certain wavelengths can fit. Waves only come in whole numbers—you can't have half a wave—so only wavelengths that can fit entirely within the space between the plates count. The plates restructure the vacuum, leaving the vacuum outside the plates different from the vacuum inside. The difference creates a force: the stronger vacuum outside pushes on the plates, and the weaker vacuum inside can't handle the pressure. It's like a child trying to hold a door shut while an entire army pushes it open. Physicists have watched this battle of the nothings play out in the lab, and it's true—the plates snap together like magnets, only there are no magnetic forces to be found. There's just nothing.

I had always thought the Casimir effect was amazing, because it made the inner workings of the vacuum visible to the naked eye. When you're talking about vacuum modes it all sounds very esoteric and theoretical, but when you have two plates crashing together like cymbals it suddenly becomes very real.

In the case of the black hole, it was becoming clear that, just like the metal plate, the event horizon restructures the vacuum. Stick an event horizon in some region of spacetime and it restricts the wavelengths of zero-point energy that can fit there. This changes the energy of the vacuum and, with it, its fluctuations, spawning particles that wouldn't otherwise be there. Those are Hawking particles, and they come in every variety, from photons to quarks. They are particles born as shifts in the nothingness, carved out by the boundaries of space and time.

But when I thought about it I realized that it was kind of an insane analogy. I mean, a horizon is different from a metal plate in a major way: you can't sail right through a metal plate. But you can fall right through an event horizon. How could a horizon be real enough to restructure the vacuum but ephemeral enough to allow stuff to pass through?

That question kept me up at night for weeks. That and the scurrying of quantum rats. During the days, seated proudly at my new editorial desk at the magazine, I plowed ahead with my research, itching to find an answer. And I quickly realized that the most powerful research tool at my disposal was my official *New Scientist* email address. That thing worked wonders. When I couldn't figure something out on my own, I would just fire off a casual email to a physicist. *Oh, hello there, superfamous physicist, I'm an editor at* New Scientist *and I'm looking into a possible story about this randomly specific aspect of black hole mechanics or quantum field theory or whatnot, and I'd love to get a better handle on it—could I bother you to explain a few things?* I'd click "Send" and within a day or two I'd have detailed answers to whatever questions had been bugging me. It was like magic. Being a "freelance science journalist" with an AOL address had worked pretty well, but being an "editor" at a major science magazine was on a whole other level. Some days I could hardly believe my luck.

Little by little, I pieced together the strange properties of event horizons. Einstein had shown that in terms of reference frames, there are two types of observers: those in uniform, inertial motion and those who are accelerating. When there's a black hole involved it's pretty crucial to know which type of observer you are. The accelerated observer is the guy who manages to safely skirt the black hole's pull and remain outside the horizon. He's accelerating because as gravity tugs at him, he has to run faster and faster just to get the hell out of there. In my mind I named him Safe. An inertial observer isn't quite so lucky. He falls in, plunging through the horizon and down into the dark depths toward whatever lurks in there. If you don't accelerate, you'll never outrun gravity. The inertial observer's fate is sealed. I named him Screwed.

According to Safe, the horizon has some rather extreme physical properties. It comes equipped with all the strange effects of relativity—near the horizon, light is stretched to huge proportions and time slows down so much that it grinds to a halt at the horizon's edge. Not only does time stop at the horizon, but space does, too. The horizon marks the edge of reality. And since its area is entropy, the horizon is hot—hot enough to vaporize anything that gets too close, leaving nothing but ashes scattered among the Hawking radiation.

Screwed, however, doesn't see any of this. According to him, the horizon doesn't even exist. Assuming the black hole is big enough, he sails right through without noticing a thing. He doesn't see light stretch, time slow, or space stop. He doesn't feel heat. He doesn't see Hawking radiation. Screwed sees nothing other than ordinary, empty space.

Two guys look at the exact same patch of universe and one sees empty space while the other sees particles? It was so bizarre, I could barely wrap my head around it. Something had gone wrong with reality. And suddenly I understood exactly what it was.

From: Amanda Gefter
To: Warren Gefter
Subject: HOLY SHIT!!!

The particles of Hawking radiation are observer-dependent! They're not invariant! Accelerated/outside observers see them; inertial/infalling observers don't. It's the horizon that creates the particles, and for the observer falling into the black hole, the horizon doesn't exist. If it did, they wouldn't be able to fall through! They have no horizon and no hidden information, therefore no entropy, no temperature, no Hawking particles. They see an entirely different vacuum state and the two vacua aren't related by Lorentz transformations—they're incommensurable. I've attached like ten papers for you to read—enjoy!

Observer-dependent matter! It's mind-blowing, right? That
doesn't happen in ordinary physics . . . like with relativity or
quantum mechanics you might have observers that disagree on
certain properties of particles, but they all agree on their existence. The horizon undoes all that. Some observers see empty
space, others see particles. Some observers see nothing, others
see something. It's insane. And why does no one ever talk about
that? Whenever you hear about Hawking radiation, it's like,
ooooh, black holes aren't really black. As if that's the interesting
part. How about the fact that *matter isn't really real*?!

From: Warren Gefter
To: Amanda Gefter
Subject: RE: HOLY SHIT!!!

Well, I think you've figured out what to say in your thesis! That is
truly amazing. I never quite realized how deep Hawking's discovery was. But isn't black hole radiation a very specific and rare situation? Does it really apply to matter in general?

I had never realized how deep Hawking's discovery had been, either. I had always harbored suspicions that his fame had been over-
hyped thanks to his illness—there's just something about a man
speaking with a robot voice that makes it seem extraordinarily profound. But as the implications of Hawking radiation sank in, I realized
that if anything, Hawking is totally underrated. I mean, everyone
knows who he is, but how many people know what he did or, more
important, why it matters? Particles can be observer-dependent. Particles aren't invariant. *Particles aren't ultimately real.*

The particles of Hawking radiation are literal embodiments of the
observer-dependence of the vacuum. In a flat, boundless space, all
observers agree on the lowest energy state, the state devoid of particles, the vacuum—which is basically to say that all observers agree on
what nothing is. An event horizon undermines that agreement. The
horizon places a boundary on the space; it restructures the vacuum.
But only accelerated observers see the boundary; inertial observers see

flat, boundless space. What looks like something to one observer looks like nothing to another.

At first glance, a black hole's event horizon doesn't seem to be observer-dependent. After all, a black hole is a concrete, localized object; there's one sitting in the center of our galaxy right now. It seems like we'd all agree on the objective, observer-independent existence of its horizon, but that's only because every one of us is walking comfortably in Safe's shoes. If we bothered to think about Screwed plunging into that omphalic abyss, we'd realize that the horizon doesn't exist for everyone—it just exists for enough of us to trick us into thinking of it as an objective feature of the world. Once we realize that it's not—that it's observer-dependent—we suddenly see that the Hawking particles, whose existence is chained to the horizon's, are ultimately observer-dependent, too.

My father was right: I had found my thesis. The philosophical reverberations of this thing could bust the Richter scale. Ever since the atomists in ancient Greece, particles had been considered the basic units of material reality—solid, objective, indisputable. Relativity taught us that observers might disagree on the location of a particle in space or time, but they'd all agree that there was a particle *somewhere*. And sure, quantum mechanics made particles fuzzier, but again their very existence remained safe and sound. The idea that different observers can disagree on whether or not a particle exists *at all* is far weirder than anything relativity or quantum theory alone ever cooked up. Particles are the so-called building blocks of reality—so if their very existence depends on whom you ask, what happens to reality?

I set to work on my thesis. My life was all Hawking radiation all the time—all day every day at the office, and all night every night in my Planck-scale flat. At work, I found ways to turn what I was researching into articles so that I could continue reading about horizons and entropy and particle ontology without raising any suspicion. At night, the tenuous turning of pages, the soft clicks of the keyboard, and the occasional rustle of an invisible rat served as a satisfying soundtrack to my quiet pursuit.

There were times when I wished my father was there: when I discovered an amazing fact, when I was tortured by a question I couldn't answer, when I was convinced that I was really on to something big. I came to know the sound of one hand high-fiving. But through frequent emails I kept him abreast of everything I was learning, and, faintly echoing through the dark sky over the Atlantic and down a sparsely lit cul-de-sac in Notting Hill, I could hear him cheering me on.

I could also hear his question echoing annoyingly in my mind: *Isn't black hole radiation a rare situation? Does it really apply to matter in general?*

It was a fair point. Even my mother, who worries more than anyone I know, doesn't worry about black holes. If black holes are so far removed from daily life, did it even matter if Hawking particles weren't real? Were they nothing more than theoretical curiosities?

In the throes of my research I quickly discovered that black holes aren't the only source of event horizons. In fact, there was a far more prosaic source: acceleration. If an observer is accelerating, light from certain far corners of the universe will never, no matter how much time lapses, be able to reach him, so long as he keeps accelerating. This sounded hard to believe until I thought back to those spacetime diagrams my dad had drawn years earlier on his yellow legal pad. The path of a light beam through spacetime is a straight line, but the path of an accelerated observer is a curve. Just as some light beams are about to collide with the observer, he swerves off along his curved path, successfully avoiding the light which has no choice but to continue along its rigid trajectory. Thus, there's a whole region of the universe from which light will never reach the accelerated observer. A whole region that's fundamentally out of touch. Dark. Like a black hole.

Actually, it really is like a black hole. The boundary between the inaccessible region and the rest of the universe marks an event horizon. Known as a Rindler horizon, it has all the same features as a black hole's. It has all the weird quirks of relativity: wavelengths of light stretch to huge proportions, and time slows down and halts at its edge. It's got an entropy proportional to one-quarter of its area—the same formula Hawking had discovered for black holes. With entropy comes temperature. With temperature comes heat. With heat comes particles.

These particles were variously described as "Rindler particles," "Unruh radiation," "Unruh-Davies radiation," or "Hawking-Unruh radiation," but they all amounted to the same thing: observer-dependent particles produced by an observer-dependent horizon. In fact, black hole horizons and Rindler horizons are totally identical at the level of equations. They might seem like very different physical situations, but as far as the math was concerned they were indistinguishable. And if you think about it, there's an obvious reason for this: the equivalence principle. Einstein said that gravity and acceleration are equivalent. Not just similar or analogous, but *equivalent*. Two ways of looking at the same thing. If gravity could create an event horizon, so could acceleration.

I imagined Safe and Screwed hanging out in ordinary flat space free from black holes. Safe was my accelerated observer, so as he accelerates through flat space, he carves out an event horizon. If he whips out a thermometer on the run, he'll measure a nonzero temperature all around him, courtesy of the Rindler-Unruh-Davies-Hawking particles. But ask Screwed to do the same and his thermometer won't register a thing. No matter how much I stewed over it, I couldn't get past how insane that was: two observers occupying the *same exact space,* and one is undeniably surrounded by particles while the other unequivocally sees empty space. And the only difference between them is that Screwed doesn't have an event horizon. Safe physically restructures the vacuum and creates actual measurable particles just by having a particular point of view. Particles that exist, objectively, for him alone.

For years I had suspected that the secret ingredient for turning my father's nothing—the infinite, boundless homogeneous state—into something was a boundary. And Fotini Markopoulou had me wondering if perhaps an observer's internal perspective, which was invariably bounded by his light cone, was somehow enough to do the trick. Still, I had been skeptical that a light cone was capable of physically trans-

forming nothing into anything. After all, a light cone is merely a reference frame; it's not a *thing* that exists out there in the universe. But
maybe my skepticism had been shortsighted. Here I was learning
about observer-dependent boundaries that create particles using nothing more "physical" than an observer's point of view. Granted, these
horizons were a little different. Unlike light cones, they were time-
dependent; they formed dynamically. But it was an intriguing prospect
all the same, one that I wrote down, and then underlined, in my notebook: *Horizons show how an observer's point of view can physically restructure the universe. Or maybe the H-state.*

The whole thing was downright freaky. And the key point was that
neither Safe's nor Screwed's vacuum state is the "real" one. Relativity
had shown that space and time were different for different observers.
They weren't invariant. They weren't real. Now it was clear that vacuum states, and with them particles, were the next to go. Particles
weren't real. Their existence was observer-dependent.

It all tied back in to that definition of particles: irreducible representations of the Poincaré symmetry group. Poincaré symmetry is a
global symmetry of flat spacetime, but global symmetries are useless in
the face of horizons. Horizons require us to define everything locally, to
slice up the global view into individual observers' patches. The problem is that there's no unique, preferred way to slice it, and different
slices give you different vacuums—just a series of incommensurable
partial views, no one any truer than the next. Curved spacetime—the
kind with gravity, with event horizons—isn't Poincaré symmetric. Take
away the symmetry and you lose any clear definition of a "particle."
When you make the geometry of the spacetime observer-dependent—
whether it's flat, as Screwed sees it, or curved, as it is for Safe—you
bring the ambiguity to a whole new level. It no longer made sense to
ask,"Is there a particle there?" Now we have to specify: "Is there a particle there *according to Safe*?" And as if that weren't enough to blow my
mind, I discovered a third kind of event horizon, one that literally
marked the edge of the universe.

* * *

If you have an accelerated observer in a flat space, you get a Rindler horizon. But I soon discovered that you could swap the situation around and have space itself accelerate while an observer like Screwed stays still in an inertial frame. With space expanding at an accelerated rate, light can only travel a finite distance even given infinite time—no matter how far the light gets, more distance just keeps opening up ahead of it, like a light beam on a treadmill. Some light beams will never be able to reach Screwed. Thus, for him, part of the universe is forever dark. The darkness is encircled by an event horizon—a de Sitter horizon.

Willem de Sitter was the first physicist to spot an accelerating universe buried in Einstein's equations—one totally devoid of matter, emptier than the emptiest stretches of cold interstellar space. Just a vast, barren, swelling nothing.

Only it wasn't quite nothing. Woven into the fabric of space was a strange form of energy that exerted a kind of antigravitational effect, a force that pushed outward, causing the space to inflate. Its source was a seemingly innocuous term in the equations of general relativity: the cosmological constant. Because it was a feature of space itself, and because it was constant, the strange antigravitational energy wasn't diluted by expansion: the more space, the more cosmological constant. This produced a runaway effect, accelerating the universe's expansion faster and faster the bigger it grew. It was the opposite of gravitational collapse—the formation of a black hole in reverse.

When de Sitter proposed his model in 1917, Einstein was convinced it had to be wrong. It flew in the face of two of Einstein's deeply held philosophical beliefs: first, that spacetime without matter was impossible, and second, that the universe was static. Eternal. In fact, looking at his equations, Einstein believed that it was the cosmological constant that would anchor the universe in place and keep it from expanding or collapsing.

Unfortunately for Einstein, philosophy wasn't strong enough to hold the universe steady. In 1929 the American boxer-turned-astronomer Edwin Hubble made a world-changing discovery: all the galaxies in the sky were moving away from us at a velocity proportional to their distance. Exactly what you'd expect to see if you lived in an expanding universe.

I have no idea how he reacted to Hubble's news, but I'd be willing to bet that Einstein punched a wall. I'm sure it took not even a moment for him to realize that he had missed out on the chance to make what would have gone down in history as one of the greatest scientific predictions of all time. And from cosmic expansion the big bang would have been two thoughts away. It had all been right there in front of his nose, in the very equations that *he* discovered, but he hadn't wanted to see it. Okay, sure, he had won a Nobel Prize seven years earlier and it wasn't like people would be walking around saying, "Man, that Einstein was an *idiot*." But still, he must have been pretty pissed.

There's a photograph of Einstein peering through Hubble's telescope atop Mount Wilson to see the expansion of the cosmos; every time I looked at it, it sent a shiver down my spine. The idea that what a man armed with nothing more than philosophical principles had worked out with pencil and paper was actually happening in this huge way, playing itself out in the enormous reality of the world, underlined the power of the mind and the exquisite potential of science. Einstein wrote, "I take it to be true that pure thought can grasp the real, as the ancients had dreamed." I couldn't help wondering how an antirealist could bear to look at that photo. Could such a person honestly pass it all off as pure coincidence? A cosmic miracle? Was the universe expanding because we all agreed it was? I could just imagine the girl in my class: *Expansion? Is there not a certain male organ that's known for that?*

After Hubble's discovery of the expanding universe, Einstein was forced to acknowledge that there were legit nonstatic solutions to general relativity. Solutions like de Sitter's. But de Sitter's model remained a mere theoretical curiosity until 1998, when those teams of astronomers went supernova hunting and discovered that the expansion rate of the universe is accelerating. Later studies pinned down exactly when the acceleration began: 5 billion years ago the cosmic expansion stopped slowing down from its initial burst of inflation and suddenly started speeding up. It was as if some strange force had lain dormant, crouching on its haunches in the stillness of space, waiting for the right moment to pounce and overtake gravity. If it wasn't Einstein's cosmological constant, it was a damn good impersonator. Physicists named it dark energy, just in case.

Today's accelerated expansion showed no signs of slowing down. As space continues to swell, it will dilute the density of matter, thinning out the universe, separating any two objects by ever-longer stretches of hopeless, starless space. As the distances between galaxies grow, the sky will darken. Eventually space will accelerate so quickly that light from distant stars will never again be able to reach us. Swept off by cosmic expansion, they will disappear, leaving only darkness, our Milky Way a dim beacon in an inky sea of empty, expanding nothing. A lonely island in the void, surrounded by an event horizon. It slowly dawned on me now: in a universe ruled by dark energy, we're all Screwed.

Living in a universe permeated by dark energy means that an event horizon awaits us. It means that *this* is a de Sitter universe. And it means that all the disconcerting effects that come with horizons aren't confined to the safety of distant black holes. It means that they are all around us.

So there I was, at a miniature desk in a miniature flat in a cul-de-sac in London in a vast and expanding de Sitter universe surrounded by an event horizon. Being surrounded by a de Sitter horizon is like being surrounded by a black hole—galaxies accelerate toward the horizon as if being pulled by gravity, then plunge straight out of the universe. This time, with space itself doing the accelerating, it's Screwed, the inertial observer, who sees the horizon. From Screwed's frame we see the galaxy's light stretched and slowed as extreme relativistic effects take hold near the horizon. By the time the galaxy has crossed into the dark region of no return, it doesn't matter whether you call it a black hole or a de Sitter horizon; either way, that galaxy is gone.

As galaxies disappear behind the horizon, the horizon grows in area and entropy. Just two years after his discovery of black hole radiation, Hawking and fellow Cambridge physicist Gary Gibbons proved that, just like black hole horizons, de Sitter horizons had an entropy proportional to a quarter of their area. With entropy came temperature; with temperature, particles. Observers in de Sitter space find themselves surrounded by heat. I began wondering why, in a de Sitter universe, London always seemed so cold. It turned out the de Sitter temperature is next to nothing, just a whisper above absolute zero. Virtually undetectable. But someday in our cosmic future, the microwave background will redshift away, leaving the de Sitter radiation as the sole source of constant heat throughout the cosmos.

The implications of all this were beginning to sink in. Like black hole and Rindler horizons, de Sitter horizons were observer-dependent. The objective acceleration of space gave rise to a horizon by hiding a region of spacetime from a given observer. It's one horizon per observer, each at a slightly different spot. No two observers will agree on the location of the universe's edge. Sitting there in London, I was in an entirely different de Sitter universe than my father back in Philadelphia. *We each have our own universe.* Markopoulou had been talking about light cones—and the thing I had learned about light cones is that they grow with time. Wait long enough and you'll see more of the universe. Wait an infinite amount of time and you'll see the whole thing. Not so in a de Sitter spacetime. A de Sitter horizon guarantees that the longer you wait, the less you'll see. In a de Sitter universe, no observer can see the whole thing. *Ever.*

Of course, if you start accelerating, like Safe, the horizon disappears. Now you're in the same reference frame as the expanding space. Nothing is hidden from you, so long as you continue to accelerate. From Screwed's perspective, you'll come to a grinding halt at the de Sitter horizon and burn to a crisp in the radiation. But as far as you're concerned, everything's just fine. There's no horizon. Just more universe.

Unfortunately, you can't accelerate forever—light's speed limit makes sure of that. Spacetime, however, can. Spacetime has no speed limit—it can expand faster than light, just as it did during inflation. If you're racing spacetime, spacetime will always win. Eventually you'll have to stop accelerating and resign yourself to life behind a horizon, stuck in a de Sitter universe. Forever.

I was realizing now that cosmology in a de Sitter universe was a

whole new ballgame. How do you begin to talk about *the* universe
when all observers have their own? In my search for answers, I came
across a talk that physicist Raphael Bousso, a former student of Hawking's, had given at a Cambridge symposium in honor of Hawking's six-
tieth birthday. Bousso's name rang a bell: he had tied for first place
with Markopoulou in the young researchers' competition at the
Wheeler symposium back in Princeton. At the Hawking conference,
Bousso gave a talk called "Adventures in de Sitter Space." He explained
how Hawking and Gibbons had discovered that de Sitter horizons had
all the same quantum properties as black hole horizons, including entropy and temperature. Noting that de Sitter horizons are observer-
dependent, Bousso said that Hawking and Gibbons had "interpreted
their results as an indication that quantum gravity may not admit a
single, objective, and complete description of the universe. Rather, its
laws may have to be formulated with reference to an observer—no
more than one at a time."

No single objective description of the universe? Markopoulou had
suggested we need some kind of observer-dependent logic to deal with
the fact that each of us occupies a different portion of the universe.
Now Hawking was suggesting that we might need an observer-
dependent theory of everything?

I pulled the crumpled IHOP napkin from where it was tucked inside my notebook and scanned the ultimate reality ingredient list. *Par-
ticles/fields/vacuum. Spacetime. Dimensionality. Strings. The universe.
The multiverse. The speed of light.*

From: Amanda Gefter
To: Warren Gefter
Subject: Another one bites the dust

Well, we can officially cross the vacuum, fields, and particles off
the reality list. The de Sitter horizon renders them all observer-
dependent. How strange . . . that this tiny table is just a matter of
perspective thanks to a horizon at the edge of the universe. I
shouldn't say *the* universe. *My* universe. Could the universe itself
really be observer-dependent? We need to talk to Raphael

Bousso. He's suggested "that quantum gravity may not admit a single, objective, and complete description of the universe."
Jesus.

This relationship between the horizon and the vacuum is still bugging me. It's as if one can be mapped onto the other despite the different number of dimensions. The entropy of the de Sitter horizon counts the number of quantum states in each of our universes—but it's finite and scales with the horizon's area. It's funny—everyone says that the big question in cosmology is, what is dark energy? But it seems like you could flip it around. You could say, dark energy is just evidence that we live in a de Sitter universe, so the more interesting question is, *why* do we live in a de Sitter universe? And how does it change what it means to do cosmology?

From: Warren Gefter
To: Amanda Gefter
Subject: RE: Another one bites the dust

How do you fit such big thoughts in such a tiny apartment? The notion of an observer-dependent universe is a bit hard to comprehend. Couldn't there still be one real, invariant universe despite the fact that each observer only has access to a limited piece of it? The dark energy/de Sitter horizon question is fascinating. Are you thinking that we have to have a de Sitter horizon in order to have something instead of nothing? Looks like we still have plenty more work to do! How's your thesis coming?

Shit. My thesis. I had spent months completely bogged down in my research. The discovery that particles weren't ultimately real was a major breakthrough in our quest, and in all my excitement I hadn't actually gotten down to the business of writing. It was due in two days.

I collected all of my notes into a huge pile and sat down to write my thesis: "The Observer-Dependence of Horizons and the Ontology of Matter."

When I had first arrived at the London School of Economics, one of my professors had pulled me aside after class to issue a strange warning. "I know you are used to writing for magazines," he said, to my surprise. "But when you write an academic paper, it shouldn't sound like it's written for a popular audience."

"I know that," I said. "I would never make an academic paper that understandable."

A few days later, a second professor stopped me in the hallway. "Be careful your papers don't sound like magazine articles," he cautioned, looking preemptively disappointed in me.

"They won't," I promised. "I am going to make them as dry and humorless as possible."

So when I finally sat down to write a paper, I conjured what I imagined to be the voice of an old British man wearing a brown tweed jacket with suede elbow patches sitting next to a weathered globe and puffing haughtily on a pipe. *Indeed, as I shall argue in the passages that follow, Duhem's holism thesis does in fact undermine the hypothetico-deductive method of confirmation. Puff, puff. Cheerio.* I littered my sentences with conjunctions and jargon; I combed them for any stray morsels of color or joy. I outlined and reiterated my arguments at every stage. *Here is a summary of what I just said. Here is what I am now saying. Here is what I'm about to say.* I used words like *normative* and *aforementioned*. My professors were duly impressed.

Now that it was time to write my thesis, I conjured the old man and wrote for forty-eight hours. Straight.

In the world of classical Newtonian physics, matter was an objective entity that existed independent of its observers, I began. *Its ontology seemed firmly rooted in the world of absolute space and time. Both quantum theory and relativity shook those ontological foundations, challenging our conceptions of matter. In quantum theory it became clear that, for instance, an electron exists in a superposition of possible states defined by probabilities, and only "collapses" into one particular state when an observer makes a measurement. Thus the observer became a participant in defining the properties of matter. In relativity, it became clear that our perception of matter is determined in part by our frame of reference, and that two observers will not always agree on, for instance, the length of an object. Again, observers were made participants of some kind, and one could no longer provide a description of matter without first establishing a particular reference frame. Nonetheless, neither of these theories entirely undermined the classical view of the ontology of matter. In both relativity and quantum theory, the properties of matter may be in some sense observer-dependent, but the very existence of matter is not. Here we will argue that the quantum properties of horizons in general relativistic spacetimes take that final step in rendering matter entirely observer-dependent, not only in its observational properties but in its very existence. This has profound implications for ontology, and is a direct product of the cutting-edge physics that have begun to unite general relativity, thermodynamics, and quantum theory. Blam.*

I continued, explaining the physics of black hole, Rindler, and de Sitter horizons. I defined their temperatures and entropies, detailing the equivalences between horizon mechanics and thermodynamics. Then I paused to consider whether the three types of horizons were equivalent or merely analogous. *Physically they seem like significantly different situations,* I wrote. *In one you've got a black hole sitting there, the heap of bones of a dead star. In another you've got an observer bolting through empty space. In another the entire universe is ballooning outward. How could they possibly be the same?*

From the standpoint of a structural realist, however, those stories didn't matter. What mattered was structure. Math. And from the math's point of view, there was no difference among the three horizons.

The physics of horizons—the radiation they produce, for instance—results from the relationship between the frame of reference of the observer and the geometry of the spacetime, I wrote. *This relationship holds equally for all three cases under consideration: in the black hole case, gravity defines the geometry of the spacetime; in the de Sitter case, the cosmological constant defines the geometry of the spacetime; and in the Rindler case, the observer's acceleration defines the geometry of the spacetime. But the physics stem not from gravity, nor the cosmological constant, nor acceleration in and of themselves, but from the relationship between geometry and observer. This relationship defines the "structures" of interest, and they are all equivalent. In adopting this position we follow suit with Einstein, whose equivalence principle states that gravity and acceleration are not merely analogous, but equivalent.*

Good, I thought; who could argue with Einstein? Or with a narrator who speaks in a royal "we"?

I explained the meaning of entropy, the derivation of Hawking radiation, and the inequivalent metrics of the observer-dependent vacuum states, finally concluding, *The presence of a horizon indicates a degeneracy of vacuum states, with no preferred field modes definable. The ontological consequences of this are fucking crazy!*

Reading it back to myself, I suspected that the last sentence might not be looked upon as "academic" enough. I tapped the delete key, then tried again. *The ontological consequences of this are dramatic shifts in our concepts of "particles" and "fields," which are shown to be intrinsically characterized by reference frames.*

Better.

Save, print, submit, sleep.

8

Making History

School had ended and my six-month editorship at *New Scientist* was up. I was itching to get back to the States. To see my family and friends and the Sun. To live in a Newtonian apartment. At the same time, my *New Scientist* gig had opened up a new universe of possibilities. It had given me steady access to physicists and the ultimate alibi for pursuing the nature of reality. I wasn't about to give that up. So I convinced my superiors to keep me on as an editor back in the United States, where I'd work in their satellite office in Cambridge, Massachusetts.

Heading back across the Atlantic, I knew that I needed to find out more about event horizons. So many questions were still bugging me. How was it possible that you can map the entropy of the vacuum onto the area of an event horizon, even though the horizon has one less dimension? What did it mean for cosmology that we live in a de Sitter universe, an impassable event horizon lurking in our future? And why had the de Sitter horizon led Hawking and Gibbons to believe, as Bousso put it, that "quantum gravity may not admit a single, objective, and complete description of the universe," but that "its laws may have to be formulated with reference to an observer—no more than one at a time"? I was sure that the answers would help us figure out what was invariant. What was ultimately real.

Back in the United States, I settled into life in Cambridge. The *New Scientist* office was located in Kendall Square, which was basically like physics Disneyland. Within a few blocks of the office were a street called Galileo Way, a restaurant called MC^2, a bookstore called Quantum Books, and a bar called The Miracle of Science. I got an apartment on the edge of the MIT campus overlooking the Charles River.

The only thing missing was Cassidy. I was thankful that my parents had taken her in while I was in London, but now that I was back in the States I couldn't wait to get her back. My mother, however, was holding her hostage. The very same woman who had reeled at the thought of having an animal in her house was now refusing to return her to her rightful owner. And Cassidy, whom I had raised to live the fast-paced life of a New York City dog, had apparently grown accustomed to suburbia, with all its "space" and "grass." What's more, the little traitor adored my parents now. She still wiggled with delight at the sight of me, but she gazed at them like they were home.

Portrait of an editor: my dream job at New Scientist's office in Cambridge, Massachusetts
W. Gefter

To help fill the void, I adopted a stray kitten. "You've been hired to deal with any rodent intruders," I told him when I brought him home. "Quantum or otherwise." He purred.

The next order of business was to go to Santa Barbara. Just before leaving London I had received an advance copy of *The Cosmic Landscape* by Leonard Susskind. Susskind, a physicist at Stanford, was one of the originators of string theory.

Over the past few years I had been piecing together a basic understanding of string theory. The premise was simple: every particle—every electron, photon, quark, and all the rest—is a different vibration of the same tiny, undulating strings. Rather than the diverse zoo of particles, the theory said, the world was made up of just one animal: the string. About 10^{-33} centimeters in length, strings vibrate like the strings of a guitar, playing the various particles like musical notes, including one that sounded just like gravity.

When I first heard about the theory, the whole idea of strings seemed pretty random. I mean, why strings? Why not stars or spirals? But, looking into it further, I learned that strings weren't random at all. In fact, they had been spotted in experimental data.

Before physicists knew that hadrons such as protons and neutrons were composed of quarks, they had been confused by the results pouring out of particle accelerators in the 1960s. At the time, an approach to particle physics called the S-matrix was becoming increasingly popular. The idea was that rather than trying to describe how two particles interact when they collide at a given point in spacetime, you can describe them solely by the starting and ending points of their travels. The S stood for *scattering,* and it went something like this. Step one: from their starting positions and velocities, two particles head toward each other. Step two: they collide, and the energy produced in their collision spawns new particles that decay into other new particles that interact to form still more particles, all of which are surrounded by swarms of virtual particles, which in turn interact with other virtual particles, which in turn . . . ad infinitum. Step three: somehow just a few particles emerge from the mess.

The S-matrix—which Wheeler invented in 1937 and Heisenberg reinvented independently a few years later—lets you skip step two altogether. It's a probability table: plug in the initial state of the colliding particles and the S-matrix spits out probabilities for which particles will emerge at the other end.

Physicists, however, were having a hard time constructing an S-matrix that would account for the outcomes of the hadron collisions they observed in accelerators. It wasn't until 1968 that physicist Gabriele Veneziano solved the problem: he discovered an equation that governed the S-matrix for those hadrons. But why did the equation work? No one was sure. What exactly was it describing?

After months spent holed up in his attic staring at Veneziano's equation, Susskind had an epiphany.

It was describing a vibrating string.

Susskind—and, independently, Yoichiro Nambu and Holger Bech Nielsen—proposed that instead of dimensionless point particles, hadrons had to be composed of tiny one-dimensional strings. You could think of them, Susskind had said, as two point particles, which you might call quarks, connected by a string of point particles, which you might call gluons.

It was a cool idea, but within a few years the successful development of QCD had wiped out all interest in strings. Almost. A few lonely physicists, including John Schwarz and Michael Green, refused to give up on string theory, working in virtual solitude for more than a decade.

Unfortunately, the math wasn't panning out. The string vibrations were far too energetic, which made them too small to be hadrons, and the values of their spins were off. Hadrons are matter particles—fermions—which means their spins come in half-integer values. But the strings' spins were whole integers, like force particles. Bosons. No matter how they looked at it, Schwarz and Green kept finding strings that looked like massless, spin-2 particles, which definitely weren't hadrons. They were something else entirely. They were gravitons.

Schwarz and Green realized that string theory wasn't a theory of hadrons at all. It was a quantum theory of *gravity*. Who cared that string theory didn't work out for hadrons? It was the holy grail!

The only problem was those missing fermions. The strings naturally produced bosons, but string theory could only be a theory of everything if it could account for fermions, too. If string theorists could find a way to transform bosons into fermions—whole-integer spins into half-integer spins—they'd be all set. Luckily, there was a way: supersymmetry.

In supersymmetric spaces, you can find reference frames in which a particle with whole-integer spin becomes a particle with half-integer spin. It was that observer-dependence that had inspired my father and me to cross spin off the IHOP napkin. What one observer calls a boson, another sees as a fermion. That means that you can take the bosons that string theory produces from the start and then view them from different reference frames to get all the fermions, too. Now you've derived all the forces of nature and all of matter from just one basic ingredient: a tiny, vibrating string. Supersymmetry turns string theory into a viable theory of everything.

What made this theory of everything particularly useful was that it did away with the dangerous infinities that occur when particles collide at a single point—the ad infinitum in step two, the same kind of infinity that causes spacetime to cannibalize and dissolve into quantum foam. Strings are one-dimensional; they extend. They can't interact at singular points. Which in some sense means that spacetime doesn't *have* singular points.

"Spacetime itself may be reinterpreted as an approximate, derived concept," the physicist Ed Witten wrote in an article entitled "Reflections on the Fate of Spacetime." String theory's success, Witten explained, came down to this: "There is no longer an invariant notion of when and where interactions occur."

That line kept bugging me, long after I first read it. What exactly did Witten mean? Usually physicists described string theory's taming of infinity as owing to the fact that strings stretch out over a finite distance, smearing out the otherwise singular point of collision. But Witten, the high priest of string theory, seemed to be attributing the smearing not to strings themselves but to observers. If observers can't agree that a collision occurs at a singular point—that is, if the collision's location in spacetime changes from one reference frame to another—then singular points don't exist. They're not ultimately real. Did that mean spacetime itself wasn't real? That it was observer-dependent?

Particles, the theory goes, are made of strings. But ask physicists what a string is made of, and they'll tell you it's an irrelevant question. Strings are fundamental. They're the basic building blocks. They're not made of anything other than themselves. It was hardly a satisfying answer. Was Witten offering a different one? He seemed to be saying that strings stretched across all the different points an observer might identify as the location of a collision. As if the strings themselves are maps of our potential to observe them. As if they are made of reference frames. As if they are, in some sense, made of *us*.

I couldn't help but think back to Wheeler: "One therefore suspects it is wrong to think that as one penetrates deeper and deeper into the structure of physics he will find it terminating at some *n*th level. One fears it is also wrong to think of the structure going on and on, layer after layer, ad infinitum. One finds himself in desperation asking if the structure, rather than terminating in some smallest object or in some most basic field, or going on and on, does not lead back in the end to the observer himself, in some kind of closed circuit of interdependences."

Perhaps string theory could pave the way toward Wheeler's vision, but you have to be willing to make some pretty radical changes to the universe. You have to add more dimensions of space.

According to string theory, different vibrations correspond to different particles. Exactly which vibrations are possible is determined by the shape and dimensionality of the space around the strings. In a one-dimensional world things can only move back and forth; add another dimension and now they can move up and down as well. The more dimensions you have, the more ways a string can vibrate. For a string's vibrations to produce all known particles, you need nine spatial dimensions, in addition to time. The problem, obviously, is that we only see three. The extra six, physicists figured, had to be curled up and compactified—tiny, complex pieces of origami that sit at every point in our three ordinary dimensions, too small for us to see but big enough to accommodate the strings, which are less than a trillionth of a trillionth of an atom, sixteen orders of magnitude smaller than the best microscopes of the future will ever resolve.

This would have worked out perfectly had there been just one way to fold up the extra six dimensions. No such luck. In 2000, Bousso and Joe Polchinski discovered 10^{500} ways, a number that's not quite as big as infinity but might as well be. There was no good reason that nature would choose to implement one folding over another, and each piece of origami created a different vacuum, which meant different particles, different constants, different physics.

The situation was a huge disappointment among string theorists, but Susskind saw an anthropic lining. In his book, he argued that string theory's failure to deliver on the promise of a single, unique "theory of everything" was a blessing in disguise, because it allows the dreaded *A*-word to explain, among other things, that infuriatingly tiny but nonzero value of the cosmological constant, the one that required some insanely improbable mechanism to cancel the infinite vacuum energy out to 120 decimal places and then stop, leaving the perfect crumbs behind to match the observed strength of dark energy. String theory describes a vast number of universes. Eternal inflation produces a vast number of universes. Put the two together and you've got a wildly diverse multiverse in which the value of the cosmological constant changes from universe to universe—rendering ours nothing short of inevitable.

Intriguingly, Susskind's book ended with a chapter about event horizons. Susskind sympathized with the multiverse critics who claim that if other universes are walled off by our cosmic horizon, forever unknowable to us, then anthropic explanations are nothing more than empty metaphysics. But it's possible, Susskind suggested, that the Hawking radiation coming from our cosmic horizon could encode information about what's on the other side. About the multiverse. In that case, we could discuss multiverse anthropics without invoking any unmeasurable physics outside our horizon. The whole enterprise would be legit.

I knew that Susskind and Hawking had battled for decades over the possibility that information can hide in Hawking radiation, and that Susskind, famously, had won. But how did the information encode what was on the other side of the horizon? The answer was all bound up in that still mysterious AdS/CFT conjecture—*something to do with string theory . . . explains liquid fireball?*—and with what Susskind called "horizon complementarity," an idea he described as "a new and stronger relativity principle." I wasn't sure what any of it meant, but in my gut I knew these were the clues to follow. I had to talk to Susskind.

I figured the easiest way to talk to him would be to interview him for *New Scientist* and run it as a Q&A pegged to the upcoming publication of his book. But I was curious to hear the arguments on both sides, pro- and anti-multiverse. David Gross, a fellow string theorist and Nobel Prize winner, was about as anti-multiverse as they came. Maybe

I could convince them to have a debate, I thought. We had never run anything like it in the magazine, but I figured I'd give it a shot.

Amazingly, both Susskind and Gross were enthusiastic about the idea, and Susskind agreed to fly down to Santa Barbara, where Gross was the director of the Kavli Institute for Theoretical Physics. They figured the whole thing might get too rowdy if we opened the debate up to public viewing, so they decided we ought to conduct it in private, just the three of us. Yup. Just the three of us. Just the inventor of string theory, a Nobel laureate, . . . and me.

Heading out to Santa Barbara, I was nervous. Not only was I going to be moderating a heated debate between two intellectual giants, but I had heard from many a fellow journalist that Susskind and Gross were notoriously intimidating. Journalists' reactions to their names alone had left me imagining an interview with either physicist to go something like this: Cowering journalist asks mildly ignorant question about string theory. Physicist stares journalist down, a ten-dimensional fire raging in his eyes. The stare pulverizes the weeping writer, leaving nothing but a puff of dust and a reporter's notebook spinning on the ground. Yeah, I was a little nervous. But when I arrived at my hotel and checked my email, I found one from Susskind. He had just arrived from Palo Alto and was wondering if I'd like to get some dinner.

We met at a quintessential Santa Barbara restaurant on the beach. In his mid-sixties, Susskind was tall and slim with a white beard and a friendly smile. I cautiously checked his eyes for any ten-dimensional flames—all clear. We sat at an outdoor table, stars shining overhead, soft surf of the Pacific sounding all around us. We chatted easily about *New Scientist* and about physics. He had this amazing voice, this thick, old-school New York accent, the kind where every syllable counts, every vowel is drawled, every consonant hammered. He had started out as a plumber in the South Bronx, and after all these years his Bronx accent hadn't faded in the California sun. Everything he said sounded tough and wise. I found myself wishing I could hire him to narrate the inside of my brain.

Susskind put me at ease, but I couldn't quite forget the fact that I was a twenty-five-year-old girl who had never taken a physics class chatting away with one of the most brilliant creative geniuses in science. I was trying my best to come across as older and professional—a plan that failed horribly the minute the waitress arrived. When Susskind ordered a bottle of wine, the waitress asked to see my ID. I blushed, then burned even redder when I realized I hadn't even thought to bring my driver's license with me. I was ready to crawl into the ocean, but Susskind just smiled. "I can vouch for her," he offered. "I'm a physicist."

The rest of the dinner went swimmingly. We drank our hard-won wine and talked about string theory, horizons, and Hawking radiation.

"I'm really interested to learn more about the ideas you brought up in the last chapter of your book," I told him. "About black hole information loss and your idea of horizon complementarity."

"That's good to hear, because I'm writing a whole book about it," Susskind told me. "John Brockman, my agent, talked me into it."

So that's how you get Brockman to be your agent, I thought. Just invent string theory. Or wage intellectual warfare on Stephen Hawking. And win.

I smiled. "I can't wait to read it."

The next morning I met Susskind at the Kavli Institute on the UC Santa Barbara campus. We were early, so we hung out and drank coffee in one of the institute's common areas, which was flanked by chalkboards covered in cryptic equations. Sitting in a physics institute with Lenny Susskind was like sitting in a coffee shop with John Lennon. Passersby did a double take, then walked over to introduce themselves to him, flustered with awe and reverence. Everyone was buzzing with excitement about the debate, bummed that they wouldn't be able to listen in. It was hard to believe that I was the one with the backstage pass. I was thankful that no bouncers would be carding.

When the time arrived, we headed upstairs. Walking into David Gross's office was like walking into the captain's quarters aboard a luxury cruise ship. It was huge, with absurdly large round windows that framed

See those round
windows? That's
David Gross's
office—at the Kavli
Institute for
Theoretical Physics
A. Gefter

the blue waters of the Pacific on all sides like portholes. I half expected
a pod of dolphins to come leaping by the office. Winning a Nobel Prize
did not look half bad. After being momentarily mesmerized by the view,
I remembered to introduce myself; Gross acknowledged me gruffly.
Something about him sent chills of terror down my spine. I checked
his eyes for ten-dimensional flames. Yup. There they were.

Gross had decided that each physicist should begin by trying to present the opponent's view, a strategy I found strange and essentially un-
usable for the magazine, but I was curious to see it play out and too
terrified of him to argue.

Gross went first, presenting the usual arguments in favor of anthropic reasoning: string theory has 10^{500} vacua, inflation grants each
one of them physical existence, and values of things like the cosmological constant vary from universe to universe—therefore things that

we once hoped to deduce as unique answers derived from simple principles can be explained away as completely random. He tried to present the case fairly, but eventually he got fed up. "That's it! I can't do any
more before I go crazy."

Now it was Susskind's turn to present the anti-landscape view.
"What are the objections to it? Jesus, it beats me!" he began. He noted
emotional objections ("It's like fingernails on a blackboard") and philosophical objections (the falsifiability requirement demanded by the
"Popperazzi"). He mentioned historical objections ("There are lots of
examples of strange, mysterious numbers—the period of the Moon,
the height of the tides—that were subject to ignorant, superstitious
explanations before science explained them."). He agreed with the objection that anthropic arguments inherently assume that our kind of
life—low-entropy, carbon-based, water-dependent—is the only kind
possible.

But there was only one objection that Susskind found worrisome.
Because eternal inflation and the string landscape produce an infinite
number of universes, everything that can happen does happen—an
infinite number of times. In such a world, probability loses all meaning. If I wanted to calculate the probability that John Brockman would
be our agent, I'd have to divide the number of universes in which he
becomes our agent by the total number of universes. In other words,
I'd have to divide infinity by infinity, which yields a depressing "unde-
fined."

You'd think you could get around the problem by taking a finite
slice of the multiverse at some particular time, counting up the number of Brockman-is-our-agent universes and the total number of universes in the slice, and then taking the limit of that ratio as the size of
the slice goes to infinity. But thanks to Einstein, that won't work.
There's no global meaning to the phrase "at some particular time," no
clock sitting outside the multiverse and defining a singular time that's
the same for everyone. Time is observer-dependent. Once you've sliced
the multiverse into coordinates of space and time you've broken general covariance and selected a preferred frame. Worse, the outcomes
you'll get for various probabilities change drastically from one time
slice to the next, yet no slice is "truer" than any other. What you need

is a probability measure that tells you how much weight to assign to each kind of universe. Without a measure, there's no way to say things like "The observed value of dark energy in our universe is probable in the multiverse," which, of course, was the whole point of the thing.

The infinity crisis was known among cosmologists as "the measure problem"—a weirdly innocuous name for something that threatened to take down the whole multiverse and the standard model of cosmology with it. The reason cosmologists believed in inflation and its subsequent multiverse in the first place was because the theory was able to predict things such as the smoothness of the CMB and the flat geometry of the space within our horizon. But under the weight of the measure problem, inflation was beginning to implode. The theory was forced to retract all of its successful predictions. It *didn't* predict a smooth CMB or a flat geometry—it predicted *every* CMB and *every* geometry, and offered no way for us to predict which ones we were most likely to observe.

"Eternal inflation is full of infinities, infinities over infinities, and it's possible that once we learn to deal with these infinities the whole idea of eternal inflation will collapse," Susskind said. "To me, this is the real Achilles' heel. It's the one I worry about most."

Finally Susskind offered up a mock rendition of Gross's famous objection. "There's a cultural danger and a danger to science itself that doesn't have to do with the truth or falsity of the proposition: if we don't resist the temptation of the anthropic principle, young people will give up looking for the mathematical, rational reasons for things. The danger that the real explanation will be overlooked is so strong and the anthropic principle is so seductive that we had better discourage thinking about it," he said, his voice dripping with sarcasm.

"What is at stake here?" I asked.

"Everything," Gross said. "There's a recent article by [Steven] Weinberg in which he claims that this is one of the great sea changes in fundamental science since Einstein. That may be carried away, but this is a very radical change in what I regard as the direction of science. . . . And my general feeling is that the arguments aren't by any means strong enough yet; they have holes everywhere you look. . . . Enormously powerful statements are made about the inevitability of this conclusion, the lack of principles that distinguish the state of the universe in string theory, but everyone admits that we don't know what string theory is. . . . We don't have the equations, the principles, the theory yet, and to base very far-reaching conclusions on a theory which everyone admits doesn't yet exist seems dangerous to me."

String theory doesn't yet exist? But Gross is a string theorist.

"And the thing that bothers me most," Gross continued, "is the absence of any consistent cosmology. Since Einstein, it has become the goal of fundamental physics to predict not just the present state, but rather the whole spacetime manifold. The business of physics, we've learned from general relativity, is the whole damn thing! The present is an illusion anyway. One of the disappointing aspects of string theory, which doesn't yet exist as a theory, is that so far we've made no insight into constructing consistent cosmologies. . . . To imagine this totally radical change in the scope of physics based on ignoring the initial conditions, the big bang, the construction of a consistent cosmology . . . okay, we have 10^{500} possible metastable states in the landscape, and on the other hand we have zero cosmologies."

I scribbled a reminder to myself in my notebook: *Why can't string theory deal with cosmology?*

"Normally in science, a scientific principle grows more powerful the more we know," Gross said. "But the anthropic arguments are the opposite—the more we know, the less force the anthropic principle has. It thrives on ignorance. And I don't think that's the way science should be done. I think the whole thing is premature and slightly dangerous. Why do I think it's dangerous? Because it's giving up the traditional route, and it can divert people from what might be the right directions. . . . In the face of all these issues, I draw a different conclusion. We're missing an important principle. We're missing the principles that allow us to construct a cosmology, to discuss eternal inflation. I would take all of the arguments for the present situation not as arguments for being stuck with anthropic reasoning but as suggestions that something fundamental is missing."

"David, you've reminded me of a few sayings," Susskind said. "'Old men are doomed to forever relive their past.' And, 'The less you have to say, the longer it takes.' As far as being dangerous, I think the shoe can be put on the other side of the foot, if that's the expression. David is very beloved in physics, he's extremely admired, and he's feared."

You think? He's freaking terrifying.

"By whom?" Gross demanded.

"Oh, by a young man we had an occasion to talk to earlier today. But also I've had many occasions to talk to young men—women seem immune to this—who when I discussed with them the possibility that the world might be anthropic were very reluctant to discuss it; they became embarrassed. Young people are being intimidated by a very powerful hostility toward these ideas."

"That people like myself have strong opinions? I'm not worried about that. What worries me is when these young people get up and talk with utter authority as if this set of ideas is based on firm knowledge."

"Boy, is the pot calling the kettle black! Wow! David, David, David. Do you remember the days shortly after heterotic string theory?"

Oh, snap!

"So? Fine. It's a little different. When these young people talk I have to remind the audience that here speaketh someone who is talking as if we know what the hell we're talking about. String theory is not something we know what we're talking about, right?"

"Yes, I completely agree."

He agrees? But he invented it. . . .

Gross continued. "So when someone gets up and says, 'String theory says . . . ,' it is part of the reaction you have aroused with your exuberance for this idea. And I can easily understand about my excitement about the heterotic string revolution of 1984—I can understand and see it happening again now with some of my colleagues, the allure of the random universe. I see it happening with poor Steve Weinberg. Steve is a man who is driven by atheism. And he is exuberant in the end of his paper because the Catholic Church has come out against the landscape idea."

"We all like that. We all find it amusing."

I chuckled.

Gross was referring to Weinberg's paper "Living in the Multiverse." In it the Nobelist wrote, "Just as Darwin and Wallace explained how

the wonderful adaptations of living forms could arise without supernatural intervention, so the string landscape may explain how the constants of nature that we observe can take values suitable for life without
being fine-tuned by a benevolent creator."

Weinberg went on to quote a *New York Times* op-ed by Christoph
Schönborn, a cardinal and the archbishop of Vienna, who wrote,
"Faced with scientific claims like neo-Darwinism and the multiverse
hypothesis in cosmology invented to avoid the overwhelming evidence
for purpose and design in modern science, the Catholic Church will
again defend human nature by proclaiming that the immanent design
evident in nature is real." Weinberg then noted, rather exuberantly,
"Martin Rees said that he was sufficiently confident about the multiverse to bet his dog's life on it, while Andrei Linde said he would bet
his own life. As for me, I have just enough confidence about the multiverse to bet the lives of both Andrei Linde *and* Martin Rees's dog."

"What I'm saying," Gross continued, "is that some of the reaction
is exactly like the reaction I got for exuberance in 1984, when we believed the answer was around the corner and we got carried away with
that position. And, Lenny, you are carried away with this position. The
stakes are damn big. So you are open to severe criticism."

"But make it scientific criticism," Susskind insisted. "Scientific
criticism would be a theorem that says you can't have metastable de
Sitter space in string theory, or a demonstration that eternal inflation is
internally inconsistent. That is science, David."

"It is incumbent upon those who build on these shaky things to
make them less shaky," Gross retorted. "I don't have to prove that
something that's ill defined doesn't make sense."

"David, do you see a way out of the landscape?" I asked. "Do you
see any particular route?"

"Do I see a particular route? Obviously not. If I did I would be in a
much better position. The anthropic line of reasoning can only be
killed by science increasing its power. It's very unlikely that over the
age of sixty I'm going to come up with the necessary insight. But where
is it going to lie? It's going to lie in the question 'What is string theory?'
It's going to lie in the question 'How does one construct one consistent

cosmology? One universe that makes sense?' We don't have one universe that makes sense!"

I turned to Susskind. "Even if you accept the landscape and the multiverse, and even if you accept that certain local physical laws are anthropically determined, don't you still need a metatheory? Don't you need something unique? Doesn't it just push the question back?"

"Yes. That's certainly true. Absolutely. The bottom line is that we need to describe the whole thing, the whole universe or multiverse. And it's a scientific question, not an ideological one."

The debate lasted a couple of hours, and afterward the three of us went to a beachside restaurant for a seafood lunch, where we chatted happily about physics. I felt at ease, as if we had bonded somehow, like soldiers returned from battle. Having heard extensive arguments on both sides, I wasn't totally persuaded by either one. Gross's view was more appealing: I, too, would breathe easier knowing that physics was determined by unique, necessary, elegant equations, and not the luck of the draw. On the other hand, I wasn't sure the universe cared much about what helped me breathe easy. And the confluence of the string landscape with eternal inflation's physical multiverse did seem to point in one obvious direction. Still, it was too soon to say anything for sure. There were too many questions hanging everywhere.

Like, why was string theory so useless in the face of cosmology? Why, as Gross said, was it incapable of describing a single universe that made sense? Perhaps because the theory doesn't exist, I thought. What did they mean by that? And how dangerous were those rampant infinities? Susskind had said that they were eternal inflation's Achilles' heel, preventing physicists from calculating probabilities, undermining the entire appeal of the landscape in the first place. Gross had said that a fundamental principle was missing from string theory, from cosmology. But what kind? If the anthropic argument thrived on ignorance, what brand of wisdom was going to take it down?

As I flew back to the East Coast, I felt overwhelmed by how much more I needed to learn, but certain that I had found the right leads to follow. I needed to delve deeper into string theory, even if it didn't exist. I was itching to get back to the issues of horizons, invariance, de Sitter space, and observer-dependence. And I couldn't wait to figure out what Susskind's horizon complementarity was all about.

I was also struck by a sudden, disturbing thought. If we really did live in an infinite multiverse, then the number of computer-simulated universes would skyrocket exponentially, and with it, the odds of our being real—whatever that meant—became infinitely negligible. In the face of a multiverse, Bostrom's little comedy routine was even more terrifying. Then again, I still wasn't convinced that there was any fundamental difference between simulation and reality, since the only view of reality that would reveal it to be a simulation would be a God's-eye view from outside reality, and reality, whatever it is, does not have an outside. Besides, the multiverse hypothesis arose as a direct consequence of the laws of physics—of eternal inflation and the string landscape—which themselves were designed to explain the universe we see around us. If the universe we see around us is a fake, then the laws of physics don't tell us anything about the "real" world beyond our simulation, the world in which the hardware lives, which might easily not include a multiverse at all, driving down our odds of being in a simulation in the first place.

My brain was starting to hurt. Were these thoughts getting me closer to ultimate reality or was I spinning in circles? And if the multiverse did exist, were there infinite copies of me thinking the same infinitely circular thoughts an infinite number of times?

Jesus, that was depressing. It was too much pressure. I couldn't stand the idea that every trivial thing that came out of my mouth was being broadcast over and over, echoing stupidly in the vast and repetitive multiverse. I suddenly understood Borges's fear of mirrors, "that horror of the spectral duplication or multiplication of reality." The multiverse rendered me even less authentic than did the Bostrom nightmare, because at least there I could imagine myself to be a unique, one-of-a-kind simulation, a simulation of myself. In the multiverse, I couldn't even take credit for a single word I said or wrote. It wasn't as though I would be the one true, original version of me, and the rest just carbon copies. If the multiverse was real, then I was a carbon copy, too, my thoughts mere facsimiles, my words as empty as their echoes. In an infinite multiverse, everything I did or thought or said would bear infinite weight and at the same time mean nothing at all. "I" would be "we," and "we" would be a dime a dozen.

I was back in the *New Scientist* office browsing the latest physics preprints on the arXiv when I spotted it, that one thing that can make any girl giggle and blush with delight: a new paper by Stephen Hawking. Written with Thomas Hertog, a young physicist at CERN, the paper promised to introduce a new approach to cosmology "in which the histories of the universe depend on the precise question asked."

Intrigued, I dove in. String theory, Hawking began, offers a vast landscape of universes, "but it has remained unclear what is the correct framework for cosmology in the string landscape." *We don't have one universe that makes sense!* The problem, Hawking explained, is that string theory grew out of the S-matrix, out of the need to make sense of those weird hadron collisions. When physicists model particle collisions, they do so from the perspective of an observer standing outside the accelerator, sending two particles racing toward each other, and recording what comes out while happily ignoring all the convoluted crap in the middle. It was a bottom-up approach, one in which you know the exact initial state of the system (step one) and from that you can evolve it forward in time to predict the outcome (step three). That works great for laboratory experiments, Hawking said, "but cosmology poses questions of a very different character. . . . Clearly it is not an S-matrix that is the relevant observable for these predictions, since we live in the middle of this particular experiment." In other words, when it comes to cosmology, we *are* the convoluted crap in the middle.

From here inside step two, how are we supposed to get a handle on step one? How do we find the initial state of the universe? According to Hawking, we don't. The newborn universe's near-infinite energies and densities are fundamentally quantum mechanical. The early universe, according to Hawking, was a quantum superposition of all possible states, with no reality to call its own. So not only do we not know the exact initial state, *there is no initial state to know.* THERE IS NO STATE.

These two facts—that we are stuck in the middle of the experiment and that the universe had a quantum origin—render the S-matrix and its bottom-up philosophy useless for cosmology.

It was time to rethink the universe, Hawking said—and that meant working from the top down. It meant starting with the observer and working *backward* toward the origin of time. The top-down approach, Hawking wrote, "leads to a profoundly different view of cosmology, and the relation between cause and effect." In this approach, "histories of the universe . . . depend on what is being observed, contrary to the usual idea that the universe has a unique, observer independent history." My mind flashed back to the Davis conference, when Hawking had rained on the inflation parade, and to the clue I had scribbled in my notebook: *No observer-independent history.*

I needed to find out more. I called a features editor in the London office, who, of course, had already heard the buzz about the paper. He agreed that we ought to run a big story, and, to my relief, told me to get to work.

I knew it wouldn't be easy, or even possible, to get an interview with Hawking, so I figured I'd start by calling Hertog.

"The top-down approach is a mixture of theory and our position within the universe," Hertog told me over the phone. "It is very much a framework for cosmology from the perspective of an observer inside the universe. It is, in contrast with the bottom-up approach, most certainly not a framework from a God's-eye view."

I asked him to take me through the details. He explained that their theory combined two key ingredients: Feynman's sum-over-histories approach to quantum mechanics and the Hartle-Hawking no-boundary proposal.

Feynman's sum-over-histories approach explained, in quantum mechanical terms, how a particle gets from here to there. That bat-shit double-slit experiment had already revealed that, when no one's looking, a photon travels multiple, mutually exclusive paths simultaneously. If I turn on a lamp, a photon travels from the bulb to my eye. Common sense tells me that it travels in a straight line, but, as usual in physics, common sense is wrong. If I were to run the lamp-to-eye experiment many times, somehow recording the interference pattern that results on my retina, I could reconstruct the photon's travels and, according to Feynman, I would find that en route to my eye the photon had simultaneously navigated infinite paths throughout the entire universe, no matter how unlikely those paths may seem. In one it bounces off the Moon, circles the tower of London, and skims John Brockman's hat before hitting my retina. In another, it flies by the Great Pyramid, hitches a ride on an elephant, and skirts the horizon of a black hole. In still others, it laps the universe. Once. Twice. It ricochets off every mirror. It does the hokey pokey. It turns itself around.

The probability for each path takes the form of a wavefunction, and the waves' phases interfere constructively in some places and destructively in others. The most absurd routes are easily canceled out by equally absurd but oppositely aligned routes. When all interference is accounted for, the last wave standing gives a high probability for the most reasonable path: the straight line, lamp to eye.

There's just one weird mathematical trick you have to employ to get Feynman's procedure to work: you have to add the waves in imaginary time. Imaginary time isn't "pretend time"—it's a time coordinate written with an imaginary number; that is, a number multiplied by i, where $i^2 = -1$. And the point is, it works. Using imaginary time instead of real time yields the right probabilities for the outcomes of experiments.

In the early 1980s, Hawking and physicist Jim Hartle decided to apply Feynman's sum-over-histories approach to the universe as a whole. It was Wheeler who had first emphasized the need to treat the universe as quantum mechanical, and Hawking was one of the brave few who followed in his footsteps. Instead of adding wavefunctions that represent the paths of particles *through* the universe, Hawking and Hartle needed to add wavefunctions that represent *the universe itself*, entire cosmic histories encoded in spacetime geometry.

Here, too, the procedure required the use of imaginary time—but now it had some rather profound consequences. When it comes to cosmic time, there are usually only two options: either the universe has always existed, and so time stretches back into the eternal past, or the universe has a beginning, and time begins at a singularity. As far as Hawking was concerned, they were both terrible options. If time is eternal, you're just stuck with it, with no hope of explaining where it came from, since it never *came* from anywhere—it just *was*. If it begins in a singularity, however, you're still just stuck with it, since the laws of physics break down there and lose their explanatory power.

In imaginary time, those two terrible options remain—imaginary time can extend into the eternal past or it can begin in a singularity—but there's also a third option. Imaginary time is indistinguishable from space, so it's possible that as you look back toward the universe's origin to just a Planck second after the big bang, what would have been the time dimension transforms into a spatial dimension, leaving the universe with four dimensions of space and no time at all. Where time was presumed to have begun at the singularity, a new dimension of space appears instead, and the singularity vanishes. Spacetime has no edge; it's more like the surface of a sphere: finite but unbounded. Hence the name "no-boundary."

Hawking and Hartle realized that no-boundary cosmology was our only hope of explaining the origin of the universe *from the inside*. In a no-boundary universe, Hawking wrote, "the universe would be completely self-contained and not affected by anything outside of itself." No blank spots on the map, no breakdown of physics, no rift in spacetime through which something else—something external—could reach in. Just a one-sided coin, all inside, no outside.

Of course if you viewed the universe in ordinary time, the singularity would still be there, that blank spot on the map, that quantum dragon. But switch over to imaginary time and the singularity disappears, the rift heals, the world is whole again.

"This might suggest that the so-called imaginary time is really the real time, and that what we call real time is just a figment of our imaginations," Hawking wrote in *A Brief History of Time*. "In real time, the universe has a beginning and an end at singularities that form a boundary to spacetime and at which the laws of science break down. But in imaginary time, there are no singularities or boundaries. So maybe what we call imaginary time is really more basic, and what we call real is just an idea that we invent to help us describe what we think the universe is like. But . . . it is meaningless to ask: which is real, 'real' or 'imaginary' time? It is simply a matter of which is the more useful description."

A PRAGMATIC APPROACH

Here's the reality test. If you can find one frame of reference in which the thing disappears, then it's not invariant. It's not real. Imaginary time was the reference frame in which the singularity disappears. Did that mean the singularity was never real to begin with? That it's merely an artifact of perspective?

Hawking and Hartle proposed that because the universe has no outside, and therefore must be causally closed, only histories that make use of imaginary time's third option—the one in which the singularity disappears—should be included in the quantum sum. But their no-boundary proposal was still a bottom-up approach: it collected every possible history that begins in a no-boundary state and summed them to find the most probable universe.

Now Hawking and Hertog were advocating a top-down twist. Rather than starting from step one, they started from step two. They started from *today*. They started with *us*.

To define step two, Hertog explained, you choose some measurements as your input—say, the universe is nearly flat, is expanding, and has a small cosmological constant. Then you work backward and consider every possible history of the universe—every history without a past boundary, that is—that could have led to our current observations. "The universe doesn't have a single history, but every possible history, each with its own probability," Hertog said. Sum those together, letting their probability waves interfere until only one wave remains, and you've determined the history of the universe.

As Hertog talked, it began to sink in just how strange this idea was. It wasn't like they were reverse-engineering the universe to uncover its actual history. No. They were saying that the universe *had* no history— history is created the minute we make a measurement. In the present. Now. "As observers, we play an active role," Hertog said. By making a measurement, we select out a subset of all possible histories, and from those histories a single past unfurls.

DETERMINATE BEING HISTORY ONLY - NO ABSOLUTE BEING BEGINNING

Of course, Hawking and Hertog weren't the first to claim that observations made today could determine the past. Wheeler was. I thought back to his delayed-choice experiment, that twisted take on the double slit in which observers choose whether a photon takes two paths or just one, even if the choice is made billions of years *after* the photon begins its journey. "The past has no existence except as it is recorded in the present," Wheeler wrote. "By deciding what questions our quantum registering equipment shall put in the present we have an undeniable choice in what we have the right to say about the past. What we call reality consists of a few iron posts of observation between which we fill in by an elaborate papier-mâché construction of imagination and theory."

Now Hawking and Hertog were applying the delayed choice not to a handful of photons but to the entire universe. The big bang, inflation, 13.7 billion years of cosmic evolution . . . it's not that these things didn't happen; it's that they happened *right now*. The universe's past is out there for us to peer into with our telescopes—but it starts with us. Like a choose-your-own-adventure story in reverse, we choose the history of the universe.

WE STRUCTURE IT.

As a teenager, I had once wondered, What if I was just born, right now, only my brain came complete with a full set of false memories, so that in reality this is the very first moment of my life and yet I feel as though I have been alive for fifteen years? Of course, I couldn't remember all the way back to my birth. The first several years were more or less blank, the occasional fragments provided by photographs and home videos. As far as I could tell, I had woken up in the middle of my life and was stuck trying to make sense of a story without a beginning or an end. But the possibility that this very moment contained within it all other moments meant, for me, that the existence of the past would never be so puzzling as the existence of the present, which itself was a kind of origin, slippery and mysterious and unaccounted for.

Still, when it came down to it, I had always conceded that time was at least approximately real and linear, a kind of tether, an umbilical cord, binding us to our own births and before them to cosmic history, stretching back through empty, interstellar space, past the stars where the atoms in our bodies were forged, through webs of galaxies, tracing the slopes and swells of space until it reached a beginning. A big bang. But the no-boundary proposal told a dramatically different story:

there's no beginning to anchor the cord. As it reaches back toward the big bang, time becomes space and the cord turns a corner, following the curve of the new spatial dimension and looping back on itself, then continuing on its way and ending up back where it started. According to the no-boundary proposal, time is an umbilical cord that tethers us to ourselves.

And really, wasn't that what Wheeler's U-diagram was all about? That giant eyeball was ogling the past. The universe gives rise to the observer, then the observer looks back in time and gives rise to the universe. A self-excited circuit that seems to turn that relentless chronological march on its head: after, then before.

Was all of this top-down business a flagrant violation of causality? Of the inviolable laws of physics? I was dying to ask Hawking. Hertog agreed to forward an email to him on my behalf. *He might respond,* Hertog told me, *but don't get your hopes up.*

A few days later, Hawking responded.

"Is there really a kind of backward causation taking place?" I had asked him.

"Observation of final states . . . determine[s] different histories of the universe," Hawking replied. "However, this backward causality is an angel's-eye view from outside the universe. A worm's-eye view from inside the universe would have the normal causality."

At first glance, that made sense. From outside the universe, where you could see the tangled superposition of possible histories, you could watch an observer in the present select a single past. To the observer here on the inside, though, the past just seems to be sitting there, as if it always had been there.

But the more I thought about it, the weirder that was. Why would the laws of physics break down in the God's-eye (angel's-eye) view? If anything, you'd expect it to work the other way around—that from a God's-eye view, where you could take in the whole of nature, everything would finally make sense, all the disparate pieces of the puzzle snapping into place, leaving the laws of physics pristine and complete. Break that symmetry with a single observer's (worm's) limited viewpoint and you'd expect to see some kind of violations, the broken puzzle pieces' jagged edges. I scribbled my confusion in my notebook. *Laws of physics intact only when viewed inside a single light cone?*

If nature's laws hold up only within a given observer's light cone, then somehow physics really is tied to observers, as Wheeler had envisioned. Did that mean that an observer really is a radioactive jewel, "the flashing purpose that lights up the whole dark universe"? I was already aware of the many disturbing questions the whole jewel thing raised.

Of course, Hawking and Hertog weren't actually singling out individual observers. They weren't suggesting that the history of the universe was different for me than it was for my father. But that's only because the measurements my father and I would use as input at step two—measurements of the universe's geometry or the expansion rate—would be exactly the same, given how close we were to each other, astronomically speaking. But if there were some observers off in a distant galaxy whose light cones barely overlapped with ours, their measurements could feasibly be quite different. If so, their whole cosmic history could be different. It's not merely that they would calculate a different history; they would literally live in a universe with an objectively different past.

Did that mean they would live in a different *universe*? That the universe itself wasn't invariant? Wasn't real?

I wasn't sure. Hawking seemed to be leaning in that direction. Still, the no-boundary proposal was just that: a proposal. It left all singular cosmologies out of the sum of possible histories out of conviction that the universe had to contain its own explanation on the inside. It seemed like solid reasoning to me—what exactly was the other option? Still, in physics, assumptions are never enough. You have to derive them from something deeper.

In the meantime, Hawking and Hertog looked at the cosmic history they had spun out from the top down, at the wavefunction of the universe that had emerged from the sum of all possible histories that began with no boundary and ended with the universe we see today, a wavefunction containing probabilities for any measurement we want to make. Interestingly, the most probable history was one in which the universe underwent a brief period of early inflation. This top-down inflation didn't require the usual fine-tuning, nor did it go eternal and produce anything beyond our observable universe. *History of the universe begins right now,* I wrote in my notebook, *and ends at the cosmic horizon. Nonetheless, it looks like it began 13.7 billion years ago and underwent a brief period of inflation. Observer looks back in time and gives rise to the history of the universe, sees exactly the kind of history needed for the observer to exist in the first place.*

I knew that proponents of eternal inflation were not going to be fans of Hawking's new theory. After all, it denied eternal inflation's origin in a well-defined inflaton field and stymied its attempts to create anything beyond our horizon, let alone an infinite multiverse. What's more, it turned inflation itself into a kind of observer-dependent illusion. But just for kicks, I picked up the phone and called Andrei Linde, the inflationary evangelist who had yelled at me back in Davis when I had suggested that the mysterious low quadrupole might cause some physicists to abandon inflation. I asked him what he thought of Hawking's theory and braced myself for an impassioned tirade. But this time his answer was short and sweet: "I don't buy it."

Hartle, understandably, was more sympathetic to Hawking's cause. As strange as their theory sounds, he said on the phone, it's really the only way forward given that we are stuck inside the universe. "It's a different viewpoint, but it's sort of inevitable. Cosmologists certainly should be paying attention to this work."

The no-boundary proposal offered to slay the quantum dragon, removing the singularity from the universe's origin and allowing us to explain the universe from the inside. But there was another reason for cosmologists to pay attention to Hawking and Hertog's work. Top-down cosmology provided a possible answer to Wheeler's question, one that had been resounding in my head for years: *If an anthropic principle, why an anthropic principle?*

The absurd value of the observed dark energy wasn't the only thing that seemed inexplicably fine-tuned for the existence of biological life, although it was clearly the worst offender. Fine-tuning is rampant— change one or two physical values by even the slightest amount and our existence is rendered impossible. Had the distribution of matter in the early universe been just a little lumpier, black holes would have formed in place of stars and galaxies. Had it been just a little smoother, we'd see no structure at all. If the weak nuclear force had been a little less weak, the only element around today would be hydrogen; a little weaker and we'd have nothing but helium. Either way, no stars could form. Without stars there's no carbon; without carbon, no life. The strength of gravity, too, was just right—a little stronger and our Sun would have burned out after some ten thousand years, far too quickly to sustain biological evolution. Had the mass difference between the proton and neutron been slightly off, atoms themselves would be unstable. And then there was the cosmological constant. That fine-tuned son of a bitch.

Many physicists were following Susskind's lead in explaining those coincidences with the infinitely varied landscape of string vacua, physically realized in the infinitely large multiverse produced by eternal inflation. It was the argument that fanned the flames in David Gross's eyes.

Hawking and Hertog bought into the string landscape—but they didn't accept eternal inflation as the means for populating it with physical universes. Instead, they saw the varied worlds described by string theory as possible histories that exist not in physical space but in the mathematical superposition from which our universe's history is derived. You can still use anthropic arguments to explain all the apparent fine-tuning without ever having to reference anything outside our own cosmic horizon. Hawking and Hertog had turned the landscape on its head: instead of multiple universes with a single history, you have a single universe with multiple histories.

When you think about it, all that fine-tuning makes sense in a top-down universe. If the history of the universe, along with all of its physical properties, is determined by our observations, then of course it will be a universe perfectly suited for us—how else would we be here to make those observations? Anthropic coincidences are problematic for bottom-up cosmology because you are starting with some initial state that's completely independent of observers; the universe evolves forward in time until observers like us just happen to arise, a fluky byproduct of physics and happenstance. Given random initial conditions some 14 billion years ago, of course we're scratching our heads and asking, what were the odds that the universe would just happen to have every minute ingredient to cook up the fragile stew of life? Top-down cosmology, on the other hand, doesn't raise the question. Instead of starting with a cosmic history and evolving observers, top-down cosmology starts with observers and evolves a history. And if you start with life, you're bound to end up with a life-friendly universe.

Why an anthropic principle? I wrote in my notebook. *Because the universe is observer-dependent?*

Such jewel-toned thoughts about life made me nervous—any theory that relied on humans or consciousness as being some kind of "special" ingredient instantly struck me as crackpot. But Hawking and Hertog weren't suggesting that observers' consciousness magically collapses the wavefunction of the universe. In their model, there's no collapse, just the interference of the universe's many paths through history. It wasn't like you *needed* life to create a universe; it was simply that there *is* life in this one, therefore the only histories relevant to the quantum sum will inevitably be life-supporting ones. Which of course was all a bit circular in its logic. Like a self-excited circuit.

I opened up a blank document on my computer, rolled up my sleeves, and prepared to write my article. It was quite a story. Here was Stephen Hawking telling us to give up all our old ideas about cosmology—the ones that told us there was some origin of time independent of us, some 14-billion-year evolution that had nothing to do with our observations, some independent reality, some way the past "really was." By combining Wheeler's self-excited circuit with the landscape of string theory, Hawking had me thinking that it might not be long before we crossed the universe off the ultimate reality list.

When I finished my article, I sent it to Hertog to check for inaccuracies and told him to feel free to pass it along to Hawking. A few days later, an email popped up in my inbox. I opened it, and there inside were the best words a reality-hunting pseudo-journalist could imagine: *The article is remarkably good and clear. Stephen.*

On a brisk April morning in 2008 I wandered happily into the *New Scientist* office, ready to kick off another day of tracking down reality, but a fellow editor greeted me with a somber expression. My smile dropped. "What's wrong?"

"John Wheeler died."

I was stunned. Speechless. It's not that I couldn't have seen it coming—he was ninety-six years old. But it felt like a blow all the same.

"You were really interested in his work, weren't you?" my colleague asked. "Do you want to write the obit?"

I nodded silently. I couldn't help feeling heartbroken. I had spent years imagining that when my dad and I finally did write our book about the nature of reality, Wheeler would read it. In a weird way, I had always assumed that we would be writing it *for him*. In my mind we would personally deliver a freshly printed copy, which he would immediately and eagerly read. As he turned the last page and let the back cover thump closed, he'd look up at us, remembering the cryptic words he had offered us years ago in Princeton, and with a twinkle in his eye say, "I see you've figured it out. Well done."

As I settled into my desk chair to begin composing the obituary, I texted my dad the bad news: *Wheeler died*.

My phone buzzed back right away: *It's a sad day for reality.*

9

A Hint of How the Universe Is Built

It didn't seem fair that time had run out on Wheeler. That the universe had blinked out of existence before he ever had a chance to solve its riddle. That his four questions—*Why the quantum? It from bit? A participatory universe? How come existence?*—still hung in the air like stunned raindrops.

They weren't the only ones hanging there. What about the self-excited circuit and the boundary of a boundary? How would we decipher their meanings now?

As I filed Wheeler's obituary, I thought back to that day in Princeton. To the awe we had felt as we consulted the oracle, and to the aftershocks that ran through us while we stood speechless on Einstein's lawn. That day had been the start of something. Something that had led me here, to Cambridge, to *New Scientist,* to day after day spent masquerading as a journalist.

Masquerading? Was I still a fake after all these years? How long would I have to do this before I'd stop feeling like it was some kind of prank? Maybe everyone feels like a fraud in their own life, I thought. Maybe everyone is.

Or maybe the problem was that I wasn't a fraud anymore. This job was supposed to be a cover for something. Had I gotten so swept up in the means that I had forgotten the end? Worse, had I forgotten the beginning?

As I mourned Wheeler's death, I realized I was mourning something else. I spun slowly in my desk chair, surveying the lifeless cubicle. Where was my father? Where had he been when I was mulling over horizons in my quantum flat in Notting Hill? Where had he been when I was discussing the multiverse on the Santa Barbara beach or non-Boolean logic at the Tribeca Grand? In Bloomsbury? At the Holiday Inn? Where had he been then? Where was he now?

My plan to score a permanent press pass and gain us access to the inside world of physics was working, but somewhere along the way I had veered off course, started speaking in singular pronouns, trying to create a life for myself. I had to, I knew, but still. Five years earlier, on a spring day just like this one, when I decided to go to the Davis conference without him, I had unknowingly set off a chain of events that had ended here, in an office, in a desk chair, alone. Was that past now written in permanent ink? If I viewed things differently now, would I compute a new wavefunction? Select a new history? Enact some quantum process of top-down regret?

And what did the old history look like through my father's eyes? Was he proudly watching his daughter from afar as she lived out a shared dream, crashing the physics party vicariously through her as life happily carried on in the suburbs? Or was he stuck there, trapped in a life that ran on its own momentum, watching his idea slip away from him, his own invention speeding off into the distance like a missed—or hijacked—train? Maybe in some other universe we had run giddily from Einstein's lawn and just kept on running. Maybe we were still hunting down reality together there. Maybe Wheeler was alive and the rain was falling. It didn't matter, though, because I was stuck in this one.

"It's a shame we never went to talk to him a second time," my father said.

I nodded my regret, then took a swig of cold soda.

It was the height of summer in Boston, and my parents had come up from Philadelphia to visit. The three of us were lounging on my balcony, overlooking a sun-drenched Memorial Drive and the Charles River, the Boston skyline rising from the far riverbank and reflecting in the sparkling water. My mother sat happily knitting while my father and I talked physics as usual. It was a perfect simulation of a perfect day.

"Why didn't you?" my mother asked, her eyes on her knitting.

"We couldn't," I said. "About a year ago I contacted Ken Ford, who worked with him, to see if we could set up an interview, but he told me Wheeler was in a nursing home. He didn't say we couldn't go, but it seemed too disrespectful."

"Since when are you worried about being disrespectful?" my mother laughed.

"We could have pretended to be doctors and gone into his room," my father said. "We'd ask him procedural medical questions, but then we'd slip in some questions about the nature of reality. *Are you having difficulty breathing? Are you experiencing any nausea? What does it mean that the boundary of a boundary is zero?*"

"You *are* a doctor!" I said.

He chuckled. "Oh, yeah!"

"Anyway, he was so hard of hearing at that point, it would have been impossible," I said.

"So now what?" my mother asked.

"I don't know."

It was depressing. With Wheeler gone, the mystery of his cryptic phrases had grown even deeper and more profound in my mind. Of course, I knew he hadn't come up with *the* answer to the universe—if he had, we would have heard about it by now. Wheeler would be a household name, the Nobel committee might have sat up and taken notice, fundamental physics would be over and done with, and we wouldn't have been on this quest in the first place. And yet somehow, despite all that, I was convinced that those phrases were the keys to reality's riddle.

Maybe I just wanted to believe that, because it was romantic and exciting and it made me feel important, as if I were somehow carrying Wheeler's torch, as if he had intended to hand it to some brilliant physicist that day back in Princeton but his eyesight was failing and he mistakenly handed it to me. To *us*. Now, years later, that flame was flickering, threatening to extinguish, and all I could see of it was a wisp of smoke, cutting the air like a rat's tail.

"What about people who knew him?" my father asked. "Like his former students? They would have some idea what Wheeler had been thinking about."

I shot up in my chair. "That's true. We could go talk to people who might know what he meant by those phrases. We could say we're writing an article about it."

"Why would I be writing an article?" my father asked.

Good point. "Okay, so we'll say we're coauthoring a book."

"You *are* coauthoring a book," my mother said.

I nodded. "Right."

That was true—kind of. We had been talking about it for years, the book my father had suggested we ought to write together when we found the answer to the universe, the book I dreamed would someday be represented by John Brockman. Every time I visited my parents' house, my father and I would excuse ourselves to go upstairs to the physics library by telling my mother we had to "work on our book." Every time I told my friends I couldn't go to a party because I was busy learning physics, I told them I was "working on our book." "Our book" had become the central character of our lives, and yet it didn't exist as anything more concrete than a distant dream. Working on the book had never included putting a word on paper. Working on the book meant working on the universe. It meant doing research. It meant living my life. And when I really thought about it, I suppose I never actually distinguished between those things: my life, our book, the universe. If I was really honest with myself, I had to admit that although I didn't buy into Bostrom's simulation argument, I had always had the acute sense that I was living *inside* our book, a book that would only become a tangible object when we found the answer to the universe. Writing had always meant searching for the answer, and the answer had always borne the shape of a book destined to sit alone in our empty library.

It all hung together, if barely. We'd tell people we were coauthoring a book in order to ask them about Wheeler's riddles, and in turn solving

Wheeler's riddles might allow us to finally write our book. It was life imitating imaginary art. A self-excited delusion.

The Cahill Center for Astronomy and Astrophysics at Caltech was a strange and fitting feat of modern architecture, a building that looked like it had been built by gauge forces, by the patching together of several mismatched reference frames.

My father and I had come to Pasadena, California, to speak with physicist Kip Thorne, one of Wheeler's former students. I could tell my father was excited to meet the celebrity physicist. "I always think of him as being straight out of *Star Trek!*" he said giddily as we approached the building.

Finding our way from the lobby to Thorne's office took some effort, as my father and I maneuvered the odd and disorienting angles.

Thorne himself was tall and bald, with a pointed goatee, and did bear a slight resemblance to Captain Picard. He kindly invited us to sit down.

"We met John Wheeler very briefly at a conference years ago, and he said two things to us that were rather cryptic," I explained. "He said that the universe is a self-excited circuit, and then, in response to a question about the universe coming from nothing, he said, 'The bound-

My dad and me
with Kip Thorne
(center)
Jo Ann Boyd

ary of a boundary is zero.' Can you tell us anything about what those phrases mean?"

"Mathematically, the boundary of a boundary is a basic principle from which you can derive certain properties of spacetime curvature. How you use that to explain the birth of the universe, I don't know. I've never found the idea terribly useful. I have the opposite view that Johnny did about how fundamental it is."

Thorne made it clear that he didn't have any more to say on the matter. From his reference frame, he was looking at two copies of the same disappointed face.

"Okay," I said. "What about the self-excited circuit?"

"From a certain point of view, which Wheeler adopts, systems can become classical only when observed. They behave quantum mechanically and indeterminately until observed, and the observation collapses the wavefunction. So Wheeler conceives of the universe as having been born and having evolved quantum mechanically until it naturally generates life. Then that life performs the observation that collapses the state of the universe to make it classical. It's self-excited in the sense that the observation comes from within the universe, not from the outside. When I describe it, it sounds fairly simple, but my impression is that it's a lot deeper than it sounds."

I nodded. "Does it have to be biological life that makes the observation?"

"I think that was his view," Thorne said. "Intelligent beings that arise naturally as the universe evolves. The person who can give you deeper insight into this is Wojciech Zurek. He worked with Wheeler at Texas when Wheeler was developing these ideas. I spent very little time with Wheeler during those years, but I think it is a deep idea. Zurek is the best living expert on that idea."

"What are your views on observers?" I asked, hoping Thorne might say more.

"I tend toward the view that they are not important at all," he said.

Not important *at all*? Without considering their perspectives, how could you ever get at ultimate reality?

"We've come to the conclusion that for something to be ultimately real, it has to be invariant and observer-independent," I said. "But we don't think observers have to be conscious or biological or anything like that. Just reference frames."

"So you don't think space is real? Or that time is real?" Thorne asked, seeming shocked and annoyed.

"Not ultimately real," I said, surprised at his reaction. Don't get annoyed at me, I thought. Tell it to Einstein! "Do you disagree?"

"As physicists, we have been tremendously successful at building pictures and mathematics that are very predictive, but we have never developed any set of tools or criteria to tell us what is ultimate reality. I think we are less in the position to probe those issues than philosophers are. But the only philosophers who have a prayer of making progress are those that understand the physics. I steer clear of asking what is ultimate reality."

Fair enough, I thought, though it was hard to imagine that kind of aversion to the big questions coming from a student of Wheeler's. "Did Wheeler influence your thinking on physics?" I asked.

"Wheeler had a tremendous ability to use physical intuition to guess how things behave. The recognition that there's enormous power in that—that had the biggest influence on me. Wheeler made great discoveries using intuition, though ultimately they had to be tested against the mathematics. In my generation, the person who has been most effective in the Wheeler approach is Stephen Hawking. Out of necessity he was unable to do complex mathematics once he lost the use of his hands, so he functions through enormous physical intuition, plus solving problems geometrically and topologically in his head."

"Do you have any good stories about Wheeler?" my father asked.

"I'll tell you one," Thorne offered. "Today there's a lot of discussion in string theory about the idea of the landscape of vacua. The particular version of the laws of quantum fields we have in our universe might be different in other universes." Wheeler, ahead of his time as always, had thought a lot about this issue, Thorne told us. Wheeler called it "mutability," the idea that the laws of physics don't exist in some Platonic realm outside the universe, but come into being with the universe at its birth and eventually die with the universe at its death. "In 1971 Wheeler was visiting, and Wheeler, Feynman, and I went to lunch at the Burger Continental restaurant here at Caltech. Wheeler was talking about this idea of mutability and asking, 'What determines which laws are in our universe?' Feynman turned to me and said, 'This guy sounds crazy. But he has *always* sounded crazy.'"

We all laughed. "What are you working on these days?" I asked.

"I'm exploring ways to be creative in other areas," Thorne said. "I'm working on two science fiction movies in Hollywood and writing an article for *Playboy*."

My father chuckled loudly and then, realizing that maybe he shouldn't, cleared his throat, furrowed his brow, and tried to be serious. "What inspired you to make that change?"

"Based on my genetic heritage, I'll probably live into my hundreds," Thorne said. "But I can't continue doing really great theoretical physics for a long time. I decided that this was the appropriate time to move into directions that I can continue with for a few decades. Also, I'm bored."

"Well, that was kind of a bummer," I said, as we walked back toward our hotel.

We had been hoping to get some answers, but all we'd gotten were a few verbal shrugs. Thorne didn't see any profound meaning in the boundary of a boundary; he pretty much said that the idea was useless.

Maybe it was. No matter how intriguing it sounded, there was no guarantee that the phrase held any shining truth. Maybe it was nothing more than the desperate, incoherent cry of an aging physicist who knew he was running out of time, or an aging man who didn't know he was running out of wits. Then again, as Feynman had said, Wheeler had always sounded crazy. And more often than not, he had been right.

"At least he told us about Zurek," my dad said. "That's useful."

That was true. Thorne had said that Wojciech Zurek, a physicist at Los Alamos, was the world's best living expert on Wheeler's self-excited circuit.

I nodded. "I guess we're going to New Mexico."

* * *

We checked in to a Pueblo-style bed-and-breakfast surrounded by white adobe walls and hanging ristras of fire-red chile peppers, then spent the day visiting art galleries on Canyon Road and discussing the nature of reality.

The next morning, we drove forty-five minutes to Los Alamos, winding our way up the mountainside to the Pajarito Plateau, seven thousand feet above sea level, to the "town that never was." Seven decades earlier the government had overtaken the mesa and set up Los Alamos National Laboratory as the top-secret headquarters for the Manhattan Project. Physicists from around the country had left their respective universities and come here to build the atomic bomb in the hopes of putting an end to World War II. Wheeler, who had first developed the theoretical underpinnings of the bomb in his work on nuclear fission with Bohr, was stationed in Hanford, Washington, at the time, working on a nuclear reactor that fed plutonium to Los Alamos. He would come to New Mexico now and then to work and to discuss electrodynamics with Feynman.

In 1944, at the start of his time in Hanford, Wheeler received a postcard from his younger brother, Joe, who was fighting on the front lines in Italy. It contained only two words: *Hurry up*. But it wasn't until the following July, nearly a year after Joe had been killed, that the Manhattan Project completed construction of the bomb. Some two hundred miles south of Los Alamos, in the Jornada del Muerto desert, they tested their plutonium "gadget," detonating the first nuclear bomb in history. The physicists watched the Trinity explosion from the safety of a base camp ten miles away as the bomb produced a blinding light, intense heat, grumbling shock waves, and a mushroom cloud that swelled more than seven miles overhead, turning one thousand feet of desert sand below to glass. J. Robert Oppenheimer, the lab's director, solemnly quoted the *Bhagavad Gita*: "Now I am become Death, the destroyer of worlds."

While his fellow physicists were still reeling from the horrors wrought by their involvement with the bomb, Wheeler was living with the guilt of his brother's death and the regret of not having gotten the job done quicker. "It does little good to second-guess history," he wrote in 1998. "But I cannot avoid reflecting on my own role. I could have understood the gravity of the German threat sooner than I did. I could—probably—have influenced the decision makers if I had tried. For more than fifty years I have lived with the fact of my brother's death. I cannot easily untangle all of the influences of that event on my life, but one is clear: my obligation to accept government service when called upon to render it." So in 1950, when he was asked to work on the hydrogen bomb, Wheeler agreed. He moved here, to Los Alamos, and lived for one year in Oppenheimer's former home.

Driving across the mesa, I found it strange to be steeped in all that tragic history. Strange to think that obscure, abstract ideas like relativity and quantum mechanics—ideas that my father and I had been discussing for more than ten years as nothing more tangible than intellectual puzzles—had such unimaginably real consequences. Not real as in invariant and observer-independent. Real as in blood and fire and grief.

We found our way to the residential neighborhood where Zurek lived. Zurek was a major figure in the science of quantum information. With Bill Wootters, another student of Wheeler's, Zurek had proven what's known as the no-cloning theorem, which says that an unmeasured bit of quantum information can never be perfectly copied. He had also made crucial inroads to understanding the process known as quantum decoherence, which helps explain why the everyday, macroscopic world doesn't seem all that quantum.

Even though you can pretend, as Bohr and his Copenhagen crew did, to draw a distinction between observer and observed, calling half of the world "macroscopic" or "classical" and the other half "microscopic" or "quantum," you can always push the boundary to larger and larger scales, the observer becoming the observed, the outside engulfed by the inside, the classical swallowed up by the growing gulp of the quantum. Why, then, don't we see the remnants of superpositions— those stripes of interference that show up in the double-slit experiment—when we measure the length of a couch or the height of a child or the position of the Moon? Why, in the world of big stuff, do classical probabilities, which assume that something always has only one position or another, work so well even though things ought to be described by quantum probabilities, which assume that things are suspended in multiple states simultaneously before we measure them?

The answer, thanks largely to Zurek, is decoherence. The idea was simple enough. Interference patterns form when the wavefunctions describing the two possible states of a system—say, the component of a wavefunction that says an electron went through slit A and the component of a wavefunction that says the electron went through slit B—add together. As the photographic plate registers one electron after the next, each lands at a random spot allowed by the probability distribution encoded in the summed wave, the superposition. The resulting pattern of stripes depends on the relative phases of the waves: dark bands appear where the waves are out of phase and cancel out, the bright bands where they are in phase and amplify. Because the phase difference between the waves remains fixed electron after electron, the superposition is coherent. If, however, the electrons are immersed in a larger environment, like air, they'll end up getting knocked around by the billions of molecules bouncing around as they travel from the slits to the photographic plate. Each time an electron is fired through the slits, its path is thrown off course and the relative phase difference between the two components of its wavefunctions changes from one detection to the next. As the electrons build up on the plate, there's no single, coherent superposition to encode the kind of probabilities that would produce light and dark stripes. Instead, the measurements reflect the probabilities of each individual wavefunction—the exact probabilities you'd expect if the particle were traveling through only a single slit at a time, and not through both simultaneously. The kind of probabilities you'd expect if the particle *wasn't quantum*.

By smearing out the coherence of superpositions and rendering quantum probability distributions classical, environmental decoherence makes it look like quantum wavefunctions collapse, transforming hosts of possibilities into single actualities. In reality, the wavefunctions haven't collapsed at all. In reality, the electron gets entangled with every air molecule it hits, its wavefunction superposing with the wavefunction of each molecule. In reality, things are getting *more*

Wojciech
Zurek and me
in Los Alamos,
New Mexico
W. Gefter

quantum. We just don't notice because we're not measuring the air molecules. If we measured not only the electrons and the detector but also the larger environment, we'd see more bat-shit interference than ever.

Zurek greeted us at the door. He was warm and wild-looking, with bushy orange hair, an equally full beard, and a thick Polish accent. We followed him into a large living room, which had a stylish southwestern flair—a stone hearth framed a fireplace at one end of the room, and at the other, floor-to-ceiling windows showcased a sweeping panoramic view of the mountains and the canyons below.

"How did you get to know Wheeler?" I asked as the three of us settled onto the couches.

"I became a graduate student at the University of Texas in 1975 and John Wheeler arrived there a year later," Zurek said. "I took his class on electrodynamics. One thing that made a lasting impression on me was when John tried to derive something on the board. Sometimes it didn't work and instead of being apologetic he would cross out the attempted derivation and write 'wrong' in big letters. That freedom to go the wrong way was one of the most important lessons I learned from him. A year or two later I took his seminar on quantum measurement. There were wild ideas being explored, but also being torn apart. It was like writing 'wrong.' You can explore wild ideas, but at some point you have to evaluate them honestly. After that, I was thoroughly hooked on Wheeler's way of doing physics and on quantum mechanics. Not just quantum mechanics—something broader. The fascinating thing that comes through in quantum measurement, but is bigger than that, is understanding how we, as observers, as beings that are alive, fit in the universe. How does our existence fit in with physical laws?"

"A lot of your work since then has been focused on understanding how the classical world emerges from the quantum," I said.

"The superposition principle tells you that if you have two quantum states you can put them together into new states in any proportions," said Zurek. "Before decoherence, every state—every superposition of every superposition—is legal. And yet the Moon is in one place; cats are either alive or dead. As Einstein pointed out, quantum mechanics of closed systems doesn't provide a reason for that. Decoherence does."

"Recently you proposed a theory you call quantum Darwinism," I prompted. I had seen some mention of it in my research but wasn't sure what it was all about.

"Quantum Darwinism goes beyond decoherence. It recognizes that we don't measure anything directly," Zurek said. "We just find out stuff from the environment. Right now you are looking at me. We are a couple of meters apart. The only reason you know where I am and what I look like is because you intercept a tiny fraction of photons—of the photon environment—that scattered from me. It's clear that there are many more copies of that very same information about me all over the place. For decoherence it's enough that the environment poses a question once. But in real life the environment is boringly asking the same darn question and disseminating the same boring answer all over the place. We grab a small piece of the environment and find out."

That was interesting. After all, what makes quantum information different from classical information is that when you acquire a bit of quantum information you actually change it. You become entangled with it. It's not like you can look at a quantum state, get some information about it, and then walk away, leaving it undisturbed for the next guy to look at. That was the root of the no-cloning law, which many physicists saw as the future of security and encryption. You can't eavesdrop on a quantum message, not without changing the message in the process. If I happen upon one unique quantum state—one bit of information—no other observer can ever see it the same way. It can't be invariant. Was that right? Confused, I scribbled a note for myself in my notebook: *Single, unique quantum state can't be real? Or reality is first come, first served?*

"So objective reality emerges when there are enough copies for us to all agree on what's there?" I asked.

"Yes, that's precisely the idea," Zurek said, nodding. "That's the whole point. To understand objectivity. In a quantum universe we do not measure anything directly. If I were to make a direct measurement of a system, I could disturb its state. But I never do that, because usually the environment does the measuring for me. It decides on the set of states that get found out and disseminated, and I never interact with the system directly. I just use the environment as a witness. The observer gets hold of the information that is already advertised widely all over the place."

"Years ago, we had a conversation with Wheeler, and in it he used two rather cryptic phrases. We're hoping you can tell us what they mean. The first was that the universe is a self-excited circuit."

"I think it's like the Taoists. It's meant to inspire you. It's not like he worked out something specific. He drew this picture of the *U* with the eye looking back at itself—I love this picture. But I don't know how to put equations on it. If you talk to people in cosmology these days, the anthropic principle is gaining more and more credence, and that goes back to some extent to Johnny Wheeler. My instinctive reaction to that is to shudder because, you know, it's like a shortcut that should somehow be illegal. But I think people are recognizing that somehow we observe the universe because it can host observers, and that if you have a huge set of possible universes with different laws then the small subset of universes that provide environments in which observers can live will be selected by the fact that observers can live there. I think that falls short of Wheeler's vision. I think he envisioned a uniqueness."

Suddenly it occurred to me that decoherence seemed to undermine the role of the observer that Wheeler felt was so essential to reality. "He believed that it's a participatory universe because quantum mechanics requires observers to measure things and bring them into existence," I said. "Does decoherence make that unnecessary?"

"Usually the measurement is done for you by the environment. But there are situations in which you deal with quantum systems hands-on. In that case, the choice is up to you how you want to set up your apparatus and decide what you're going to measure. Wheeler's delayed choice is a fantastic example where it's actually up to you. Generally we don't have to worry about that because of decoherence. But our universe *allows* us to do things that smack of direct intervention or something."

"Ultimately there can't be a real distinction between the observer and the environment," my father chimed in. "From some other perspective, the observer is the system or is part of the photons and molecules that make up the environment. So when you set up the formalism—"

"What do you mean by *ultimate*?" Zurek interrupted. "The point of view where you are outside the universe and looking at the whole wavefunction evolve? Fine. But that's God's point of view, that's not ours. We're stuck on the inside. 'Ultimately' is a big carpet under which you can sweep a lot of important things. But I do think there are important questions along these lines. One question that ought to be looked at more carefully is, why systems? That's probably your line of thought, right? Why not the observer and the environment and the apparatus all in one? My preliminary answer is that if you don't subdivide the universe into systems, you don't have a measurement problem. So you don't have to apologize for having systems when you're trying to solve it."

That made sense. From our point of view inside the universe, we always seem to divide it into systems using Bohr's dividing line, observer and observed, or observer, observed, and environment. As my father was pointing out, the categories themselves can't be ontologically distinct, and were, in some sense, frame-dependent. But as Zurek said, maybe the deeper question was, why divide things at all? I wondered if Fotini Markopoulou had already given us the answer: light cones. Thanks to the finite speed of light, we're all stuck with limited perspectives, with horizons that carve the supposedly singular world into pieces. Into systems. The existence of dark energy and the de Sitter horizon it creates only made those horizons all the more permanent, a quantum universe forever divided.

"The other phrase Wheeler used was 'The boundary of a boundary is zero,'" I said.

"Well, that's simply true," Zurek said. "If you put a boundary on something, you don't have to bind it anymore. It's closed. It's an observation, a regularity, that's strikingly simple. One should strive to have similarly strikingly simple explanations of other things."

Put a boundary on something and you don't have to bind it anymore, I wrote in my notebook. *It's that simple?*

I waited for Zurek to say more, but he didn't.

"That's it?" I asked. "Wheeler seemed to think it was so profound."

"I think he liked the idea of getting a lot from a very little. Charlie Misner, Kip Thorne, and I wrote an article about Wheeler and we included a photograph of notes he had written on a blackboard. One was a quote in Latin. We struggled with translating it because our Latin was rusty, but it turned out to be a quote from Leibniz, which he exclaimed after he realized that if you have just two elements, zero and one, you can construct complete mathematics. Adding one to zero gives you everything. I think that's the spirit."

It reminded me of set theory, of building the whole number line from the empty set, just by putting it inside some brackets.

"Does the fact that the boundary of a boundary is zero actually give you a prescription for doing that? For getting everything from nothing?" I asked, desperate to hear something more significant. "Wheeler made it sound like a clue."

In my mind, Zurek sighed deeply, the kind of sigh that said, *I knew one day someone would come asking for this. But I hadn't suspected that day would be today.* Then he walked over to the fireplace and pushed on a brick in the wall. The brick slid back into the hollowed-out wall, revealing a secret compartment, and in that secret compartment lay a black velvet box. Zurek carried the box back over to the couch, where my father and I sat perched on the edge, wide-eyed. He carried it in two hands, as if the world would end should it fall. I spotted Wheeler's U-diagram embossed in gold on one side. Zurek stood before us and lifted the lid. From inside came a blinding white light. When our eyes adjusted we saw, in the tiny center, which appeared somehow infinitely far away and infinitely visible, the answer to the universe.

In reality, Zurek shrugged. "I have no idea what to say."

I sighed and decided to let it go. "We're trying to figure out what reality is," I said. "We've defined something as being real if it's invariant, or observer-independent. What's your view of reality?"

"We have confidence that our language is good enough to describe the world we live in, but it's not going to work," Zurek said. "It's been developed for very specific purposes that don't have to do with analyzing fundamental physics. Philosophers try to get you to commit to a bunch of words, and all of those words dissolve when you start thinking about physics in a deep way. My view of reality has to do with what philosophers call intersubjectivity. That's what quantum Darwinism is about. Reality is what we agree on. In that sense it's what's invariant. But that invariance—and hence, quantum reality—is not fundamental, it's emergent and approximate. Big words are seductive, but if you look at them closer, you don't know what they mean."

I nodded, though I was pretty sure I knew what reality meant.

"You and Wheeler seemed to work a good deal on the role of information in physics," I said.

"I would very much like to understand the reason why information is around," Zurek said. "John was much more bold in trying to use it as a foundation. The connection between information and reality—to use a word I just outlawed—is extremely interesting. In classical physics, information is completely unreal. You have objects, and information just *describes* them; you have information *about* them. It's completely subjective. In quantum mechanics, information is much more fundamental. Wheeler had that picture where the observer breaks through the boundary. Information was outside Newtonian physics, but it's part of quantum physics, it's *physical*. That's absolutely crucial. My aim is to understand how it comes out of quantum mechanics. But a connection like that is often a two-way street. You might understand how information comes from quantum mechanics, and then you can turn around and understand how quantum mechanics comes out of a deeper understanding of information. Johnny Wheeler gave us courage to take that idea seriously."

"Why is information more real in quantum physics?" I asked. "Because things are binary, so you can describe them in bits?"

"It's more than that," Zurek said. "In classical physics you can find out the state of a system and then someone else can come along and find out the state of the same system and you'll both agree. In quantum mechanics, that's generally impossible. Acquiring the information does not reveal some preexisting reality. Acquiring the information somehow *defines* reality. That's closer to Wheeler's participatory universe. And yes, we don't do that normally—normally there is decoherence and information spreads by quantum Darwinism, but still, the laws of the universe allow that to happen. That's something we shouldn't ignore. That's a hint of how this universe is built."

As we pulled away from the curb, a coyote darted across the road, pausing briefly in front of our car to stare us down. Its pelt and bone were just illusion, I told myself, information incarnate, carnivorous and evolved and only approximately objective, a product of quantum Darwinism, an endless repetition of bits emanating redundantly into the desert.

"He looked like the flutist from Jethro Tull," my father said.

"The coyote?"

"Zurek."

"He was really interesting," I said. "But I still don't fully understand what Wheeler was trying to say."

"Maybe there's not so much to it?" my father said. "Maybe it's like he said—a kind of Zen koan or something, just meant to make you think differently?"

I shrugged. "That would be depressing. What do you want to do now?"

We decided to go visit the Bradbury Science Museum, which featured exhibits about the history of Los Alamos National Lab, the Manhattan Project, and nuclear weaponry.

"When I was a kid, maybe eleven, I had this rock and mineral collection," my dad told me as we drove into town. "And someone gave me these rocks of trinitite. I can't remember who. They were like these glass rocks that were formed in the Trinity test. The package came with this little piece of cardboard that had a picture of the explosion. I was totally fascinated by it. I think in my mind I conflated it with kryptonite."

As we wandered around the museum, I thought about what Zurek had said. Even if decoherence does most of the participatory work for us, there's still a kind of ambiguity that requires observers to make choices, choices that bring information into the world, constructing the universe bit by bit, as Wheeler had envisioned. *That's a hint of how this universe is built.* Decoherence, as Zurek had said, effectively solves the problem of why we don't see interference effects—the trademark signs of quantum superposition—in our everyday, macroscopic lives. But a deeper question remained: why is there any interference to get rid of in the first place?

"What Zurek said about the differences between classical and quantum information was fascinating," my dad said, pretending to talk to a statue of Oppenheimer. The destroyer of worlds was wearing a Brockman hat. "That classical information is *about* something, but quantum information *is* the thing."

"Hey, look!" I said, spotting a small case mounted on the wall off to Oppenheimer's side.

My dad talking to J. Robert Oppenheimer at the Bradbury Science Museum in Los Alamos
A. Gefter

It was filled with small, grayish rocks, "a new, manmade mineral, christened 'trinitite,'" the plaque explained. "These are the rocks you were
talking about!"

"Oh, yeah!" My father peered into the case, smiling like he had just
rediscovered a long-lost childhood toy. But as he read the description,
his smile quickly turned to a look of confusion and concern. "Seriously,
though, who would have given that to me? They're radioactive. My
parents just let me sit around holding those things. . . ."

I was glad we had come here. I felt like we had been nudged just a
little closer to understanding Wheeler's words. According to Zurek, the
self-excited circuit meant that observers have the power to create information—a result, presumably, of the fact that the world is divided
into pieces, itself a result of our being stuck inside the universe with
no access to a God's-eye view. The boundary of a boundary—the simple geometric fact that "when you put a boundary on something, you
don't have to bind it anymore"—somehow suggested a way to bring
that information into existence *from nothingness,* but I still wasn't sure
how. It from bit, bit from nothing. Deep questions remained unanswered. Decoherence was great for hiding reality's quantumness, but
why was reality quantum at all? Wheeler's question remained as prob-
ing as ever, hanging like a mushroom cloud over the New Mexico desert. *Why the quantum?*

Back at home, I was hunting around online for more information on
Wheeler's former students, trying to find more people who might be
able to help us decipher those phrases, when I came across a profile of
Wheeler in a 1978 issue of the *Alcade,* the University of Texas at Aus-
tin's alumni magazine, written one year after he had moved to Texas
from Princeton. One particular paragraph caught my attention.

"And when he gets that new idea, what does he do with it?" the
article asked. "He records it in a hardbound, handwritten notebook.
Dr. Wheeler has filled almost forty of these notebooks since he started
keeping them during the war. . . . When he talks with a colleague, he
records what the person told him and what he thought about it afterward. When he is asked to give a lecture, he organizes what he will say

in the notebook, writing it out in black ink, in beautiful, legible script. When a neighborhood child constructs a paper greeting for his birthday, when he buys a postcard in a foreign country, when he comes across a cartoon that makes him laugh, he pastes these items into his notebooks, saving them for his own future reference and for the future delight of science historians."

Holy shit.

Wheeler kept journals? *Forty* journals? Recording his every thought and conversation? That was in 1978. How many more had he filled in the three decades that followed? What had become of them? How were we going to get our hands on them?

All I knew was that we needed to see those journals.

10

That Alice-in-Wonderland Shit

At Arizona State University in Tempe, physicist Lawrence Krauss was launching the Origins Initiative, an ongoing project to explore "the origin of the universe, stars, planets, life, consciousness, culture, and human institutions." I wondered what exactly that wouldn't include. To kick it off, they were throwing a huge symposium, and *New Scientist* sent me to cover the event. Over the course of three days, it featured talks by some of the biggest names in science. But by the end of the first night, I had only one name in mind: Brockman.

The first day had featured panels of physicists discussing the knowability of the universe's origin. Andrei Linde evangelized the merits of the multiverse, while David Gross sat shaking his head and looking like he was itching to stab someone. Guth and Vilenkin optimistically addressed the difficulties of making predictions in an infinite multiverse. Then Gross took an axe to them all. We don't know what string theory is, it doesn't offer a consistent cosmology, the technical and conceptual underpinnings of eternal inflation are shaky at best, and we simply don't know the rules of the game, he said. Resorting to the multiverse is a cop-out. We ought to be searching for real answers. The crowd cheered.

Oddly, no one mentioned what seemed to me to be the biggest

(Left to right) Sheldon Glashow, David Gross, Andrei Linde, Paul Davies, Alex Vilenkin, and Alan Guth at the ASU Origins Symposium Dan Falk

cosmological challenges—like what it means to talk about "the universe" when each of our observer-dependent de Sitter horizons delineates our own universe, or how we can move beyond inflation's semiclassicality to understand the universe's quantum origin, or whether the origin lives not in the past but the present, a top-down, delayed-choice kind of birth, its umbilical cord looped back around its observer?

That evening there was a cocktail reception at the university's art museum. I was wandering around the outdoor plaza, attempting small talk, when I saw it. The Panama hat. I didn't need to see the head it sat on.

My mouth went dry and my heart began to race. I pulled out my phone and texted my father: *Brockman is here! What do I do?*

A minute later my phone buzzed with his reply: *Panic?!*

Strategy, I told myself. *I just need a strategy.*

I came up with a good one: slink pathetically around the party, keep a safe distance, and rehearse what I might say in case I muster up the guts.

I didn't.

Milan Kundera says that every action is a self-portrait of the one who acts. Today's painting looked something like this: a girl, pressed against a wall, conspicuously inconspicuous, peering round a corner at a man in a Panama hat. Title: *Pull It Together*.

The next day's lectures were held at the Boulders, a posh resort in Scottsdale that sat against a backdrop of 12-million-year-old granite rock formations and proud saguaros in the foothills of the Sonoran desert. During a break between sessions, everyone filed out into the hallway where coffee and snacks were being served. I stood chatting with Dan Falk, a freelance journalist whom I had met back at the Davis conference and had gotten to know at several physics conferences since, discussing the lectures we had just seen and trading notes about whom we were hoping to interview. "Honestly," I confessed, "I really just want to talk to John Brockman. But he scares the crap out of me."

"Here's your chance," Falk said, pointing his chin toward the far end of the hall behind me. I turned around.

There was Brockman in his white linen suit and Panama, looking like a meaner, tougher Tom Wolfe and talking to a pack of Nobel laureates. It wasn't the kind of conversation you interrupt. But eventually the Nobelists headed back toward the lecture hall and for a brief minute Brockman was alone. I had to introduce myself. I couldn't pass up the opportunity, even if every biological instinct in my body was screaming, *Flight! Flight!*

Falk laughed as I took a deep breath, pushed my shoulders back, and walked down the hallway toward Brockman. Then I panicked. At the last second I veered off to the side and began waving at some imaginary colleague I'd apparently just spotted. I headed back into the lecture hall, defeated.

The next round of talks was followed by a break for lunch, which was being served in the main dining area of the resort, a short walk away. I was heading outside when I saw Brockman hanging out by the door. I mustered up all the nerve I could find. He caught my eye as I approached; there was no chickening out now.

"Hi, John? I just wanted to introduce myself. My name is Amanda Gefter. I work for *New Scientist*." I offered my hand for a shake, but Brockman just stood there.

With a stern expression, he looked me up and down, then in a husky voice said flatly, "I know who you are."

I hadn't seen that coming. I wasn't sure how to respond, so I went with a bewildered, "You do?"

"Roger's talked about you," he said.

Roger, I assumed, was Roger Highfield, the British science journalist who had recently taken over as editor of *New Scientist*. I knew that Roger had written several science books, but I hadn't realized that he was one of Brockman's clients. The idea that Roger Highfield and John Brockman had had any kind of conversation about me was exhilarating, if surreal, but I had a good suspicion that it had probably gone something like this.

Brockman: *How's life since you've taken over* New Scientist?

Roger: *It would be spectacular if it weren't for Amanda Gefter, who is probably going to get us sued and bankrupt the bloody magazine.*

I had recently written an opinion piece that provoked the threat of a libel suit.

I cringed sheepishly. "I've been causing a bit of trouble for Roger."

Brockman stared me down with stony approval. "It's good."

I smiled. It didn't surprise me that Brockman would approve of a little troublemaking. I opened my mouth to respond, but apparently he had decided that three sentences were the most he was willing to waste on me, and he abruptly walked away to talk to someone more important.

Back in Cambridge, I was eager to delve deeper into Susskind's horizon complementarity. Brockman had convinced him to write a book about it, so I knew it had to be important. I also knew that if I could write an article about it for the magazine, I'd have the perfect excuse to talk more with Susskind, to finish the conversation we had started on the Santa Barbara beach.

"He says it's a new and stronger form of relativity," I told one of the features editors, knowing full well that no editor can resist a story that messes with Einstein. It worked like a charm. I got the green light and immediately contacted Susskind.

It all began with a paradox, he told me over the phone, one born of Hawking's monumental discovery, and rendered graver still in Susskind's Bronx inflection. When black holes radiate, they evaporate, ever-shrinking spheres that eventually up and vanish from the universe, taking with them whatever had fallen in. That's what Hawking had believed, anyway: if an elephant falls into a black hole and the black hole radiates away, it takes the elephant with it, leaving no trace of it behind, not a single bit of information to betray its strange extinction.

For Susskind, the scenario was nothing short of a crisis. "We have a principle that in physics information is never lost," he told me. "In quantum mechanics it means that the initial state can be recovered from the final state. That's very, very fundamental. Quantum states should mean something. All of physics as we know it is conditioned on the fact that information is conserved, even if it's badly scrambled."

If a physical law like the conservation of information could fail at the edge of a black hole, it could fail anywhere. Either the world is described by quantum mechanics or it isn't—find one scenario in which it breaks down and the whole thing is totally useless. If black holes could lose information, Susskind said, the entire edifice of quantum mechanics would come crumbling down. The Schrödinger equation, which describes the evolution of quantum systems in time, would be meaningless. Wavefunctions would go limp and wither. Any semblance of continuity from past to future would melt away. Predictions based on quantum mechanics would be rendered absurd, as probabilities would sum to less than and greater than 1.

On the other hand, if black holes *couldn't* lose information, then general relativity was doomed. That's because there was only one realistic way to save the information from evaporating into oblivion. It couldn't climb up out of the black hole's interior and escape, because crossing back over the horizon would require moving faster than light. The only hope was that the information never fell into the black hole in the first place, that the horizon somehow prevents its passage into the dark.

That scenario, however, violates the equivalence principle, the cornerstone of general relativity. Einstein's happiest thought was that a free-falling observer always finds himself in an inertial frame free from gravity, a conviction that any physical experiment will inevitably confirm. Like a man falling off a roof, an elephant falling into a black hole feels no gravity. As far as physics is concerned, the elephant may as well be at rest. "Gravity" is the fictitious force we introduce when we view the elephant from some other reference frame in which it appears to be inexplicably accelerating. It is our way of patching up two misaligned frames to save the semblance of a single reality.

If the elephant is at rest in its own reference frame, some kind of impenetrable wall isn't going to suddenly materialize in front of it. Information-blocking walls don't just appear out of nowhere—not without violating the laws of physics.

"The equivalence principle says that if you're in a neighborhood of spacetime where the curvature is small, then you should experience no strange or violent behavior," Susskind explained. "The curvature is very small near a horizon, so someone falling through shouldn't experience anything strange. If no information is to be lost, then it must never pass through the horizon. On the other hand, the principle of equivalence says that the horizon is no place special, so information should be able to pass right through."

At first that didn't sound right—the curvature near a black hole horizon was *small*? You'd think it would be pretty big, considering a black hole has the strongest gravitational pull of any object in the universe. But if the black hole was big enough, Susskind explained, gravity's tidal forces at the horizon would be negligible. Given any horizon size, really, you could always look at some segment that was small enough to make space appear flat and ordinary, not capable of blocking information or betraying Einstein.

It was a perfect paradox: information couldn't be lost without violating quantum mechanics, and it couldn't be saved without violating general relativity. Hawking took Einstein's side, saving relativity by sacrificing elephants and quantum mechanics. But Susskind was convinced that quantum mechanics couldn't just fall apart, not without the world around us falling apart, too. He had a nagging hunch that information never crosses the horizon in the first place, but he had to find a way to keep the equivalence principle intact.

Actually, it's not hard to show that the information never crosses the horizon from the point of view of an accelerated observer outside the black hole. In studying Hawking radiation I had already seen how Safe, my accelerated observer, would see light stretched to perverse proportions and time slow down until it freezes at the horizon's edge. Safe sees nothing cross to the other side because as far he's concerned, *there is no other side*. For him, the horizon marks the edge of reality, the end of the world. Safe can't lose any information, because there's nowhere for it to go.

But that story becomes tricky to uphold when you think about what's happening to Screwed. Screwed sails right through the horizon, because, thanks to the equivalence principle, the horizon doesn't exist for Screwed. As far as he's concerned, information, such as the vast amount of it contained in his own body, can easily cross into the black hole, even if it can't climb back out. Safe says the information remains outside the horizon; Screwed says it's inside the black hole. Susskind knew that if somehow *both* stories were true, then neither quantum mechanics nor general relativity would be violated and all would be right in the universe.

The problem was that for both stories to be true, it seemed like the information had to be in two places at once, as if there were two identical clones of every bit of information. Unfortunately, such a scenario was expressly forbidden by Zurek's no-cloning theorem. The reason is simple: if you could clone a quantum particle, you could outsmart the uncertainty principle. You could measure the position of one clone and the momentum of the other and now you have precise knowledge of two aspects of a conjugate pair, uncertainty be damned. But the uncertainty principle is a principle—there's no outsmarting it. Information can't be cloned. Once again, Susskind was stuck with a paradox: both stories have to be true and both stories couldn't be true.

When the solution came to him, even Susskind realized just how crazy it sounded. "Every other option had been eliminated, leaving one possibility," he said. "It seemed totally absurd but I knew it had to be right." He had first announced his answer at a conference back in 1993. "I don't care if you agree with what I say," he told the audience. "I only want you to remember that I said it."

"I feel like we should make a move while Brockman still remembers who I am," I told my father over the phone. "I think we should write a book proposal."

"But the idea was to write a book once we had found the answer to the universe," my father said.

"Yes," I said, "but if you build it, it will come."

"What?"

"*Field of Dreams*. Ghosts playing baseball? If we have a book deal, the answer will come."

"I'm not sure that's how it works," my father said.

"I think if Brockman's your agent it is."

My father was right—the book had always lived somewhere off in the vague future, an event horizon that receded as quickly as we approached, and I think we both wanted it that way, because we knew that no actual book would ever measure up to the idea of Our Book, and because we knew that the day the book became a physical object would be the day our journey was over.

But Wheeler's death had made the clock tick a little louder. I didn't want to wake up ten or twenty or thirty years from now a magazine editor who once had talked about the meaning of nothing with her father. I wanted something tangible to keep us on track. To keep us together. Something like a book contract.

"Okay," my father said, sounding apprehensive but excited. "If you think now is the time to go for Brockman, let's do it."

"To be? To be? What does it mean to be?" Niels Bohr demanded when asked where a particle can be said to be prior to being observed.

Susskind was following not only in Bohr's footsteps but in Einstein's when he proposed his radical solution to the black hole information-loss paradox: there is no observer-independent meaning to the location of a bit of information. If you want to ask where the information is, you have to answer the question "According to whom?"

In order to uphold quantum mechanics and its conservation of information, Safe has to see information remain outside the black hole's horizon. In order to uphold general relativity and its equivalence principle, Screwed has to see *that same information* inside the black hole. The quantum no-cloning theorem forbids the duplication of information. But that doesn't matter, Susskind said. After all, who can see the information in both places? No one can be inside and outside an event horizon at the same time.

The key to solving the paradox, Susskind discovered, was realizing that there's no frame of reference in which the information has been cloned. As long as you stick to what a given observer can actually see, there's either Safe's story or Screwed's, but never both. It was sort of a mindfuck: both stories are equally true, but you can only talk about one at a time. You have to pick a reference frame and stick to it. Within any given reference frame, no observer ever witnesses a violation of the laws of physics. Violations only show up in the God's-eye view, a view that luckily no observer can ever actually have. The two descriptions—inside and outside the horizon—are complementary, Susskind said, in the same way that wave and particle are mutually exclusive but complementary descriptions of, say, an electron. He called the principle black hole complementarity, or, more generally, horizon complementarity.

Physicists were intrigued by Susskind's argument, but Hawking stubbornly maintained that information really did disappear behind the horizon and evaporate into oblivion, and many physicists followed Hawking's lead, living in a state of shared denial about the fate of quantum mechanics. Susskind, however, saw the problem all too clearly. The black hole information-loss paradox loomed like a dark cloud over everything. Cumulus chaos.

Then, in 1997, came a game changer. Physicist Juan Maldacena had been working on a model of string theory in anti–de Sitter, or AdS, space. Unlike our de Sitter space, which is defined by the positive value of its cosmological constant, AdS space has a negative cosmological constant. Our positive cosmological constant pushes outward on space, causing the universe's expansion to accelerate. Flip the sign and instead of pushing it pulls, causing space to bend in on itself, curving like a saddle at every point, distorting space and time in ways that only M. C. Escher could imagine and making possible the seemingly impossible, such as the ability of a light beam to travel to infinity and back in a finite amount of time. To make things more complicated, the AdS space in Maldacena's model was ten-dimensional, five of its dimensions curled up like origami at every point. To make life easier, Susskind told me, just picture it as a sphere, one with five large dimensions (plus time) surrounded by a four-dimensional boundary.

Through some genius intuition and complicated mathematics, Maldacena had discovered that the string theory operating in the ten-dimensional interior of the AdS sphere was mathematically equivalent to an ordinary quantum theory of particles operating on the four-dimensional boundary. The quantum theory of particles, he found, was remarkably similar to QCD, the theory that describes the interactions of quarks and gluons here in our universe. The only difference was that Maldacena's quantum theory was a conformal field theory—a CFT—which meant that it was the same at every scale, unlike QCD, in which the strong force grows weaker as you look to smaller distances. This equivalence between string theory in AdS and the CFT on its boundary became known as the AdS/CFT duality.

It all sounded pretty abstruse, but the more I thought about it, the more amazing I realized it was. First of all, it showed that string theory, a theory *with* gravity, was completely equivalent to an ordinary theory of quantum particles *without* gravity. Everyone had been trying to unite quantum mechanics and general relativity by shoehorning them into a single "theory of everything," but AdS/CFT suggested that maybe gravity is what quantum mechanics looks like when viewed through a different geometry. No wonder the world's leading physicists got up and danced the Macarena when they first heard the idea. ("Ehhh, Maldacena!") Second, there was that weird issue of dimensionality again. A theory with five large dimensions could be perfectly mapped onto a different theory in four.

Susskind had been thinking about the dimension issue ever since

Bekenstein had discovered that a black hole's entropy scaled with the area of its horizon, and not with its volume. If entropy counts the amount of information hidden within the black hole's three-dimensional interior, why would its value be determined by the two-dimensional area of its boundary? It was as if the three-dimensional black hole was somehow also two-dimensional. It had bugged me when I had first learned about it, and I was glad to hear it had bugged Susskind, too.

Susskind realized that the curious relationship between entropy and area wasn't confined to black holes—it applied to any region of space. After all, any region of space can be made into a black hole if you stick enough mass in it. Black holes are the highest-entropy objects around, so if *their* entropy can fit on a lower-dimensional surface, so can the entropy of anything else.

It was insane, counterintuitive, and undeniable: the total amount of information in any region of three-dimensional space scales with the area of its two-dimensional boundary. Susskind called the idea the holographic principle, reminiscent as it was of holograms, two-dimensional films that encode all the information necessary to reconstruct a three-dimensional image.

I was looking around the *New Scientist* office as he explained this to me over the phone one afternoon, and it dawned on me just how impossible it seemed. Every chair, every journalist, every air molecule from floor to ceiling could be precisely mapped, *with no loss of resolution,* onto the surfaces of the walls. A three-dimensional volume of space is far bigger than the surface area of its boundaries, yet the information content is the same? It was as if one of those three dimensions is just totally useless. As if we'd been thinking about dimensionality all wrong.

Susskind had suggested that the world itself was a kind of hologram, a projection of some lower-dimensional gravityless theory encoded on the edge of the universe. I wondered which was weirder, the idea that I was nothing but a computer simulation or the idea that I was a holographic projection from the edge of the world. Probably the hologram. In any case, Maldacena's AdS/CFT duality was the perfect embodiment of Susskind's holographic principle. It convinced the doubting physicists, including Hawking, that information couldn't be lost in black holes.

In AdS/CFT, there's a one-to-one mathematical mapping between the five-dimensional interior of the space and the four-dimensional boundary, so given any object or physical process in the higher-dimensional bulk spacetime you can follow the math to find its precise counterpart on the boundary. That raised a fascinating question: what's the lower-dimensional counterpart of a black hole? Black holes are made of gravity, but in Maldacena's model, there's no gravity on the boundary. What could a gravityless black hole possibly look like? Maldacena calculated the answer. It would look like a hot gas of ordinary particles. In fact, it would look like a quark-gluon plasma.

Quark-gluon plasma? Suddenly I remembered the reminder I had written for myself in my notebook back when I wrote my article on the plasma observed at RHIC, the one that, to everyone's surprise and confusion, had a viscosity that made it the most ideal liquid ever observed, up to twenty times more liquid than water, a fact that ordinary physics couldn't explain. *Look into AdS/CFT . . . explains liquid fireball?*

"The quark-gluon plasma is dual to a black hole?" I asked Susskind, amazed. "I read somewhere that AdS/CFT can explain the RHIC measurements."

That's right, Susskind said. The plasma is dual to a black hole, and a black hole's event horizon has a calculable viscosity. As it turns out, the viscosity of Maldacena's ten-dimensional black hole is nearly the exact value measured for the quark-gluon plasma at RHIC.

"So wait," I said. "Is the idea that we can use the mathematics of ten-dimensional black holes to calculate the viscosity of a four-dimensional quark-gluon plasma? Or is it that when we measure the quark-gluon plasma we are literally looking at a ten-dimensional black hole through four-dimensional glasses?" As an ontic structural realist, I had a pretty strong inkling of what the answer ought to be.

"It depends who you ask," Susskind said. "Maybe the quark-gluon plasma is analogous to a ten-dimensional black hole. But the connection may be deeper. A lot of us think it's probably deeper."

"So much for claims that string theory is untestable," I said.

"Is that testing string theory?" Susskind asked. "I think so."

It had to be a little vindicating for him, I thought, how it had all come full circle. After all, Susskind had originally invented string theory to describe hadrons, particles made of quarks and gluons. Later everyone realized the theory was actually describing things with much higher energies. But as it turned out, Susskind was right after all. String theory *does* describe hadrons—only in ten dimensions and a radically different spacetime geometry.

"And this duality between black holes and the quark-gluon plasma convinced physicists that information can't be lost?" I asked.

Yes, Susskind said. Everyone knew that information couldn't be lost in a hot gas of ordinary particles—that was just basic quantum mechanics. If a quark-gluon plasma was dual to a black hole—if they were two different descriptions of the very same thing—then a black hole couldn't lose information, either. Hawking admitted he had been wrong. Susskind won their thirty-year battle.

"But we don't live in an AdS universe," I said to Susskind. "Our universe is de Sitter. AdS/CFT was enough to change Hawking's mind?"

"It was," Susskind said. "The opposition, including Hawking, had to give up. It was so mathematically precise that for most practical purposes all theoretical physicists came to the conclusion that the holographic principle, complementarity, and the conservation of information would have to be true. It was the nail in the coffin of information loss."

It was also the nail in the coffin of dimensionality's invariance. I could finally put my finger on what it was that had bugged me about the dimensional reduction of black hole entropy: it meant dimensionality wasn't ultimately real. The holographic principle, and specifically AdS/CFT, showed that two descriptions of the same exact physics could have different numbers of dimensions. The two descriptions were mathematically equivalent. As an ontic structural realist, I knew that neither description could be considered the "real" one; the only thing that was real was their mathematical relationship. Dimensionality wasn't invariant. It wasn't an ingredient of ultimate reality.

Neither were strings. AdS/CFT showed that strings were nothing more than ordinary particles viewed in a higher-dimensional warped space. If the particles on the boundary and the strings in the bulk could be perfectly mapped to one another, there was no genuine difference between them. Particles, strings . . . they were just two ways of looking at the same thing.

Once the AdS/CFT duality had convinced physicists that information couldn't be lost in black holes, they all jumped on board Susskind's horizon complementarity train. As radical as it was, it was the only way to save information without violating quantum mechanics or general relativity in the process. But its implications were profound. *Really* profound. I realized just how freaking profound they were when I thought more about Safe watching an elephant fall toward a black hole.

It's a gruesome scene. As it nears the horizon, the elephant gets stretched from trunk to tail, twisted and deformed as it moves ever more slowly toward the looming abyss. Slower still, it approaches the point of no return, the space around it growing dangerously hotter. But before the elephant crosses the horizon, it's torched by the Hawking radiation, reduced to nothing but a sad mess of burning ash.

Screwed, true to form, is riding the elephant. From his point of view, he and the elephant sail smoothly into the black hole, encountering nothing remarkable at the point in space where Safe sees a horizon. No twisting, no burning. Just easy, empty space. If the black hole is big enough, Screwed and the elephant will happily live out the rest of their lives before hitting the singularity.

So the elephant is dead outside the black hole and alive and well on the inside. It was a pretty serious discrepancy. It was as if Schrödinger's cat was both dead and alive in a box and the box was both floating in empty space and bursting into flames a billion light-years away. There seemed to be two copies of a single elephant, but quantum mechanics forbids cloning and a single elephant can't be in two places at once. But Susskind's point was this: *no observer can see both elephants.*

"Traditionally people thought there was stuff behind the horizon and stuff in front of the horizon and they were different things, different bits of information," Susskind said. "You wouldn't confuse the two.

But what we've discovered is that you cannot speak of what is behind the horizon and what is in front of the horizon."

The confusion, he explained, could be pegged to the misuse of the word *and*. "The operating word is *or*, not *and*," he said. "Complementarity in quantum mechanics always involves replacing *and* with *or*. Light is waves *or* light is particles, depending on the experiment that you do. An electron has a position *or* it has a velocity, depending on what you measure. In each case there are complementary descriptions which are incompatible if you use them both at the same time. The same thing is happening with black holes. Either we describe the stuff that fell in in terms of things behind the horizon *or* we describe it in terms of the Hawking radiation that comes out. What's so surprising is that this kind of confusion or redundancy is occurring on such a huge scale. If we have a black hole that's a billion light-years in diameter, then we have a description of information a billion light-years deep into the black hole. People always thought quantum ambiguity was a small-scale phenomenon. We're learning that the more quantum gravity becomes important, the more large scales, even *huge* scales, come into play."

The coolest part was that any experiment you could imagine to try to see both elephants would fail perfectly. For example, there seems to be a quick instant when both versions of the elephant are outside the horizon, accessible to the eyes of a single observer. That's because the elephant gets toasted by the Hawking radiation *before* it hits the horizon, say within a Planck's distance of it. At that exact moment, Safe sees the elephant reduced to ashes while Screwed sees it happily unharmed, both versions of the elephant still outside the black hole. You'd think that some third observer—Sucker—could try to catch a peek of both elephants. But, technically, catching a peek of something means bouncing light off it and hoping some of the rebounding photons hit your retina, and the light's wavelength has to be smaller than the object it's trying to resolve. The smaller the wavelength, the higher the energy. To glimpse the elephant at a Planck's distance from the horizon, Sucker would need to use photons with energies larger than the Planck energy—which is either downright impossible or would create another black hole on the spot, cloaking the elephant with another horizon. Either way, Sucker's plan to see Safe's elephant and Screwed's elephant simultaneously will never work.

What if Safe watches the elephant burn and then jumps into the black hole to see its healthy doppelgänger? Again physics conspires against it. In order to measure even a single bit of information about the toasted elephant in the Hawking radiation, Safe has to wait until half of the black hole's mass has evaporated away. By that time, thanks to some simple geometric rules, it's guaranteed that Screwed and his elephant would be destroyed by the singularity. There was no way around it. No observer can see both elephants.

As I thought about it, it occurred to me that the phrase "both elephants" was totally misleading. There's only one elephant. *Or*, not *and*. There's Safe's elephant *or* there's Screwed's elephant. End of story. Any talk of two elephants automatically breaches the quantum no-cloning rule. It violates the laws of physics.

My mind was officially blown. Top-down cosmology had suggested that you violate the law of causality when you try to take a God's-eye view of the universe, leaving me to wonder if you would violate other laws, too. *Laws of physics intact only when viewed inside a single light cone?* Now, horizon complementarity answered with a resounding yes. All the laws of physics—both relativity and quantum mechanics—only hold true within a single light cone, finite and limited as it may be.

For years my father and I had talked about the impossibility of a God's-eye view. After all, that was Einstein's lesson: you can't talk about the universe without asking, from whose point of view? Reference frames make all the difference. *There is nothing outside the universe.* Thanks to the finite speed of light, any given observer can see only a piece of it. But we had been talking about it as a philosophical point: if no one can ever see the universe from a God's-eye view, it makes good pragmatic sense to avoid using that view in your description of the universe. Horizon complementarity was saying something way more powerful: it isn't just a matter of philosophy anymore; it's a matter of physics. Try to describe the universe from an impossible reference frame that can see across horizons, and *you'll get the wrong answers*. You'll count two elephants instead of one. The laws of quantum physics will break down.

Horizon complementarity's message was clear: *physics makes sense only within the reference frame of a single observer.*

The idea was so radical it was hard to wrap my head around it. The situation with the elephant seemed so weird because intuition tells us that even if we can't be inside and outside the horizon simultaneously, there's still some ultimate answer to where the elephant *really* is. But "really" assumes a reality that can be described from a God's-eye view. There is no single "really." There's Safe's "really" and Screwed's "really." Nothing more.

"It's not only a new form of complementarity, it's also a new form of relativity," Susskind told me. "Relativity told us that certain things are relative to the motion of an observer—for example, whether two events are simultaneous or not. But there were other things that remained invariant. A flash bulb goes off inside my house. That's an invariant statement. But now we're saying that's not true at the level of black holes. The location of the information, whether it's behind or in front of the black hole horizon, depends on the motion of the observer. Where a basic event takes place is subject to the observer in a way that was not true in standard relativity. The location of a bit becomes ambiguous and observer-dependent when gravity becomes important."

Einstein had rendered three-dimensional space and one-dimensional time observer-dependent but left the unified four-dimensional spacetime invariant. Now, horizons made that observer-dependent, too. Spacetime was no longer invariant. It was no longer real.

"If horizon complementarity tells us that spacetime is observer-dependent, what's left invariant?" I asked Susskind.

"What's left invariant?" He paused. "That's a good question."

"I'm starting to think we should focus our book on the search for invariants, how they seem undermined at every turn," I said to my father. I had come to my parents' house in Philadelphia for a weekend, and my father and I were hanging out in our physics library, working on our book proposal. Thanks to AdS/CFT, the holographic principle, and horizon complementarity, we had now crossed dimensions, strings, and spacetime off our list. "Every time physicists think something is invariant it turns out to be observer-dependent. An illusion. It's like that Lewis Carroll poem. What's it called?"

"*Alice in Wonderland?*"

"No, the poem," I said. "The Snark?"

"I don't remember that one."

"These characters are hunting for a Snark, but no one has ever seen one. Every time they think they've caught one, it turns out to be a Boojum and it disappears."

I dug through the books in my old bedroom until I found my copy of *The Hunting of the Snark,* then read aloud to my father the story of the Bellman, the Banker, the Beaver, and their crew, who set sail in hopes of capturing a Snark.

"'What's the good of Mercator's North Poles and Equators, / Tropics, Zones, and Meridian Lines?' / So the Bellman would cry: and the crew would reply / 'They are merely conventional signs!'"

"They're observer-dependent!" my father piped in.

"'Other maps are such shapes, with their islands and capes! / But we've got our brave Captain to thank' / (So the crew would protest) 'that he's brought us the best— / A perfect and absolute blank!'"

I looked up. My father was grinning.

I read on until the fateful end, when the Baker believes he's found the elusive creature: "'"It's a Snark!" was the sound that first came to their ears, / And seemed almost too good to be true. / Then followed a torrent of laughter and cheers: / Then the ominous words "It's a Boo—"'"

"I love it," my father said.

"Oh my God, listen to this," I said, reading from Martin Gardner's preface to the book. "'The Snark is a poem about being and nonbeing. . . . The Boojum is more than death. It is the end of all searching. It is final, absolute extinction, in Auden's phrase, "the dreadful Boojum of Nothingness." In a literal sense, Carroll's Boojum means nothing at all. It is the void, the great blank emptiness out of which we miraculously emerged; by which we will ultimately be devoured; through which the absurd galaxies spiral and drift endlessly on their nonsense voyages from nowhere to nowhere.'"

"Cheery," my dad said.

"But appropriate! The Snark is invariance, ultimate reality. But every time we think we've found a Snark, it turns out to be a Boojum. It disappears."

We set to work on a proposal for a book called *Hunting the Snark: Physicists' Quest for Ultimate Reality.*

Like the characters in Carroll's surrealist poem, physicists are on the trail of their own Snark, we wrote. Their shadowy creature is ultimate reality, the objective world independent of observers. It is the world "out there," as it exists in and of itself regardless of how we perceive it. This has proven far thornier than one might think. In the early twentieth century, Albert Einstein found that space and time were not fundamentally real but are observer-dependent. Meanwhile, the founders of quantum mechanics were coming to grips with the realization that observers play a far more profound role than anyone had imagined. But what few people know is that in recent years, things have gotten a whole lot weirder. Today, cutting-edge physics is forcing us to completely rethink the nature of reality and our place in the cosmos. In studying the physics of black holes physicists have found that particles are observer-dependent; in exploring the consequences of the holographic principle they've found that even four-dimensional spacetime—which had been left intact by Einstein's theories—is also observer-dependent, introducing what Leonard Susskind has called "a new kind of relativity."

Each chapter of the book, we explained, would have physicists hunting for a different Snark—space, time, gravity, particles, spacetime, dimensions, the gauge forces, strings—which would inevitably and profoundly turn out to be observer-dependent Boojums, culminating in a total rethink of cosmology itself.

These are conclusions that boggle the mind. The things we have long believed to be the most fundamental features of reality have turned out to be nothing more than mirages. One by one we see that the seemingly solid building blocks of the universe are consequences of our own points of view—and that the universe itself is a strange kind of fiction. Snark-hunting is no task for the fainthearted. It seems the more deeply we look into the nature of reality, the more clearly we find only a reflection of ourselves. In the end, the closest thing to a Snark may turn out to be Nothing itself.

"I think this is great," my father said, "but to be honest, it doesn't feel like our book. It feels like *a* book."

I knew what he meant. Like most imaginary books, *our* book was exhaustive, encompassing every last bit of life, the universe, and everything. It was inimitable. It was infinite. It was the answer to the universe.

"At some point *our* book has to become *a* book," I said.

He nodded. "I guess that's true."

Eventually my father went to bed, but I kept working, determined to finish the thing. According to the book I had bought on how to write a book proposal, the next step was to include some biographical information about the authors.

In Hunting the Snark, *father-daughter writing duo Warren Gefter and Amanda Gefter journey to the frontiers of physics and cosmology in search of ultimate reality. Warren Gefter is . . .*

Fuck.

How was I going to explain my father's role? I had snuck him past the organizers of the Wheeler symposium under the vague guise of "plus one," but I had no idea how I was going to sneak him past Brockman. His credentials, no matter how impressive in the medical world, were going to seem seriously random here, and his inclusion in the project was sure to raise more questions than it answered. For example, why did I, a supposedly seasoned science journalist, need a coauthor in the first place? Why did said coauthor happen to be my father? And what the hell was a radiologist doing coauthoring a book about ultimate reality?

There was simply no way to explain it. My only choice was to write the proposal as if my father's inclusion were perfectly normal and hope that the book sounded too good to refuse.

For the Snark was a Boojum, you see.

When an email from John Brockman appeared in my inbox, I felt a sickening twist in my stomach. *Here we go,* I thought. *This is it.*

I clicked it open, holding my breath.

This wasn't it.

The email made no mention of our proposal. It was an invitation to a conference on moral psychology.

Morality had never been of much interest to me, but I jumped at the chance to attend a Brockman event and drove from Boston down to Washington, Connecticut—a wealthy country town in the southern foothills of the Berkshires, just swanky enough to provide country homes for rich Manhattanites and just rural enough to have no cell phone reception.

The conference was held at a luxury hotel that had the look and feel of a country manor. In true Brockman style, a select group of only nine elite scientists were invited for the roundtable event, along with a handful of press from outlets such as *The New York Times, Newsweek,* and *Scientific American.*

The conference lasted three days. On the second night, Brockman invited everyone for dinner at his country home, the legendary farmhouse, which sat on seventy-five sprawling acres in a neighboring town. As I sipped wine and tried to make small talk with the scientists, the surreality of it all had me tongue-tied. It felt as though only yesterday I had been reading about the Brockmans' farmhouse and wondering how I would ever get invited to mingle with the members of the Reality Club. Now, inexplicably, I was.

Everyone helped themselves to food and sat down to eat at the large dining table. Katinka Matson, Brockman's wife, sat down next to me. An artist, agent, and president of Brockman, Inc., Matson was strikingly attractive, with bright white hair that framed her face in sharp, modern angles. As we chatted, she looked down at my arm. "That's not real, is it?" she asked.

I followed her gaze to the tattoo on my forearm and laughed. "Yeah, it's real."

"That's permanent? What does it say?"

I held out my arm so that she could read the words that were inked there. It was a poem by Seamus Heaney, from *Station Island.*

In the book, Heaney travels to Station Island in Donegal, Ireland, the legendary site of Saint Patrick's Purgatory and a place of Catholic pilgrimage. Heaney's not searching for God, but for himself, for his voice as a writer. At the end of the pilgrimage, sleep-deprived and weak from hunger, Heaney meets the ghost of James Joyce, who tells him, "Your obligation / is not discharged by any common rite. / What you must do must be done on your own / so get back in harness. The main thing is to write / for the joy of it. Cultivate a work-lust / that imagines its haven like your hands at night / dreaming in the sunspot of a breast. / You are fasted now, light-headed, dangerous. / Take off from here. And don't be so earnest, / let others wear the sackcloth and the ashes. / Let go, let fly, forget. / You've listened long enough. Now strike your note."

In the next poem, Heaney does. Entitled "The First Gloss," it is inked in black Garamond on my writing arm. Matson read the words aloud: "Take hold of the shaft of the pen. / Subscribe to the first step taken / from a justified line / into the margin."

When I had gotten the tattoo years earlier, the tattooist had suggested that I take a photo of my arm and send it to Heaney. "He would be so honored!" he insisted.

"Sure," I had replied, "I'm sure he'd hang it right next to his Nobel Prize."

"Did John see this?" Matson asked. "John! Come look at this!"

Brockman came over to the table, and Matson pointed to my arm.

He read the poem aloud, then looked up at me. "Why did you get that?"

"It's about being true to your voice," I said. "About taking risks in your writing."

He nodded approvingly.

The following afternoon, as the conference wrapped up, I walked over to Brockman to say goodbye before heading back to Boston.

"You should come down to New York," Brockman said. "I'll take you around to meet all the publishers. See if we can find a project for you."

I started to nod, unsure what to say. *What about* Hunting the Snark?

"John, she already has a book she wants to write," said Matson, who had just walked over. I wanted to hug her.

"Okay, fine," he said. "Tell me about it. But I don't want to hear any of that Alice-in-Wonderland shit."

I explained that it was about what the latest, most cutting-edge theoretical physics was telling us about the nature of ultimate reality.

"What's the title?" Brockman demanded. "It's got to hit hard."

My mind raced. Suddenly every thought that popped into my head was some kind of Alice-in-Wonderland shit. It was like trying not to think of a pink elephant. Or a pink Cheshire cat. Finally I spat something out: "*The End of Reality.*"

"Not bad," he said. "Send me a proposal. Two pages."

"Okay," I said, enthusiastically. "I will."

I thanked them both for having me, then walked away.

"Good luck with your tattoo!" Matson shouted as I made my way out the door.

After writing my article on horizon complementarity, I couldn't help thinking back to the Santa Barbara debate about the string landscape, and I was suddenly more than a little suspicious of the whole notion of the multiverse. When we talk about the multiverse, we're talking about an infinite number of causally disconnected universes, all separated by horizons. If you violate the laws of physics with a description across one event horizon, I was pretty sure you'd obliterate them with a description across an infinite number. If the laws of physics only made sense within the reference frame of a single observer, what the hell could the multiverse even mean?

I was about to call Susskind to ask him when an email from him popped up in my inbox. He was sending me an early draft of his book about horizon complementarity and his thirty-year battle with Stephen Hawking, the one that Brockman had convinced him to write. I was thrilled to get to read it so far ahead of publication, and I dove right in.

Toward the end, Susskind described the current acceleration of the universe's expansion and wondered if horizon complementarity applied to our de Sitter horizon in the same way that it applied to the horizon of a black hole. "At the present time, we understand very little about cosmic horizons," he wrote. "The meaning of the objects behind the horizon—whether they are real and what role they play in our description of the universe—may be the deepest question of cosmology."

"We all read your proposal and we all agreed that while we'd like to work with you, this just isn't your book."

Katinka Matson had called to deliver the verdict on our proposal for *The End of Reality*. "The coauthorship structure is too confusing," she said. "It's not clear why your father is involved."

"I got that," I said. "But this has always been a joint project."

"We can't sell the coauthorship," Matson said. "There's no voice. Where are *you* in this? Your tattoo, where's *that*?"

I cringed. The irony of the words inked on my forearm was not lost on me. For years now I had not been true to my voice; I had been writing in a whisper. I could barely hear myself at all in the magazine articles I wrote, because from the start I had been pretending to be someone else, someone with a different voice, a proper voice, the voice of a journalist. I certainly couldn't hear myself in the papers I'd written for school, which I had composed under the guise of a stuffy British man. And thinking about it now, I saw that Matson was right— I couldn't hear myself in the voice I had conjured for those stillborn books, either, *The Alice-in-Wonderland Shit* and *The End of Reality*, books intended to simulate other books.

"Just think about it," Matson said kindly.

I told her I would, and hung up the phone.

Matson's words echoed: *There's no voice*. For one long decade, the only way I had ever known to get to the truth of reality was to be a fake. But what exactly was I covering up? What if behind this mask there was nothing, and even the imposter was an imposter itself? What if I was like Clark Kent ripping off his business suit only to reveal another business suit underneath? What if I was nothing more than an ordinary journalist who had somehow convinced herself that she was on some secret mission to uncover the nature of reality, because it made life exciting and meaningful, and my father, being a father, had played along for my sake, or perhaps had himself been similarly duped? We shared so many thoughts, I figured, it was hardly a stretch to think we could share a delusion.

Matson felt it was the coauthorship that had muted my voice. Maybe she was right. Maybe it was like Safe and Screwed, like *their* confusing coauthorship structure. You violate the laws of physics when you try to speak from two observers' points of view simultaneously. Maybe you violate the laws of publishing, too. Maybe our book had been an impossible object from the start. Maybe it didn't make sense to try to write a book using both of our voices, since it would add up to no voice at all.

Suddenly I understood just what had struck me as so funny about the academic use of the royal "we," back when I was writing my thesis. That pronoun, which was meant to denote pure objectivity, pure *reality*—the essence and observer-independence of science—was in fact describing something nonexistent, forbidden by physics, disappointingly voiceless, and markedly *not real*.

The only pronoun that held any hope of reality was *I*, but I wasn't sure that I knew who that was. After so many years of writing with phony voices, I was no longer convinced I had a real one to offer.

Somewhere in the world—on a shelf? encased in glass? in a box under a bed?—sat a stack of dusty hardbound notebooks, and somewhere in the handwritten pages of those notebooks hid the solutions to Wheeler's riddles.

Chasing down his former students hadn't exactly provided the enlightenment we were hoping for. No one seemed to really know what Wheeler had been thinking. The only option was to hear it straight from the source. I might have failed in my mission to score a book contract for my father and me, but I was determined to get our hands on those journals.

I was scouring the Web once again for any clues to their whereabouts when I stumbled upon a transcript of a lecture that Charles Misner, another one of Wheeler's former students at Princeton, had recently given at the University of Maryland. Halfway through the lecture, Misner said, "John did have this habit for, I guess, all of his life of having bound notebooks. . . . They were always there. When he had a group of students in the office he would sit down and take notes as the discussion went on. He would also make notes to himself about the calculations he was doing, or the work he planned to do. What were the important questions in physics? And so forth . . . Those notebooks, incidentally, have been given to the American Philosophical Society in Philadelphia."

Philadelphia?

They were in my hometown?

I immediately pulled up the American Philosophical Society's manuscripts library catalog and searched for the Wheeler journals—but nothing turned up. I tried again . . . nothing. I tried WorldCat, searching the collections of more than ten thousand libraries. Finally, a listing appeared: the Wheeler collection, twenty-eight volumes of notebooks. With it, an error message: *Sorry, no libraries with the specified item were found.*

Eventually I found a reference to the Wheeler collection on the American Institute of Physics's International Catalogue of Sources, which confirmed Misner's claim that the journals were indeed at the American Philosophical Society. It listed neither a call number nor a status. It simply said, *Contact repository.*

So I did. I told them that my father and I were coauthoring a book about Wheeler and that even though the journals were on restricted access, we desperately needed them for our research. The coauthorship structure may have violated the quantum no-cloning theorem and the laws of publishing, but I figured the librarians probably wouldn't notice, and I was right. They agreed to let us look at the notebooks.

I called my father at his office to tell him the news: "I found the journals."

11

Hope Produces Space and Time

The thing about the American Philosophical Society is that it's exactly the kind of place you'd expect to find the secret of the universe.

The APS began as the Junto, Ben Franklin's secret society. Franklin—inventor, printer, politician, Freemason—had specified that the original Junto consist of only twelve members, its proceedings kept tightly under wraps. The group met every Friday, when each member was required to "produce one or more queries on any point of morals, politics or natural philosophy," which the twelve men would then discuss "in the sincere spirit of inquiry after truth." Sixteen years later, Franklin transformed the Junto into the Philosophical Society. "That one society be formed of virtuosi, or ingenious men, residing in the several colonies, to be called the American Philosophical Society," Franklin stipulated, the aim of which would be to explore "all philosophical experiments that let light into the nature of things." Franklin was elected the group's first president. Early members included Thomas Paine, George Washington, and Thomas Jefferson. Later members included Robert Frost, Albert Einstein, and John Wheeler.

The APS building was erected on Fifth Street, in the heart of Philadelphia's Old City. The building is a classic work of Georgian architecture around the corner from Independence Hall, where Jefferson would later present the Declaration of Independence and near where the Liberty Bell sits today in all its fractured irony. Soon the APS built Library Hall, nestled in a tiny cobblestone street next door to the country's second bank, now a Revolutionary War–era portrait gallery. Today the library holds 11 million rare manuscripts and countless rare books, including first editions of Newton's *Principia* and Darwin's *Origin of Species*.

"I can't believe we're about to read his journals," my father said, sounding almost giddy, as he tugged open the library's large wooden door. It was a hot August morning in what appeared to be the eighteenth century, the sound of tourist-filled horse-drawn carriages clacking in the thick summer air.

I was awfully giddy myself. According to the manuscript librarian, only two other people had read through the journals. As per Franklin's requirement, we were prepared with our own query: what had Wheeler meant that day back in Princeton? I had no idea what to expect. I hoped it wouldn't be another dead end.

The foyer of the American Philosophical Society library in Philadelphia
A. Gefter

Inside we entered a small but impressive foyer, with a high ceiling and a striking black and white checkered marble floor. On the walls, framed by ornate molding above and colonial wainscoting below, were glass cases displaying Jefferson's handwritten copy of a late draft of the Declaration of Independence and a map of Lewis and Clark's expedition drawn in Clark's own hand.

We checked in at the front desk and were handed badges that gave us access to the library's reading room. Heading through the arched doorway, we found ourselves in a kind of holding room with lockers for us to store all the things we weren't allowed to bring near the manuscripts, including pencils and paper. Once we had stashed everything except our laptops and power cords, we entered the huge glass doors to the reading room.

The whole scene looked suspiciously perfect. Old books lined the walls; a staircase led to a balcony for reaching the higher shelves. Brass chandeliers hung over wide mahogany tables. Busts of Franklin and Jefferson watched over a handful of scholars quietly turning fragile pages. It was beautiful. I couldn't help wondering if the old books were empty props, the scholars hired extras. "It looks fake," I whispered to my dad.

He nodded. "Like a movie set."

Secret societies? Rare manuscripts? Restricted access? Masons? A treasure hunt through inscrutable symbolism to decode the mysterious phrases that had now haunted us for a decade? At stake nothing less than the secret of existence?

"I feel like I'm in a fucking Dan Brown book," I whispered.

We approached the librarian seated behind a desk and explained that we had come to look through the Wheeler journals. "Which ones?"

Standing next to
John Wheeler's
journals in
the American
Philosophical
Society library
W. Gefter

he asked, handing us a thick manila folder containing a catalog of the Wheeler materials.

I obligingly thumbed through the pages, then answered, "All of them, please."

He looked skeptical. "There are *a lot* of notebooks."

I smiled. "We know."

"Okay," he said, a hint of warning in his voice. "I'll bring out a few for you to start."

We took a seat at the center table and waited until the librarian emerged wheeling a cart full of thick bound notebooks, maroon and brown and bursting at the seams.

"Be careful with them," he said, eyeing us nervously. We must have looked like a pair of ADD-riddled kids on Christmas morning, ready to tear into our presents, waiting to pounce.

"We will," I said, perhaps a little too eagerly, and he left us alone with Wheeler's words.

"Remember," I whispered to my dad, "keep your eyes peeled for anything about a self-excited circuit or the boundary of the boundary."

I grabbed a journal off the cart at random. It was labeled "Relativity Notebook No. XIX." Carefully cracking open the cover, I discovered a fortune, the kind that comes out of a cookie, which Wheeler had glued onto the first page: *Persistence will be rewarded.* I took it as a sign of encouragement and began to read.

In the journal, I found Wheeler contemplating what he called the "paucity of laws," the unbelievable fact that all the complexities of the universe seem to be governed by a few simple laws of nature, which in turn are likely unified into something simpler still. "Closer to center we get, the fewer laws we find and the simpler they are," he wrote.

Several pages later I read, "The law of force, law of interaction should be such as automatically to guarantee conservation of the source, via the principle that the boundary of a boundary is zero."

"Look!" I shouted in a whisper. "Already!"

Wheeler continued. "Seems to make these laws of Maxwell and Einstein almost trivial manifestations of something simpler going on 'inside'—except that 'inside' presupposes idea of space—and space should be a secondary notion." A few pages later, in an entry titled "Down with Space," he had scrawled, "Space prejudices dimensionality when there ought not even to be any dimensionality."

"Feynman would say one is talking dreams here," Wheeler wrote, "that one should instead be doing calculations. My fear: if one thinks only on those things which one can calculate, he fails to think about lots of important issues. In other words, it may be more important to look for the right questions than to look for the right answers. ['Truth doesn't come flying in. It has to be dragged in by the heels.']"

I couldn't find any more clues to the boundary of the boundary in that journal, but soon my father was nudging me, excitedly pointing to a passage. I leaned over his shoulder to read. "We don't apply logic to something, logic is the whole thing. . . . No structure, no equations. *We* impose the structure. We form what forms us. [View of relativity as the boundary of a boundary consistent with this. Would be *wonderful* to show it 'coming out of nothing' . . .]"

We form what forms us. That sounded like the self-excited circuit. But how exactly was the boundary of the boundary related? And in what way did Wheeler envision it emerging from nothing?

A few turns of the page later, my dad was pointing again. "Yesterday, during unexpected stop-over at London airport because of Alitalia strike, rang Penrose, told him of view that there is no structure except pregeometry beneath it all, no dimensionality, no general relativity. . . . Once again concept of 'mere elasticity.' I noted [the] boundary of a boundary, and how this idea comes close to being as dimensionless as any idea can well be. He felt there is too much beauty in the structure that he works with for this to be just an illusion, an appearance, of four-dimensionality. I, on the other hand think of structure without structure, law without law, and physics without physics. The very nature of the laws we see derives, we conceive, from their being the grossest way we can ascribe structure to something that has no structure."

General relativity is "mere elasticity"? I was starting to wonder if the journals were going to offer more questions than answers.

Wheeler would have said that was a good thing, but I suspected that an answer or two couldn't hurt.

As I continued to read, I was amazed by how meticulously organized the journals were—not by the library staff, but by Wheeler. Every journal began with a detailed table of contents referencing the heading and page number of every entry. In the margin of each entry, Wheeler noted the date, time, and location of the writing. Had he glued into the journal a loose piece of paper—the back of an envelope, perhaps, or notes scrawled on a conference itinerary—he noted not only the date and time of the note-taking itself but also the date and time of the gluing. If it was a photograph, Wheeler had created a transparent overlay on which he indicated in colored markers the identity of each person in the photo. Had he spoken to a fellow physicist on the phone, he jotted down the precise duration of the call as well as the caller's phone number. Had he gone out to dinner with colleagues, Wheeler sketched the seating chart. Had he spilled some coffee on the page, he labeled the stain: "Coffee."

He was also on a constant search for words. I saw now that it was no accident Wheeler had coined so many terms and phrases, like "black hole," "wormhole," "quantum foam," "the self-excited circuit," and "the boundary of a boundary." His desire to find the best words for such things bordered on mania. The journals were full of pages that were nothing but lists of words, hundreds upon hundreds of words, seemingly copied out of a dictionary: "Secret seek sought senses shadow shade shape shapeless shares shatter sharp sheen shelter shield shine shock shore shot show shuffle siege sink siren song skeleton skin sky slate slide slumber sight smoke smuggle snow sober soil sojourn soldier solid solitude solution solve . . ."

My father, reading my mind as usual, leaned over and whispered, "This guy had a serious case of OCD!"

I nodded. Ya think?

He even traveled with a thesaurus, a fact I gleaned from his packing lists, which cataloged every item that ever graced his suitcase along with their respective weights. A fair amount of weight was allocated to books. On an average trip he brought two physics books, a few philosophy books, several volumes of poetry, and one thesaurus. Wheeler's parents had both been librarians. Clearly he had inherited their devotion to the written word and to cataloging.

"Spurn spyglass squeeze stable staccato stage stamp stand standing star start state stay steel steep steer steersman . . ."

My father and I read in silence, the massive journals propped up on foam wedges for support, carefully turning the brittle pages, which were all loosening from their bindings, weakened by the passing of time and by the weight of all the pictures, postcards, and papers Wheeler had glued to them. I wondered if Wheeler had developed his compulsion for gluing everything down after Eisenhower had personally scolded him for losing classified H-bomb documents on a train.

I soon came to recognize Wheeler's symbol for "the boundary of a boundary is zero": $\partial\partial \equiv 0$. It popped up everywhere: "Of all regularities, the one to me most suggestive is $\partial\partial \equiv 0$." "Everything from nothing $\partial\partial \equiv 0$." He described it as "spacetime's grip on mass," the "faint frown of space." He wrote, "The principle of algebraic geometry that 'the boundary of a boundary is zero' is dimension free. It would be difficult to find any simpler principle on which to build the laws of physics if, at bottom, these are all based on 'law without law,' a universe of 'higgledy-piggledy' construction. Therefore it is interesting to see that major portions of electromagnetism, gravitation, and the Yang-Mills theory of the quark-binding field are built exactly on this principle. However, in each case there is a part of the theory that has no equally simple structure."

How were electromagnetism, gravity, and QCD built on the boundary of a boundary? It was clear that Wheeler saw it as the unifying principle of physics and a possible way to get something from nothing. But what the hell did it mean? And what was with the triple equals sign? I leaned over to ask my father.

"It's an identity," he whispered.

Inside John Wheeler's journal (American Philosophical Society library)
A. Gefter

"Isn't that what a normal equals sign is?" I asked.

"An equals sign says these two things happen to be the same. The three lines mean they're identical, by definition." Fair enough. I read on.

"Go anywhere, see anybody, ask any question," Wheeler wrote—and he did. Many entries were written in transit as he traveled around the globe meeting with physicists, philosophers, mathematicians, and virtually anyone else who might have anything to contribute to his quest to understand the universe. At times, though, he grew frustrated. "All this traveling about trying out ideas—still it isn't clear how we're going to build up structure out of nothingness."

Wheeler knew, I soon discovered, that such lofty philosophical ambitions were looked down upon by his colleagues, a fact that clearly bothered him but seemed to bother his wife, Janette, even more. On January 31, 1976, he wrote, "Janette and I arrived in Princeton. . . . Wonderful people. But on the whole they took a skeptical point of view

toward what I had to say. Janette was mortified she told me this morning when we woke, and said that at the meeting she almost burst out
in tears. I was talking, she said, about the kind of thing college students discuss, incredibly naïve, vague comments about the nature of
reality, lots of quotes but nothing hard to get one's teeth into. A happy
manner of presentation didn't make up for lack of hard content. She
thought she saw Gi[a]n Carlo Wick making a sniggering remark to I. I.
Rabi sitting beside him, about halfway through my presentation. And I
went overtime . . . Rabi after the session said to me, 'I hear you're giving up physics for this sort of thing.' I expostulated, spoke of my paper
on scattering theory for the Borgmann Festschrift. Lewis Thomas, too,
was a bit anti-philosophical. Aage Peterson (Yeshiva) referred to my
'beautifully poetic talk' but said there was one term in it, 'quantum
principle,' that might lead to confusion."

The next day, Stephen Hawking had come to visit Wheeler in
Princeton and invited him to speak at a conference in Cambridge. A
downhearted Wheeler replied, "Maybe yes, maybe no." "I brought up
my problem about reactions I received to general ideas and desire I
have for a low profile, and yet the problem that one can't think if one
can't talk. He sympathized and says he talks about ideas 'in formation'
only in company of one or two others."

Wheeler worried, too, about the effect his quest might have on his
loyal followers. On May 25, 1979, in his hotel room at the Hyatt Re-
gency in New Orleans, he wrote, "Wheeler has been leading people
along a trail. He can't abandon them at the bottom of a cliff. He has to
show the way up. Many are staking their futures on the promise of the
road he espouses. He can't let them down. He is under a tremendous
obligation to deliver."

When the library closed for the evening, my father and I headed back
out to the cobblestone streets. It took a moment for my eyes to adjust
and my mind to emerge from inside Wheeler's head. It was that feeling
you get when you exit a movie theater and your brain has to snap out
of the fictional world and back into reality, only this movie had lasted
eight hours and featured the mind-bending ideas of a tortured genius.

I looked at my father with a grin. "Wow."

"I know," he said, looking dazed and exhilarated.

"They're not how I imagined they'd be. I always thought of him as this kind of cheery, lighthearted guy. I didn't imagine him to be so . . ."

"Relentless?"

"Yeah. Relentless. Determined. But it's not normal determination."

My father nodded. "It's obsession."

Over the next few days I discovered that Wheeler had one particular obsession: Kurt Gödel. "I argue that the Gödel business has to come into physics, and physics into the understanding of it, it is important to see the G.B. through and through," Wheeler had scrawled in his journal on July 22, 1973. "My problem, I am afraid, is this, that after I've written down the circuit I see nothing next, and infinite number of things to do next, but nothing clearly sorted out in between."

"The Gödel business," I assumed, referred to Gödel's incompleteness theorem, which said that if a mathematical system is consistent, it can't be complete. Gödel had proven the theorem by creating a mathematical sentence that said, in numerical language, something to the effect of: *This sentence is not provable by this mathematical system.* If the system could prove it, then the statement would be rendered false and the system inconsistent for producing falsities. On the other hand, if the system couldn't prove it, then the statement would be true, keeping the system consistent but incomplete, unable to prove one of its own statements. If it was true it was false; if it was false it was true. A classic paradox. *This sentence is a lie.*

But here's the thing: it's obvious that Gödel's sentence is actually true. The mathematical system can't prove it, because if it did the system would self-destruct. The sentence clearly speaks the truth. It's true but unprovable. *We* know the statement is true, even though the mathematical system making the statement doesn't. How? Because we have a vantage point the system doesn't: *we're on the outside*. From our God's-eye view outside the system, we can determine the truth or falsity of the sentence. From inside the system, it's nothing but paradox.

The whole mess came down to one word: *this*. The statement is talking about itself. Self-reference is the root of the whole inside/outside problem. You can't assess the validity of statements about the system from inside the system—to determine their truth or falsity, you have to look down on them from some higher level of reality. Gödel's incompleteness theorem said that the minute self-reference is involved, there are limits to what we can know, unless we can step outside the thing. And if I had learned one thing about the universe, about reality, it was that *you can never step outside the thing*.

A few months before writing that entry, Wheeler had approached Gödel at a cocktail party thrown by the Princeton economist Oskar Morgenstern. Wheeler and Gödel had a close friend in common: Albert Einstein. Wheeler explained to Gödel that he intuited some deep connection between Gödel's incompleteness theorem and Heisenberg's uncertainty principle, two principles that had been discovered within a few years of each other and placed sudden, disturbing limitations on what is knowable in the universe. Gödel didn't want to talk about it. When Wheeler asked why, he learned that Gödel wasn't exactly enthusiastic about quantum mechanics. As Wheeler put it, "He had walked and talked with Einstein enough to have been brainwashed out of any interest in quantum theory—to me a great tragedy, because Gödel's insight might be the key thing."

Wheeler seemed convinced that hidden inside Gödelian incompleteness sat the meaning of quantum mechanics. But why? "Idea of isomorphism between calculus of propositions and pregeometry more interesting than ever," he scribbled.

I knew that propositions were like the atoms of logic, simple statements that could be either true or false, such as "Snow is white" or "My pants are on fire," and the calculus of propositions was just a set of logical rules that related the propositions to one another. And I knew that pregeometry, another Wheeler coinage, was some mysterious thing more fundamental than spacetime, the building material of reality, one presumably based on $\partial\partial \equiv 0$. But why there would be an isomorphic mapping from the calculus of propositions to some kind of pregeometry was beyond me.

I spotted a clue when Wheeler quoted mathematician Hans Freudenthal speaking about the calculus of propositions: "Our vocabulary need not contain a single constant subject. The predicates are, as it were, floating in the air; they do not refer to anything." Opposite this Wheeler had written, "How entrancing—almost an open invitation to be foundation for quantum mechanics plus pregeometry."

Still, Wheeler was lost. "Not seeing a dramatic clear path ahead," he wrote. "Now have concluded just have to push in through the undergrowth. 'Traveler, there are no paths. Paths are made by walking.'"

The calculus of propositions gave the rules for relating two-valued, true/false binary propositions without any reference to the *meaning* of the propositions. Inherent meaning was irrelevant; all that mattered were the relationships, which held true regardless of the truth or falsity of the individual propositions. If p implies q and q is false then p is false—a rule that holds regardless of whether or not my pants are in fact on fire. In that Wheeler saw a glimpse of structure without structure, form without content, the something from nothingness to which he hoped $\partial\partial \equiv 0$ would lead.

"A certain uniqueness, naturalness, and beauty must characterize the real equations," he wrote, "and above all, simplicity. What is the simplest mathematics one knows? . . . There is nothing simpler than the + −, true-false, yes-no, up-down choice. As subsequent thought and analysis has revealed, there are many structures that can be built upon this binary element, but every one of them that has been looked into possesses some arbitrary element or number or structure with the sole exception of the calculus of propositions. It would seem to have the desired element of uniqueness and simplicity, and this in a satisfying and even truly striking way. Logic is too important to be left to the logicians."

Wheeler wasn't ready to give up on Gödel. After having been spurned at the cocktail party, Wheeler brought some of his students in to see Gödel at Princeton to ask him again about the connection between undecidability and quantum mechanics. Gödel threw them out of his office. So Wheeler tried another tactic. Glued into the journal I found a letter he had written to Gödel in December 1973. Apparently Wheeler thought that Gödel might be more likely to answer the question if it was multiple choice. The letter read:

I was quite startled a few months ago at that little party at Oskar
Morgenstern's to learn that you believe in the existence of what is
sometimes called "an objective universe out there" in contraven-
tion of the consequence of the quantum principle as currently
envisaged. Of course I may well have misunderstood. But if I understood correctly that would explain why one should have no
particular motive to try to understand the quantum principle in
new terms, whether in terms of the calculus of propositions or in
any other terms. Why try to explain something that one does not
believe to be correct! You are a very busy man, I know, and I'm
the last person that would want to impose on you to ask you to
write a letter. But could you take time enough to send me the
substitute for a letter by checking off items in the attached copy
of this letter and return that copy to me in the enclosed stamped
self-addressed envelope? With Oskar Morgenstern and many
many others I share great admiration for you and your work, and I
will be honored and very much helped to have this indication of
your present thinking. Every good wish for 1974.

Sincerely,
John Archibald Wheeler

Below was the multiple-choice question Wheeler had written for
the reluctant Gödel.

Have you published on why you disagree with the quantum
principle?
_ Question not well-defined
_ Have not published the slightest mention
_ Have published; see   _____
What is your central point? _____
_ Too long to state here and now

I laughed in admiration of Wheeler's chutzpah. *Persistence will be
rewarded.*
Within weeks he sent a second letter, this time to Paul Cohen, a

mathematician at Stanford who had won a Fields Medal for his work in logic. In analogy to Gödel, Cohen had proven that certain mathematical statements in set theory were similarly undecidable.

"I'm winding up twenty years' involvement with gravitation physics and relativity," Wheeler wrote to Cohen, "with the conclusion that the mystery of things lies still deeper, in the quantum principle; that the quantum principle is connected in some deep sense with logic and the calculus of propositions; that the structure of the universe is connected with our own existence in some deep Leibnizian sense; and that only when we recognize how strange the universe is will we be able to understand how simple it is. I have no special axe to grind. I am simply in search of deeper understanding. I can't help feeling that the marvelous things you have done must have some much deeper connection with the issues that puzzle me than most physicists would recognize. There was a time when the parallel axiom of Euclid seemed to be 'merely a matter of logic.' Then came Bolyai and Lobachevsky. Then came Riemann who opened the door to Einstein and general relativity, with the most direct possible tie to physics. Similarly today, so many people think the questions of undecidability are 'merely matters of logic.' But the good Lord, I'm afraid, didn't have the benefits of the modern university business office, with precise allocations of so much to physics, so much to mathematics, and so much to philosophy. I'm afraid he got everything all mixed up together."

Wheeler was clearly on a mission—but why? What was he trying to find? A few pages later, his agenda became clearer. Scrawled on TWA in-flight notepaper and glued into the journal was an entry entitled "Add 'Participant' to 'Undecidable Propositions' to Arrive at Physics."

"We consider the quantum principle," he said. "Of all the well-analyzed features of physics there is none more central, and none of an origin more mysterious. This note identifies its key idea as participation; and its point of origin as the 'undecidable propositions' of mathematical logic. On this view physics is not machinery. Logic is not oil occasionally applied to that machinery. Instead everything, physics included, derives from two parents, and is nothing but cathode-tube image of the interplay between them. One is the 'participant.' The other is the complex of undecidable propositions of mathematical logic. The 'participator' assigns true-false values to appropriate ones among these propositions at his own free will. As he does so, the corresponding world unrolls upon his screen. No participator, no world! . . . The propositions are not propositions about anything. They are the abstract building blocks, or 'pregeometry,' out of which 'reality' is conceived as being built."

I was so engrossed in the journal that I nearly fell out of my seat when the sudden banging of some kind of drum shattered the silence, flooding the library with the sound of—"Yankee Doodle"? I looked over to see a marching band in full Revolutionary War–era garb drumming past the library window. Now that we had been jolted out of Wheeler's brain and into, apparently, 1776, I figured it was a good time to break for lunch.

"I just don't get this whole Gödel thing," I said to my father as I spooned some lobster fried rice onto my plate. "He thinks that Gödel's undecidability makes room for quantum observer-participancy. Like you have some logical proposition, 'The electron is spin up.' Is it true or false? You need an observer outside the calculus of propositions to decide. So Wheeler's idea is that every time an observer determines the truth or falsity of a proposition, he's registering a bit of information, or rather he's *introducing* a bit of information to the world, and bit by bit we build reality. He wrote, 'Logic as building material.'"

"So determining whether the proposition is true or false is like collapsing the wavefunction?" my father asked.

"I guess. Undecidability requires an external observer, the same way quantum mechanics does."

"But when it comes to the universe, you can't be an external observer. Isn't that the basic problem in combining quantum mechanics with general relativity?"

I nodded. "Yup. I don't see how undecidability helps."

"You think Wheeler knew how it helped?"

"Not so far. So far I think he just suspects that they're two expressions of the same thing. But two expressions of a problem don't add up to a solution. Besides, I'm not convinced that Gödelian undecidability would apply to things like electron spin. Wouldn't it only apply to propositions that observers make about themselves?"

Back in the library, I immersed myself in Wheeler's world. Page after page he agonized over the same questions, circled the same ideas, desperate for some new connection to leap out from the words. He was convinced that there was more to the Gödel business, a kind of quantum logic embodied there. It was a logic that demanded external observers to decide the internally undecidable, to lift the equations from some lifeless heap so that they might "spread their wings and fly," an "intervention . . . envisaged as the quantum principle showing up in its most primitive form." Again and again he returned to the boundary of a boundary, a clue, he was certain, to how we might build physical structure from structurelessness, using the bootstraps of a self-referential loop—a self-excited circuit carved from "airy nothingness." "Physics," he wrote, is "machinery to make something out of nothing." Bit by bit, measurement by measurement, proposition by proposition, he saw that airy nothingness solidifying, and he dreamed that together we would build the world from the primordial haze from which we ourselves arose.

I was confused. The self-referential loop invoked internal observers; it was the universe looking at itself, a windowless entirety, inside with no outside, a one-sided coin. As Thorne had said, "It's self-excited in the sense that the observation comes from within the universe, not from the outside." But Wheeler's Gödelian vision required external observers who could make sense of things from some higher level. Which was it? Inside or out?

I wasn't the only one confused. In the journal, Wheeler noted a question that a student had asked during one of his lectures: "By slightly red-headed mathematically minded student, Isn't participator himself built from physics? And so don't you have a Gödel type of situation where the system makes metamathematical statements about the system. . . . I answered that I would think of the abstract element of participator as outside the system. This I said to him (or someone) is trial approach. It is not mathematical logic. It is mathematical logic plus participator. The most important test is whether it gives anything like quantum mechanics. If it does, we have a go-ahead sign; if not, we have to revise our thinking."

He seemed to be following in the footsteps of his mentor, Bohr, in drawing a stark line between observer and observed. But at the end of the day, Wheeler knew that such a line could never be maintained. "Elementary phenomena are impossible without the distinction between observing equipment and observed system," he wrote, "but the line of distinction can run like a maze, so convoluted that what appears from one standpoint to be on one side and to be identified as observing apparatus, from another point of view has to be looked at as observed system." *We form what forms us.* "Aren't we mistaken in making this separation between 'the universe' and 'life and mind'?" he wrote. "Shouldn't we seek ways to think of them as one?"

I glanced over at my father, who was reading intently, and I smiled. Everything was back to the way it was supposed to be, back to the way it had been in the beginning. Back to Wheeler. Back to the two of us sitting side by side reading, quietly searching for the answers to the universe.

Several pages later I discovered a newspaper article from London's *Daily Telegraph* that Wheeler had glued into the journal: "Days Are Getting Shorter."

The next morning we settled into our usual seats in the library and dove back into Wheeler's head. Once again I found him wrestling with the role of observers in the creation of reality. He needed to know what criteria were necessary to be an "observer" and what constituted an "observation" that would be capable of rendering some bit of the airy nothingness as real.

"What we have been accustomed to call 'physical reality' turns out to be largely a papier-mâché construction of our imagination plastered in between the solid iron pillars of our observations," Wheeler wrote in his journal. "Those observations constitute the true reality." But there were questions he couldn't seem to answer: Who was the observer? And, crucially, was there just one or were there many?

"On few issues in my life have I ever been more at sea than I am now on the relative weight of the individual and the collectivity in giving 'meaning' to existence," Wheeler wrote. "Last night before falling asleep I could not see how anyone could doubt it is the individual who gives meaning to existence—where else except in my mind is the world I seek to explain?" But soon after, on a train ride from Rhode Island to Boston, Wheeler scrawled, "How preposterous to think that each has to invent the universe afresh. Moreover, meaning comes from interaction with others, not from 'one consciousness alone,' whatever that would mean."

Why drag consciousness into it at all? I wondered. Wheeler knew it was a mystical morass, and that one gap in our understanding couldn't be plugged by another. Observers, sure—but why not stick with Einsteinian observers, just reference frames, coordinate systems, rods, and clocks? After all, in the first arc of the self-excited circuit, the universe gives rise to the observer. That meant that the observer, conscious or not, had to be built out of ordinary physics, not fairy dust.

But as I read further, Wheeler's reasons for turning to consciousness became clearer. As far as he could see, it was the only solution to the problem of the second observer.

"What happens when several observers are 'working on' the same universe?" Wheeler asked in his journal. It wasn't a minor question. It was *the* fundamental obstacle to understanding the meaning of the quantum and, in turn, existence. Wheeler was sure that the self-excited circuit was the only viable explanatory structure for existence—no tower of turtles, no fundamental ingredient left unaccounted for—but he couldn't see how to squeeze more than one observer into the circuit. It was torturing him.

Wheeler wasn't the first to point out that quantum mechanics slips into paradox the minute you introduce a second observer. The Nobel Prize–winning physicist Eugene Wigner, for one, had emphasized it with a Schrödinger's-cat-style thought experiment that became known as "Wigner's friend." It went something like this: Inside a lab, Wigner's friend sets up an experiment in which an atom will randomly emit a photon, producing a flash of light that leaves a spot on a photographic plate. Before Wigner's friend checks the plate for signs of a flash, quantum mechanics shows that the atom is in a superposition of having emitted a photon and not having emitted a photon. Once the friend looks at the plate, however, he sees a single outcome—the atom flashed or it didn't. Somehow his looking collapses the atom's wavefunction, transforming two possibilities into a single reality.

Wigner, meanwhile, is standing outside the lab. From his point of view, quantum mechanics shows that until his friend tells *him* the outcome of the experiment, the atom remains in a superposition of having emitted a photon and not having emitted a photon. What's more, his friend is now in a superposition of having seen a spot of light on the plate and not having seen a spot of light on the plate. Only Wigner, quantum theory says, can collapse the wavefunction by asking his friend what happened in there.

The two stories are contradictory. According to Wigner's friend, the atom's wavefunction collapsed when he looked at the plate. According to Wigner, it didn't. Instead, his friend entered a superposition correlated with the superposition of the atom, and it wasn't until Wigner spoke to his friend that *both* superpositions collapsed. Which story is right? Who is the true creator of reality, Wigner or his friend?

"The theory of measurement," Wigner wrote, "is logically consistent so long as I maintain my privileged position as ultimate observer." From Wigner's point of view, quantum mechanics "appears absurd because it implies that my friend was in a state of suspended animation before he answered my question." One way out of the absurdity, Wigner said, would be to contend that he himself was the only observer in the universe, a solution he just couldn't buy. "To deny the existence of the consciousness of a friend to this extent is surely unnatural attitude, approaching solipsism, and few people, in their hearts, will go along with it," he wrote. Instead, Wigner took the paradox to mean that consciousness plays some special role in physics—that while atoms and photographic plates and obsessively organized notebooks can be in superpositions, conscious people cannot.

Hugh Everett, Wheeler's student at Princeton, likewise saw the problem of the second observer as the central mystery of quantum mechanics. In his 1955 dissertation, written under Wheeler, Everett wrote, "The interpretation of quantum mechanics . . . is untenable if we are to consider a universe containing more than one observer." A possible escape route, Everett suggested, is "to postulate the existence of only one observer in the universe. This is the solipsist position, in which each of us must hold that he alone is the only valid observer, with the rest of the universe and its inhabitants obeying at all times [linear evolution of the wavefunction] except when under his observation. This view is quite consistent, but one must feel uneasy when, for example, writing textbooks on quantum mechanics, describing [the collapse of the wavefunction through observation] for the consumption of other persons to whom it does not apply."

But Everett offered a solution that was the complete opposite of Wigner's: instead of privileging conscious observers, Everett gave up the notion of wavefunction collapse altogether. Instead, he said, we should assume that the universe, along with all the observers it contains, is described by a single wavefunction that never collapses. In that case, it *is* reasonable to consider Wigner's friend to be in a superposition. Then again, so is Wigner. So are all the rest of us. With every little decision we make—hit snooze or get up, eggs or waffles, Route 76 east or Chestnut Street—we evolve into increasingly complicated superpositions of having done and having not done everything we've ever done and not done. If we could stand outside the universe—and outside ourselves—we would see this tangled mess, this infinite repetition of our every possible variation, but since we can't, we inevitably see only a limited perspective, one in which it *seems* as though quantum wavefunctions collapse, one in which it *seems* as though we live out a single reality.

In 1957, *Reviews of Modern Physics* published a shortened version of Everett's dissertation along with an article by Wheeler analyzing its merits. Everett's interpretation of quantum mechanics, Wheeler wrote, "does not introduce the idea of a super-observer; it rejects that concept from the start." That's a good thing, Wheeler explained, because aside from Everett's interpretation "no self-consistent system of ideas is at hand to explain what one shall mean by quantizing a closed system like the universe of general relativity." There, again, was the central question of quantum gravity: if there's nothing outside the universe, how do you make sense of quantum mechanics from the inside? Now I realized it was just another way of asking, what happens when you have more than one observer?

Despite his endorsement of Everett's ideas, Wheeler leaned far more toward Bohr's point of view that the distinction between observer and observed is fundamental, even if "the line of distinction can run like a maze." Reading his journals, it wasn't hard to see why. In quantum mechanics, Wheeler glimpsed a mechanism for creating reality out of nothing, bit by bit, through the posing of yes/no questions. But if observers are nothing special, just part of the system, ordinary physical objects, then there's no way they can have some special reality-building power. If the observers are stuck on the inside, they're subject to the same Gödelian uncertainty as everything else, helpless to decide the truth-values of propositions. In Everett's world, observers play no active role. With no need to go around collapsing superpositions, they're simply swept along by Schrödinger's wave. Genesis by observership short-circuits. You're left with a single, massive, universal wavefunction and no possibility of explaining where the hell it came from.

Still, Bohr's view, as Wigner made clear, held up only for a universe with one observer. So Wheeler knew he had three options. One: he could become a solipsist, rendering his propensity to go anywhere and talk to anybody just a tad insane. Two: he could accept Everett's parallel branching realities and surrender the hope of explaining existence through observership. Or three: he could maintain that consciousness was magically exempt from the laws of physics and yet mysteriously capable of affecting the physical world through wavefunction collapse. "About no feature of 'It from Bit' do I feel less comfortable than *whose* bit," he wrote.

At times Wheeler did veer toward solipsism, toward the idea that each observer was the sole enabler of wavefunction collapse in his or her own independent universe. It was a "many worlds" view, but not the kind that's been attributed to Everett. In Everett's view, each observer occupies their own branch of a single, shared wavefunction. In the solipsistic picture, there's one universe—one wavefunction—per observer. Rather than all observers sharing a single reality, each has his or her own.

"Feeling I am wrong in looking for 'one meaning' that would summarize, and be built on, the findings of all the 'observer-participators' that I have been concerned about for so long," Wheeler wrote in one of the journals. "I now cite [Walter Lippman] (which make[s] me think it is conceivable we must give up any 'one-world' view of physics): 'Man is no Aristotelian god contemplating all existence at one glance.'"

Is "'this world' a misleading simplification for 'all our worlds'?" he wondered. "What these separate worlds are and how they fit together the central problem? ... Lots of recognition circuits, not a universal one? That's honesty. And that raises the issue of linkage, how to link, fascinating in its own right. 'Big buzzing confusion' indeed!" A few days later he wrote, "Idea, surely not new, that there is not 'one world,' but as many worlds as observers, and that 'meaning' comes in the reconciliation of them, but how much reconciliation?"

"He's starting to question whether the universe itself is invariant," I whispered to my father. My father looked up from the journal he was reading and leaned over to see the passage to which I was pointing: "I'm prepared to question the very term 'universe.'"

But reading on, I saw that solipsism made Wheeler profoundly uncomfortable, and he inevitably swung back toward a more communal view. "I can't make something out of nothing," he wrote, "and you can't, but altogether we can."

I thought back to Wheeler's book At Home in the Universe, where he had argued explicitly that you need multiple observers in order to build reality. No single observer, he had said, was capable of making enough measurements to bring into being all the bits you'd need to build the whole universe. "Mice and men and all on Earth who may ever come to rank as intercommunicating meaning-establishing observer-participants will never mount a bit count sufficient to bear so great a burden," he wrote.

Wheeler was stuck. The only way to have multiple observers living in the same universe without having to give up the observer's ability to create reality was to afford some special role for consciousness, however reluctant he was do it. That opened up a host of bizarre but unavoidable questions, like "What level of consciousness?" "Does a worm qualify?" "What about household appliances?"

"For computer, is there a brain? For hard-wired worm, what difference from any gadget—such as electric dishwasher—with built-in responses to a few standard pokes? For higher brain, learning possible: strengthen favorable responses, weaken unfavorable ones. For still higher, learn from others, especially learn language, invent words like meaning, then puzzle what they mean!" At one point he even pasted in an article by E. O. Wilson on animal communication.

"Where then is the role of consciousness in giving meaning?" Wheeler wrote. "And how does one see the necessity of the quantum principle in the construction of the world? I keep going around and around the same circuit of questions, trying to find a way in to the center of the mystery."

There we were going around and around with him. I had to wonder whether we were getting anywhere, or just circling the drain.

"I can't stop thinking about the problem of the second observer," I told my father over lunch. We had been going to the same restaurant every day and the waitstaff had gotten to know us.

"Are you doing some kind of work nearby?" our waiter asked us, having witnessed several of our lunches spent poring over our notes.

"We're doing research at the American Philosophical Society library around the corner," my father told him.

"Oh, cool!" he said. "What kind of research?"

"Physics," I answered.

"Very cool," he said. "I studied philosophy of physics."

My dad gave me a smile that seemed to say, *Congratulations! Behold your future.*

"Wow," I said, nodding. "Wow."

The waiter smiled. "I'll be right back with your food."

"The second observer?" my father asked.

"Yeah, the idea that one observer can collapse a wavefunction, while a second observer sees him in a superposition correlated with the thing he's measuring. Wavefunction collapse is supposed to be, like, the thing that *creates reality,* and yet it seems to occur at different times for different observers."

"Like reality is observer-dependent?" my father asked.

"I guess. It's just another way of stating the inside/outside problem," I said. "Because according to the first observer, he's outside the system he's measuring. But then the second observer comes along and measures the first observer, which means the first observer is now *inside* the system, and the second observer is outside. And then you could have a third observer—"

"There's always some perspective where the observer becomes the observed."

"Right. And that implies that there's nothing ontologically special about observers; they're just part of the universe, on the inside. But then you can't explain the universe as a self-excited circuit, because how do observers have some special power to create the universe? Like, what counts as an observer? Wheeler thinks consciousness could be the criterion for an observer, but that's obviously bullshit. I mean, consciousness is just a physical process in the brain. It's not magic."

My father nodded. "The consciousness talk made him sound a little like a crackpot."

"Right," I said. "But you can tell he's, like, forced into it. He's backed into a corner. He can't take the Everett view without giving up the ability to explain existence, and he can't take the solipsistic view and say there's just one observer."

"Why not?"

"Because he doesn't think one observer is enough to create all the bits you need to make a universe. He said that the total number of bits in the universe has to be finite and that you have to count the contributions of all observers who had ever and would ever live."

The waiter arrived with plates of steaming food.

"You know, the holographic principle has taken care of that first part," I said. "It says that the amount of information in any region of spacetime has to be finite. But I think it makes the second part more difficult."

"The part about counting up the measurements of multiple observers?"

"Yeah, because Susskind's horizon complementarity shows that you'd end up vastly overcounting. You'd clone information, violate quantum mechanics." You couldn't count Safe's and Screwed's observations together or you'd get the wrong number of bits—not to mention elephants. "When you think about it that way, solipsism seems the most likely."

"So you're talking to yourself right now?" my father said with a laugh.

"Yup!" I said. "And it looks like I don't have to share these noodles!"

Back in the reading room, I thought more about the holographic principle. It was a shame Wheeler hadn't had a chance to appreciate its discovery. Maldacena had given a talk about AdS/CFT at the Princeton symposium, but Wheeler was ninety then and most likely couldn't hear it. Still, it was strange that there wasn't more mention of Hawking radiation and event horizons in the journals, given their radical blow to our conception of reality. Not only had Wheeler been around for those discoveries, he had been at the center of them. After all, it was because of Wheeler's riddling that Bekenstein discovered that horizons have entropy. In fact, it now occurred to me that the drawing featured on our Science and Ultimate Reality press badges—the sphere tiled with 0s and 1s—was a drawing of the entropy on a black hole horizon. Wheeler had obviously thought it was important. It showed that the universe was built of binary bits. Of information. But why hadn't he focused more on the fact that horizons rendered particles and the vacuum observer-dependent? Given how badly he wanted to show observers creating reality, you'd think he would have been psyched.

In a 1974 journal entry, Wheeler had commented on Hawking's original paper "Black Hole Explosions?," which had just been published in *Nature*. "Amazing result," Wheeler had scrawled. "Viewpoints. Conservative."

Two years later, Wheeler wrote, "Hawking, Unruh says *observer* determines number of particles around a black hole," but left it at that. A few months after that he scribbled, "Clear up: are Hawking particles a math[ematical] artifact?"

In a later journal, from 1990, I stumbled on an entry that read, "Think again about this horizon business. It seems to me I rang Bill Unruh already once about this question in the last 3 mos. If I remember his answer correctly, he feels much surer about the acceleration side of life than the horizon aspect. I remember—in perhaps important connection with this issue—Bill Unruh's Santa Fe talk re 'there's no such thing as a particle' . . . Ring him now!"

Still, Wheeler never did go too far into the horizon business—maybe, I thought, because it only served to further fracture the world into a multiplicity of incommensurable viewpoints, while Wheeler was there trying his damnedest to bring it all together. "Not a word or a thought that we consider 'ours' did not come from someone else or at someone's push," he wrote. The universe—the *one* universe—is "an enormous construction we all have a part in."

He never went for the string theory business, either. During a conference in Santa Fe in June 1989, Wheeler wrote, "Murray Gell-Mann . . . said, 'You ought to learn string theory. It has everything you want.' I said, it's nothing but turtles; no place for observer-participancy. It real, 'observer' a delusion? Just the other way around . . . All this talk about an external world—*that's* what's theory!"

That night, lying in bed in the bedroom where I had grown up, I thought more about the inside/outside problem, the problem of the second observer. The first principle of cosmology must be that there is nothing outside the universe, but a universe that contains its own observers on the inside is marred by pathological self-reference, since any observer who describes the universe has to include himself in the description, the kind of self-reference that Gödel had shown would lead to inherent uncertainty, the truth-values of propositions left hanging in the wind.

He wasn't the only one. In 1936, five years after Gödel had proven his incompleteness theorem and nine years after Heisenberg had discovered the uncertainty principle, Alfred Tarski, the Polish logician, had shown that any language capable of making self-referential statements couldn't assess the truth of its own sentences. That same year,

Alan Turing, the father of computer science, had shown that computer programs cannot determine whether they will run for a finite or infinite time—that is, computer programs, and presumably human minds, cannot calculate themselves. Amid a spate of such discoveries, scientists, philosophers, logicians, and mathematicians were confronted again and again with the same stark lesson: self-reference leads to limits that you can transcend only by stepping outside the system.

"Call observer, or data that can be got from the proposition, on one side; and 'observed' on other side," Wheeler wrote in his journal. "Can add more axioms or enlarge system, decide all in original, but then a new and larger region of undecidable, uncoordinatizable, outside the system. Always have this 'uncertainty.' Always the same—can't decide true or false values from the inside."

In this "uncertainty" Wheeler had glimpsed the essence of quantum mechanics, and it convinced him that Bohr's requirement that the observer be outside the system had to be obeyed at all costs. The problem was that Bohr's picture falls apart as soon as a second observer comes along and observes the first. When Wigner approaches, Bohr's external observer is suddenly internal to the very system he thought he was viewing from the outside. The observer becomes the observed. How can someone be simultaneously standing outside the system causing wavefunctions to collapse and internal to the system suspended in superposition? Both stories couldn't be true, and yet here we are, trapped inside the universe, where you can always find a second observer who sees us as part of the system inside. We are, as Wheeler's U-drawing showed, the universe looking at itself. It seems impossible, then, to apply quantum mechanics to the whole universe without spiraling into strange loops of self-reference and recursion.

I thought back to my teenage nights in this room, reading the existentialists in search of some justification for my angst. I remembered reading Sartre's *Being and Nothingness,* a book I had scored in the used-book store where I worked after school. Sartre had described the feeling you get the moment you realize that someone is watching you. What was it? Shame? Vertigo? Nausea? That sudden shift in reality as you are transformed from subject to object. Sartre would have been *really* nauseated if he had thought about quantum mechanics. Not only do you shift from subject to object, you also go from creating reality out of papier-mâché to hovering in some kind of feline purgatory.

It came down to the same inside/outside question I had confronted back at the Tribeca Grand, the problem of quantum gravity: where's the observer? Restated: quantum mechanics *forbids* us from being both subject and object, and being on the inside of a one-sided universe *requires* us to be both subject and object. And, really, aren't we? My father was a character in my story, and I was a character in his. Subject in one reference frame is object in another.

Then again, Susskind had taught me that physics only makes sense when you talk about one reference frame at a time—at least when event horizons are involved. Safe or Screwed. Accelerated or inertial. Did something similar apply here? Maybe quantum mechanics doesn't allow me to talk about my father's and my stories at the same time. Maybe neither of us could be subject and object simultaneously. Was that the solution to the problem of the second observer—to only talk about one observer at a time? But how then do you piece together our splintered stories into some single arena we call the universe? Moving from the single observer's reference frame to the global, God's-eye view seemed to lead inevitably to the overcounting of elephants, and to wavefunctions that both had and had not collapsed. "Properly to draw together the local and the large scale was one of the greatest of Einstein's tasks," Wheeler wrote. "Here how do we draw together the 'local' yes-no and the global 'all there is'?"

And if there is no "global all there is"? If there is no universe? "The number one issue: Is everything nothingness?" Wheeler wrote. He had said he was prepared to question the very term *universe,* and I was feeling like I ought to prepare myself, too.

"Something of an information-theoretic character is at the bottom of physics, spacetime, existence itself," Wheeler wrote on April 19, 1986. "This is a quick sentence, if anyone asks me for a last word before I leave this Earth." He was en route to the hospital for heart surgery.

I couldn't help wondering what my quick sentence would be, should I need one. *What's real is what's invariant.* No, that was too simple. *Nothingness is a state of infinite, unbounded homogeneity.* I couldn't use that; it was my father's line. *The first principle of cosmology must be . . .* Jesus, I couldn't come up with one original thing to say for my last sentence on Earth?

Wheeler survived the surgery and continued on with his "lovely, lonely search," but he was more poignantly aware than ever of those increasingly shorter days, torn between his need to solve the mystery of existence before it was too late and his fear that he was ruining his reputation in the process. "Darwin wouldn't, Bohr wouldn't, Newton wouldn't *expose* his half-baked ideas so widely as I do," he wrote on March 6, 1987. "Yet I do it *anyway*. I see no way except *dialog* to get on with the job of clarification. And I don't have forever, nor do I see who will take up the work when I'm gone."

Reading John Wheeler's journals at
the American Philosophical Society library
A. Gefter

August 16, 1988: "I need to be tortured into 'confessing' a really great and simple idea. Ha! Not *an* idea. *The* idea. And stop talking about how someone somehow sometime is *going* to find this idea. Keep quiet, talk quietly with those who can help, get busy and *find* this idea."

"Life expectancy of the order of 3 to 5 years," he noted later that month. "I consider my biggest commitment of all to be this greatest of all puzzles that we see around us every day, all that is, existence: how come? . . . I am not I unless I continue to hammer at that nut. Stop and I become a shrunken old man. Continue and I have a gleam in my eye."

Days of reading and fourteen years later, Wheeler was still at it, still gleaming. In an entry dated March 8, 2002, he wrote, "Still agonizing over what shall I say at next week's Templeton Conference."

"Oh my God, look!" I nudged my father, showing him the page. "It's the conference we crashed!"

"I'm still trying to develop the theme of it from bit," Wheeler continued. "Nothing! How inject *us* into that nothingness? A question of philosophy? But maybe philosophy is too important to be left to the philosophers."

I eagerly turned to the following entry, excited to read his postconference impressions and hoping, just a little, for some mention of an adorable young girl and her father and the penetrating questions they asked. He didn't. Strangely, he never wrote about the conference at all.

"I think I finally get it!" my father said excitedly when I walked into the kitchen. "The boundary of a boundary!"

My mother had gone to bed, so the two of us poured some coffee and set up shop at the kitchen table.

"Something to do with conservation of energy and momentum?" I asked. I had caught glimpses of it in the journals but was still struggling to understand.

"Exactly. Wheeler showed that the equations of general relativity follow from the requirement that the boundary of a boundary is zero. The local curvature of spacetime cancels out the energy and momentum of the mass that's present there. That's why mass curves spacetime."

"So it all adds up to zero?"

"In any local region. But the calculation only works out if the boundary of the region doesn't have any exposed edges. As long as the boundary of the boundary is closed, everything cancels out."

He broke out a yellow legal pad and sketched the geometry, showing me how a four-dimensional region would be bounded by three-dimensional cubes whose own boundaries were two-dimensional faces, each sharing its edges with the adjacent faces to create a closed manifold. Adding up the currents of momentum-energy passing through the faces required counting each edge twice, once for each face of which it was a part, counting once as positive and once as negative and summing to a perfect zilch. Curvature manifested as distortions in the geometry of the faces, their opposite edges veering toward or away from one another, no longer parallel, tugged on by adjacent faces, the entire boundary warped and distorted by the presence of the mass. It was all rather technical, but the point was simple enough: most of the complicated structure of four-dimensional spacetime seemed ultimately to be rooted in nothing.

"In the journal he says that not only general relativity but all the quantum field theories are based on the boundary-of-a-boundary principle," I said. "All the gauge theories."

"Right. It's a universal thing. It's the basis for the way fields respond to masses or charges."

That the boundary of a boundary is zero meant that everything you needed to know about what was going on in the interior of some region of spacetime could be read directly off its boundary without ever looking inside. "Outside reveals inside," Wheeler had written. The conservation of momentum and energy could be read from the faces, the amount of mass inside deduced from their warped edges.

"That's what he meant when he compared it to elasticity," my father said. "To calculate the elastic forces on some deformed body you only need to calculate what's going on at the surface. Everything inside cancels out."

"He wrote something about that, hang on," I said, sifting through my notes. "Here. In 1973 he wrote, 'Elasticity is the lowest form of physics one can have without having no physics at all. Deals only with "the surface of things." Want to argue that electromagnetism and gravitation are of the same breed of cats, except that they may not require any dimensionality at all.'"

No dimensionality? The surface of things? Outside reveals inside? Was the boundary-of-a-boundary principle an early sign of the holographic principle at work?

My father knew exactly what I was thinking. "Sounds a bit holographic, huh?"

In one entry Wheeler had written, "One looks forward to the 'dimensionality' of space (or better the 'average effective dimensionality') as being determined out of basic theory in terms of a simpler basic substrate, pregeometry, which itself has no such property as dimensionality." In another, "Ask if every law can be put in the $\partial\partial \equiv 0$ form. How come many dim[ensions] can be reduced to so few?"

The question reminded me of the unease I had felt after learning that all the information contained in some volume of spacetime was proportional to the area of its boundary—as if a whole dimension was redundant. Was that what Wheeler had been thinking when he drew the surface of a sphere tiled with 0s and 1s? That the ability to read out the black hole's information content from its boundary meant that the world, at bottom, was made of information

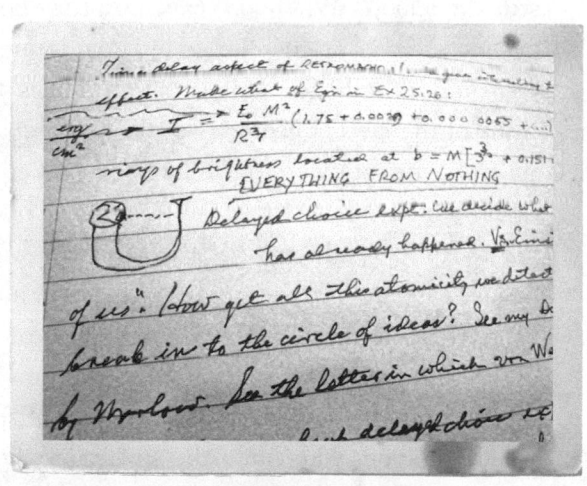

Wheeler never stopped drawing his self-excited universe (American Philosophical Society library)   A. Gefter

and that information, at bottom, was made of nothing?

January 8, 1988: "What comes closer to a circuit and to $\partial\partial \equiv 0$ than what we can build—with imagination enough—out of questions and answers?"

October 20, 1985: "More and more certain everything must be built from 'nothing.' What reasonable alternative is there!"

February 22, 2005: "Nothing! Nothing! You start with nothing to get everything."

I spent the last day in the library alone with Wheeler's journals. We had made it through nearly all of them, and my father couldn't take another day off work, so I had come to finish reading the final entries without him.

Wheeler was over ninety years old now, and the journal entries became fewer and further between, and increasingly illegible. He continued drawing the U-diagram, the swoop of the curve growing shakier each time. He frequently quoted a poem by Piet Hein, a friend of Bohr's: "I'd like to know / what this whole show / is all about / before it's out."

Eventually he began dictating his entries to his secretary, Jackie Fuschini. They grew strangely repetitive, as if Wheeler's thoughts were stuck in a loop. "Where do space and time come from?" he would ask, again and again, often followed by the refrain "It's all an illusion." There were an oddly large number of references to Charles Darwin. "We think of the laws of geometry as having been established at the beginning of time, but were they?" Wheeler asked. "This question reminds us of another How Come, the species of plants and animals that we see today 'here always' that the question was not so easy to answer became apparent to the young investigator aboard the HMS Beagle."

He also returned to that central question: one observer or many?

November 8, 2005: "It's hard to ask 'how come anything' when we take it for granted that we live in a world of space and time. . . . And if we are the ones who 'build' the spacetime, how come we don't get [as many] spacetime[s] as people? How come just one? Pursue further that one."

Wheeler was living with Janette in the Meadow Lakes nursing home, in Hightstown, New Jersey, just outside Princeton. Fuschini would visit him there and take down his thoughts in the journal.

October 11, 2006: "72 degrees, warm and sunny. Sitting on sun porch at ML, looking out picture window at beautiful flowers and shrubbery on patio. It is Indian summer. . . . John has nothing to report on space and time or Charles Darwin today."

The next entry was dated December 7, 2006, and consisted of one surreal sentence: "Hope produces space and time?"

A year passed before another entry was written. In that blank space, Janette had passed away. She was ninety-nine years old when she died. They had been married for seventy-two years.

January 2008: "Johnny has not entered anything in the journal in a long time. The next time I visit we will begin."

I turned the page.

It was blank.

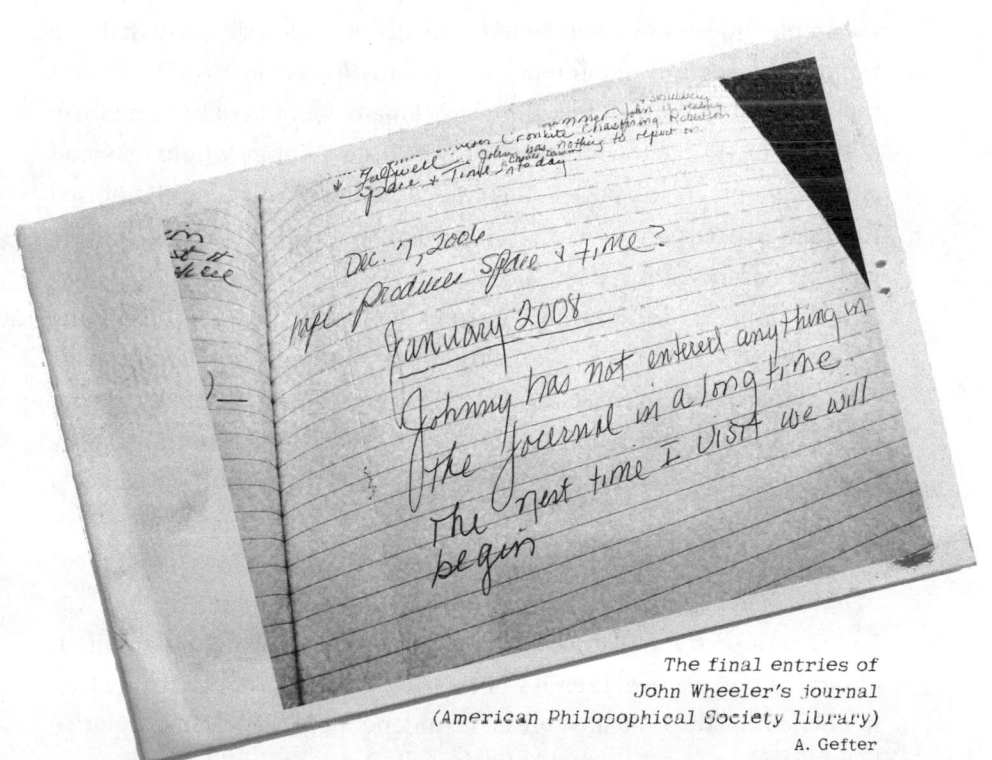

The final entries of
John Wheeler's journal
(American Philosophical Society library)
A. Gefter

12

That Hypothetical, Secret Object

I t was kind of crazy, how far this thing had come. One deceptively simple question—*How would you define nothing?*—had led me from a low-security conference center in Princeton to a quantum-rodent-infested London flat, from Brockman's Connecticut farmhouse to the New Mexico town that never was, from a library where I sobbed at the sight of a blank page to a Hertz car rental kiosk at San Francisco International Airport, where I now stood listening to refueling policies with my parents at my side.

My father and I were here to track down the answers to so many questions that remained. Like, why would nothing, the H-state, have ever changed, if it ever changed at all? What role do observers play in the universe's existence? If, as Guth said, "something *is* nothing," why does the nothing *look* like something? What is the meaning of objects behind the cosmic horizon? Will the sales guy ever stop talking so we can hit the road in search of ultimate reality?

We had learned that the criterion for ultimate reality was invariance—if some feature of the universe is merely an artifact of an observer's perspective, then it's not a true ingredient of the reality that lies beneath. "Real" means "observer-independent." And amazingly, we had found hardly anything that fit the bill. Relativity had booted space, time, and gravity off the list, and with them concepts such as mass, energy, momentum, and charge. Gauge theory had crossed off electromagnetism and the nuclear forces, leaving no fundamental forces behind. Hawking radiation spelled the end of particles, fields, and vacua, while AdS/CFT and the holographic principle had wiped out dimensionality, strings, and spacetime. The only items remaining on our now threadbare IHOP napkin were the universe, the multiverse, and the speed of light, and, thanks to horizon complementarity, the first two had fallen under deep suspicion. I was growing increasingly convinced that my father's speculation might be right: that nothing is real. Which, of course, was to say that Nothing is real.

Wheeler's journals had taught me one thing for certain: whatever the answer to ultimate reality may be, it was going to have to navigate that Möbius dichotomy of inside and out, subject and object, referencing a single observer, no more than one at a time, and avoiding at all costs any confusing coauthorship structure.

Coauthorship seemed to be the source of every problem: it was the culprit behind the black hole information-loss paradox, Brockman and Matson hated it, it made a liar out of Wigner's friend, it drove Wheeler to study worm consciousness, and it undermined the very meaning of the word *universe*. After all, what is the universe if not the sum total of every possible point of view? Yet as I had learned from Susskind, summing points of view violates the quantum no-cloning theorem and leads us to overcount information by mistaking different views of the same bit for different bits. *About no feature of "It from Bit" do I feel less comfortable than whose bit.*

No matter how many times he circled the issue, Wheeler never figured out how to fit a second observer into the self-excited circuit. Did that mean there is no second observer? "The theory of [quantum] measurement is logically consistent so long as I maintain my privileged position as ultimate observer," Wigner had said. Everett had agreed— reality makes sense so long as we're willing "to postulate the existence of only one observer in the universe."

The whole situation was bumming me out. How exactly do you toss aside the existence of a second observer when the second observer is your *father*?

Physics may have been pushing for sole authorship, but to me it still sounded a lot like betrayal. Besides, we had already coauthored something: our life, my career, a secret mission, the universe. Was it really all an illusion? I tried my hardest to shove Matson's words back down, but they kept resurfacing like Whac-A-Moles: *Just think about it*.

I sighed impatiently, then decamped for an empty seat on a nearby bench, abandoning my parents at the rental counter. Something was gnawing at my stomach. Bad pizza? Guilt?

Cosmology's standard model was rooted in coauthorship. The whole point of inflation, of having the universe temporarily expand faster than the speed of light, was to extend the big bang theory's reach beyond a single light cone. It's a theory about the space that lies on the other side of our cosmic horizon. Meanwhile, horizon complementarity forbids us from considering what exists on one side of a horizon and what exists on the other side as constituting a single spacetime geometry, replete, as it would be, with redundant information. And if that weren't bad enough, you can't stop the thing from going eternal, from producing a global multiverse, a single object that encompasses infinitely many spacetime regions separated by horizons. Cosmologists were asking for *infinite* coauthors. I had been shot down for two. And yet everywhere I looked, people were still talking about cosmology's golden age, as if nothing had ever happened at all.

Cosmology was going to have to change in a big way, but how? Could physicists rein in eternal inflation and trap it inside a single light cone, or would they have to go back to the drawing board and come up with a whole new theory? My father and I had come here, to California, to find out, and my mother had happily tagged along. "I know you'll be busy with reality," she said, "but we'll just have to make time to go to the beach."

I looked over at the counter and saw my father signing the last of the contracts and scooping up the keys to the Toyota that might lead us to some answers. My mother caught my eye and smiled. She smiled as if all was well, as if we all existed in the same, happy universe. "I can't make something out of nothing, and you can't," Wheeler had written, "but altogether we can." It was kumbaya cosmology. It was a child's dream. Maybe the problem of the second observer was nothing more than the problem of growing up.

The next morning, my father and I took the BART train from downtown San Francisco to the Berkeley campus to meet Raphael Bousso.

Outside the physics building, I stared in disbelief at a parking sign: *Reserved for Nobel laureate at all times*.

"Finally, a reason to want a Nobel Prize!" my dad said.

I nodded. It wasn't David Gross's dolphin-front office, but it was a damn good parking space.

Inside, Bousso welcomed us into his office. In his early forties, he was young and handsome, with a calm demeanor and a sweet smile. He had grown up in Germany and studied under Hawking at Cambridge before ending up at Berkeley, his accent a melodic mix of the three nations.

My father and I wanted to know how the multiverse could survive the holographic principle, and Bousso was the guy to ask. Along with Joe Polchinski, it was Bousso who had discovered the string landscape—that collection of some 10^{500} universes, each with its own physical laws and its own cosmological constant—opening the door to anthropic physics in the multiverse. But he had also discovered the most general form of the holographic principle. Known as the covariant entropy bound, it showed that given *any*

Nobel parking at the physics building, UC Berkeley
A. Gefter

spacetime region—even one that is expanding or collapsing—the amount of information that can fit in that region can't exceed one-quarter the area of its boundary, which meant that the implications of the holographic principle were inescapable in cosmology. There was no avoiding it: spacetime is observer-dependent, and physics only makes sense according to the viewpoint of a single observer. How could Bousso possibly reconcile a global multiverse cutting across 10^{500} or even infinite causal horizons with the local viewpoint demanded by holography?

"The global approach has to be wrong at some level," he told us once we had settled into our seats.

That much was evident even without horizon complementarity—just look at the measure problem. In a global multiverse, probabilities drown in infinities and physicists lose the ability to make scientific predictions—which was, after all, the point of science. Bousso suspected that those infinities were related to the duplicate information you get when you try to view the world from inside and outside a horizon.

"The lesson to take away from the black hole information-loss paradox is that you have to restrict to the viewpoint of only one observer," he said. "It seems very natural that this is a restriction we should impose on the multiverse."

So he did. And while he was at it, he looked to see if such a restriction would solve the measure problem.

"I thought, why not do something that we've been forced to do by the black hole information-loss paradox anyway and kill two birds with one stone?" Bousso said.

The usual method for coming up with a measure went something like this. Arbitrarily choose to call one dimension of the multiverse "time" and the others "space," despite the fact that time and space are observer-dependent. Look at a finite region of the multiverse at a given time and count, say, the number of universes that have a certain value of dark energy relative to the total number of universes in the sample. Take the limit of that ratio as the sample size goes to infinity. Then compare your answer with the one universe we actually see. If your measure says that the probability of a universe with a value of dark energy as small as ours is basically zero, trash it and try again.

Bousso, however, took a completely different approach: only count what's inside a single observer's light cone, otherwise known as a causal patch. That way, you don't have to come up with suspect ways to whittle down infinity because the whole scenario is finite to start. What's more, you don't have to worry about overcounting the redundancies that crop up when you cut across horizons.

But I was confused. "If you're looking at a single causal patch, why do you have to deal with probabilities at all?" I asked. If you're restricting yourself to a region of spacetime in which everything you'd ever want to observe is within your causal reach, what do you need a measure for? You don't have to calculate the odds of seeing something—just open your eyes and look around.

Unfortunately, Bousso said, it's not so simple. In a universe ruled by eternal inflation, the physics within a causal patch changes over time, since every vacuum with a positive value of dark energy is unstable and subject to decay. According to the standard story, our universe started off in a false vacuum—a temporarily stable state that is not the lowest possible energy—and then plummeted downward, inflating faster than light in the blink of an eye, until it hit the true vacuum below. That's where we've been living for the past 13.7 billion years. But as we now know, our vacuum wasn't the true vacuum, either, even if it was slightly less false. Our vacuum is permeated with a small but significant amount of dark energy, whereas the true vacuum would have no dark energy at all. Our universe is perched atop a stable plateau, but the right quantum push could send it careening over the edge and soaring down toward rock bottom, stopping at various lookout points along the way. Each stop is a big bang, each plateau a universe. This finite but wildly varied plunge through cosmic history is a kind of multiverse in its own right, despite the fact that it never refers to any reality beyond the light cone's edge.

Just like the decay of a radioactive atom, the vacuum decay of a given universe is random, ruled by nothing more certain than probability. As an observer, you don't know which universe you're in, so if you want to make testable predictions, you need a probability measure to calculate what you're most likely to see.

"So you start with some vacuum, you calculate all its probabilities for decay, and you add them all up," Bousso said. "Instead of a global multiverse you have an ensemble of vacua within a single causal patch." If you want to predict something like the value of dark energy, you don't have to look to other bubbles in the global multiverse; you just have to consider the possible histories within your own cosmic horizon. The result was what Bousso named the causal patch measure.

It seemed like a good plan: take Occam's razor to everything outside your horizon, make spacetime finite, and generate a probability measure based on a history of vacuum decay, harnessing the anthropic power of the multiverse without having to deal with the Popperazzi or the metaphysical burden of eternally inaccessible universes.

Then it got even better.

"I thought I was going off in a radically new direction by rejecting the global viewpoint," he said. "Then I learned that the local causal patch measure reproduces the exact same probabilities that you get with the global light cone measure. That came as an enormous shock to me."

The global light cone measure was more "global" than "light cone." It was one of many global measures on the market, the kind that counts across horizons, the kind that, as Bousso said, has to be wrong at some level. This particular global measure, however, had something going for it that the others didn't: it happened to give the exact same probabilities as the causal patch measure.

"I had no idea that two measures could look so different and turn out to be the same," Bousso said. "That was amazing to me."

The two views—the God's-eye view of the whole multiverse and the internal view of a single observer—were completely equivalent. Dual. So even though the global view is fundamentally flawed, the local causal patch measure offers us a way to still talk about a global view as if it actually means something. To talk about a multiverse without ever referencing anything beyond our own horizons. I jotted in my notebook: *Best of both worlds.*

"We had been thinking there was a kind of shift under way in cosmology," I said, "from a God's-eye view of the universe to the point of view of a single observer."

"Not everyone sees it this way," Bousso said. "But I would argue that it's inevitable. Lenny [Susskind] saw that in the context of black holes you get the wrong results when you think of the bird's-eye view of inside and outside the horizon—it looks like you're Xeroxing information and violating quantum mechanics. My view was that it was too radical a statement to restrict just to black holes. It must be telling us something deep about how things work in general."

"But you showed this duality between the global and local pictures," I said. "Doesn't that make the global view valid again?"

"The local view is still more fundamental," Bousso said. "It has to be. The idea is to somehow build a global multiverse out of these local patches."

I nodded. Maybe there was hope after all. Maybe you could take multiple sole authors and patch them together, though I had no idea what that book would look like. Would it alternate authors from one chapter to the next?

"Would you ascribe reality to things outside our cosmic horizon?" I asked, echoing Susskind's deepest question in cosmology.

Sharing a laugh with Raphael Bousso, UC Berkeley
W. Gefter

"I'd say there are things happening there that might be a tiny bit different from what's happening here, but I would not talk about them simultaneously. So it depends on how 'real' you want things to be," Bousso said with a laugh. "It is certainly my belief that the most fundamental way to describe cosmology is to restrict to one causal patch."

"We're trying to figure out what the ingredients of ultimate reality might be," I said. "And clearly we need quantum gravity to find them. What could they be? Are they strings?"

I was pretty sure I knew the answer to that, since AdS/CFT had knocked strings off the IHOP list, revealing them to be holographic projections of ordinary particles, neither description more "real" than the other. Still, Bousso was a string theorist, so I figured I'd ask.

"It's one of the big questions," Bousso said. "String theory wasn't discovered in the way that theories are usually discovered in physics, where you have some idea of the principles and the basic ingredients and you build it from there. We just sort of stumbled upon string theory, and through mathematical consistency we keep discovering new ingredients. As a result, we don't know which ingredients are fundamental. In one setting you may want to think of strings as fundamental, and in another you might think of D-branes as fundamental. It's not even clear that there's a unique answer to the question."

I quickly scribbled in my notebook: *D-branes?*

If strings and particles were merely two ways of looking at the same thing, I thought, what's the thing? As an ontic structural realist, I figured the thing had to be the holographic principle itself. But what was that telling us? If a holographic principle, *why* a holographic principle?

"Is the holographic principle the key clue to quantum gravity?" I asked.

"It shows a relationship between geometry and information that's completely general but of which we don't understand the origin," Bousso said. "We can say what the relationship is and we can look at nature and check over and over that it really holds true, but it's a bit like a conspiracy. It's got to have a deeper reason. And we think that reason has something to do with quantum gravity, or more precisely with a unified theory of matter and quantum gravity. On one side of the relation you have geometry, the area of surfaces in spacetime, and of course spacetime is gravity. On the other side you have the amount of information that can fit in adjacent spacetime regions, and information is really just the number of quantum states. So you have gravity on one side related to quantum theory on the other. The relationship is so universal there has to be a simple reason for it. It's like, why do things fall at the same rate? The hope is that the holographic principle can play a role analogous to the role the equivalence principle played in guiding Einstein toward general relativity."

"The fact that you get the wrong answers when you use a global, God's-eye view," I said, "what is that telling us about reality?"

Bousso thought for a moment. "There's a sense in which it's telling us that what we call reality, no matter what it ends up being, is probably an approximate concept. If a typical observer has access to a finite number of quantum states, because your past light cone has a maximum area that dictates how much information will fit in it, that means there's a limit to how sharply you can measure anything, how precisely you can describe the world. There's clearly no infinitely sharp sense in which the world is some particular way."

On the train ride back to San Francisco, I thought about what Bousso had told us. The local view—the perspective of a single observer, a light cone—has to be fundamental. If you want a multiverse, you can try to patch it together out of local views. But at the end of the day, it's an illusion. We could officially cross the multiverse off the ultimate reality list.

I tried to imagine what a multiverse cobbled together from local views would look like, but my brain couldn't piece it together. I knew it couldn't be the sea of bubble universes that cosmologists always show to illustrate the thing. Such a picture assumes that the bubbles have some definite, observer-independent existence. But it's not like the edges delineating individual bubbles are physical walls floating in some invariant space—they are internal markers of the boundary of a given observer's point of view. Viewing the multiverse from the outside would be like seeing every possible perspective simultaneously.

Whatever that would mean.

I couldn't help thinking of Borges's Aleph. In the story "The Aleph," the protagonist discovers that a man named Carlos Argentino Daneri has found, under his basement stairs, an Aleph: a point in spacetime from which every point can be seen simultaneously, from all angles. A God's-eye view.

"The Aleph was probably two or three centimeters in diameter, but universal space was contained inside it, with no diminution in size," Borges wrote.

> Each thing (the glass surface of a mirror, let us say) was infinite things, because I could clearly see it from every point in the cosmos. I saw the populous sea, saw dawn and dusk, saw the multitudes of the Americas, saw a silvery spiderweb at the center of a black pyramid, saw a broken labyrinth (it was London), saw endless eyes, all very close, studying themselves in me as though in a mirror, saw all the mirrors on the planet . . . saw the circulation of my dark blood, saw the coils and springs of love and the alterations of death, saw the Aleph from everywhere at once, saw the earth in the Aleph, and the Aleph once more in the earth and the earth in the Aleph, saw my face and my viscera, saw your face, and I felt dizzy, and I wept, because my eyes had seen that secret, hypothetical object whose name has been usurped by men but which no man has ever truly looked upon: the inconceivable universe.

Susskind had suggested we meet him at a little café in Palo Alto. My father and I arrived early and snagged a corner table on the outdoor patio. Soon, Susskind came strolling down the street looking casual and cool.

He greeted me warmly and in turn I introduced him to my father. As I watched them shake hands I caught a glimpse of something I had rarely seen in my life: my father was nervous.

Susskind and I headed inside to order some beverages while my father held the table. When we emerged with coffees and teas in hand, I looked over at my father and knew exactly what he was thinking: Holy shit. Lenny Susskind is carrying my coffee.

As we set the drinks down, the table wobbled. "Every time I'm with another physicist and we wind up with a wobbly table, we always

end up trying to figure out the laws of table mechanics," Susskind said. "Nobody's had a good idea!"

"Maybe you need eleven dimensions?" I offered, cringing at my own stupid joke.

"Well, in a one-dimensional world, this wouldn't be a problem, would it?" Susskind said. "It gets worse as the number of dimensions goes up." He laughed. "That's the kind of brilliant insight I'm capable of coming up with."

"It seems like there's a huge change under way in cosmology," I said when the conversation grew serious. "Maybe even a paradigm shift, from a God's-eye view to the perspective of a single observer. Do you think that's happening?"

"Yeah, I do," Susskind said. "I do think that idea is starting to take off. But at the same time there are times when it's useful to think about the global perspective. We do attribute a degree of reality to what's out there beyond the horizon whenever we invoke things like anthropic reasoning. We do. On the other hand, we should be able to formulate complete theories of observation and experiment without invoking anything beyond the horizon. So there's a tension there. There's a real tension there. And I think that tension is going to come to a head. And hopefully when it comes to a head we'll find a clearer picture of this relationship between the local and the global—it's something I've been thinking about for years."

The tension, I knew, was mounting in his own work. On one hand, Susskind had argued in Santa Barbara that he believed strongly in the explanatory power of the anthropic principle, power made possible by the intriguing convergence of the string theory landscape and the infinite bubble universes produced by eternal inflation. On the other hand, his own discovery of horizon complementarity gave cosmology no choice but to toss aside the unphysical God's-eye view and describe the universe in terms of what a single observer can see.

"We've been thinking a lot about the nature of reality," I said, "and defining what's real as what's invariant. But when we look at the ingredients we think constitute reality, we keep discovering that none of them is invariant. So what's ultimately real?"

Susskind shook his head. "My only guess is that there will be a big surprise and everything will be turned on its head."

"Do you suspect that string theory has the answer?" my father asked.

"No, I don't," Susskind said. "String theory is an incredible edifice that has a startling degree of internal consistency, and it does contain quantum mechanics and gravity of some kind, but it doesn't describe the universe. It doesn't describe any known cosmology except empty space."

By "empty" Susskind meant a space with no dark energy lurking in its depths—a space, that is, with no event horizons. Strings are described by the S-matrix, which computes the probability of what will come out of a string interaction given what went in, ignoring all the convoluted crap in the middle. In such a space, a string actually means something. But as Hawking said, "We live in the middle of this particular experiment." We don't know and can't measure what went in or what comes out; all we'll ever know is the middle. Here in the middle, the S-matrix is useless—and so are strings.

It suddenly occurred to me why: it's because the S-matrix loses its invariance. Strings interacting in the middle of the universe can be viewed from countless perspectives, from different directions in space and time, from reference frames in various states of motion. What looks to one observer like a string vibrating at one particular frequency and producing its associated particle will look to another like a string vibrating at a different frequency and producing a different particle. Observers can't agree on what's what, and, as Einstein emphasized, no observer's view should be more valid than another's.

It reminded me of particle handedness. For all observers to agree on the handedness of a particle, the particle has to be moving at light speed so that no observer can outrun it and see it spinning the other way. Similarly, by defining strings at future infinity, the S-matrix ensures that every observer will see them the same way. In a universe free from dark energy, every observer's light cone, given infinite time, will grow large enough to encompass the entire universe and overlap completely with everyone else's, so all observers will share a single reference frame. Strings will be observer-independent. They'll be *real*.

But in a universe with dark energy—a universe like ours—no one can access future infinity, trapped, as we are, behind observer-dependent event horizons. In our de Sitter universe, we are all Screwed, adrift in an ever-accelerating, ever-emptying space, each of us surrounded by our own steadfast horizon, forever doomed to provinciality and finitude. Our light cones will never overlap completely, no matter how long we wait. No two observers will ever see the same universe. In de Sitter space, all hope of invariance is lost. The S-matrix doesn't mean anything, and neither does string theory.

"Is there any hope of doing cosmology in de Sitter space?" I asked.

"There are three ideas that I know of," Susskind said.

The first, he explained, was to look to the boundary of our universe at future infinity and find the field theory that encodes our higher-dimensional de Sitter space as a hologram. If it worked, we'd have a dS/CFT.

Susskind, however, wasn't convinced that it was such a good idea. "That may be misguided, because it's global," he said. It encompasses every observer's light cone—Safe's, Screwed's, and infinite others.

Having a hologram at future infinity wouldn't do anyone much good, because whose world would it be describing? No observer can access the whole space; only a God's-eye view could make sense of the hologram, which would have to simultaneously describe the insides and outsides of infinite regions separated by event horizons. According to horizon complementarity, the hologram wouldn't make any sense—it would overcount elephants and violate the very laws of quantum mechanics that made the world holographic in the first place. "We learned from horizon complementarity that quantum mechanics will only apply within a single causal patch," Susskind said.

So much for dS/CFT.

"The second idea," Susskind said, "is to formulate physics in a single causal patch in de Sitter space."

That made more sense. Just place the hologram on the event horizon of a single observer's patch. If each of us is stuck with a permanent horizon, why not make use of it and find the dual physics that lives there? Of course, that horizon is observer-dependent. Which would make the cosmic hologram observer-dependent. Which would make the *universe* observer-dependent.

"That probably can't be right," Susskind said, "because the de Sitter space can decay."

As Bousso had explained, in a universe ruled by eternal inflation, every vacuum with a positive cosmological constant is unstable and bound to decay. Our vacuum, with its trace amount of dark energy, is a false vacuum, one that's temporarily stable—if "temporarily" means billions of years—but will eventually drop to a lower energy state. It's happened before. It was called the big bang.

"According to our understanding of eternal inflation," Susskind said, "the endpoint of evolution along any trajectory is an open, FRW universe." When our space decays, it will drop to a state with a smaller cosmological constant; in turn, that vacuum will most likely decay to an even lower energy state, and so on and so forth, rolling down the hills of string theory's landscape until it bottoms out in the lowest possible valley, the one with a cosmological constant of zero—an ordinary, expanding, flat space, also known as Friedmann-Robertson-Walker, or FRW, space.

"It's not clear that that helps," Susskind said, "but it might."

Well, sure, I thought. An ordinary, expanding, flat space is exactly the kind of space you need to have an invariant S-matrix, to make sense of string theory. Reality holds up there, because all observers can see the same thing. "It helps because observers will no longer be stuck in finite patches?" I ventured.

He nodded. "An open FRW universe has an infinite number of particles in it, and a late-time observer can look back and as time goes on see more and more of it. Eventually he can see an arbitrarily large amount of it. There's always more out there, so you'll never see the whole thing, but take any portion of it and you will eventually see it. That situation doesn't have the same problems that we face in de Sitter space."

That brought Susskind to the third idea for dealing with the problem of cosmology. "The idea is to take the point of view of that late-time observer," he said—the one who lives in FRW space. "I call him the Census Taker." The plan, he explained, was to construct a holographic duality between the Census Taker's patch of flat space and a quantum conformal field theory living on the patch's two-dimensional boundary: FRW/CFT.

I understood the appeal. Because it has room for invariance, FRW/CFT would allow cosmologists to save not only string theory but reality as well. At the same time, by restricting the theory to a single observer's causal patch, it would satisfy the demands of horizon complementarity and avoid any breakdowns of physics. They could even keep eternal inflation and the string landscape in play, allowing for an anthropic explanation of the value of dark energy or anything else in need of an explanation.

Then again, satisfying the demands of horizon complementarity in flat space was pretty trivial. In flat space, causal patches don't mean very much, because at the end of the day—or at the end of infinite time—every observer shares the same patch. At that point, you might as well just call it the universe. And sure, FRW/CFT respects the holographic principle, but how hard is that? The area of the observer's boundary becomes infinite. Respecting the holographic principle in flat space requires about as much constraint as a driver needs to obey an infinite speed limit.

Besides, I wasn't sure how the Census Taker's cosmology was supposed to help us here and now. Don't we want to account for *our* universe? So what if our de Sitter space decays to a flat space billions of years from now? Is that really such a comfort? Is it even fair to consider the FRW universe a continuation of our own? If our universe decays, it will be destroyed in the process, the same way that whatever universe lived uphill from us was destroyed in our big bang. *This universe isn't real now, but someday it will be*. I just wasn't sure if that was enough.

"Maybe one of these ideas will be right, maybe all of them, maybe none of them," Susskind said. "But it seems to me important to delineate what the right questions are as clearly as possible. Understanding the relationship between the local and the global views of physics, I think that's *the* really big question. There's going to be a lot of crap published about it. That's obvious. But it's a real question. It's all connected with questions of horizons, the holographic principle . . . all these things are interconnected and they haven't come together into a single, comprehensive view. I hope that will happen. This question about cosmology, about how to think about it, I'm obsessed by it. But I'm seventy-one years old. The prospects of me solving it are very small. So I am sort of involved in proselytizing, not for any viewpoint but for the questions. I'm telling people, I think these are important questions, maybe the most important questions."

His urgency reminded me of Wheeler. On first glance, they couldn't be more different. Where Wheeler was sweet and soft-spoken, Susskind was brash and tough; where Wheeler was open to wild speculation, Susskind was cautious and skeptical. But both physicists were prone to bold ideas, to great leaps of intuition, and both were ahead of their time.

"Can I ask you for one last thing?" my dad said, reaching into his briefcase.

Oh, God, I thought. What is he doing?

He pulled out Susskind's book *The Holographic Universe*. "Could you sign this?"

I blushed, embarrassed. Here I was trying to act cool and professional, and he was acting like a star-struck fan. But as Susskind scribbled in his book, I smiled. All those years ago in the Chinese restaurant, my father had chosen to confide his idea about nothing, about the H-state and the universe, in me. Now, sixteen years later, I was overwhelmed by the feeling that I was finally beginning to return the favor.

As we watched Susskind head off down the street, I turned to my father. "What was going through your head when Lenny and I came back to the table with our drinks?" I asked him.

He laughed. "I was thinking, 'Holy shit. Lenny Susskind is carrying my coffee.'"

Everything I knew about road trips had come from Jack Kerouac. *All that road going. All the people dreaming in the immensity of it.* This trip was exactly like that, except instead of wise hitchhikers there were physicists, and instead of cheap motels there were Marriotts. Instead of diners and roadhouses there were Jamba Juices and sushi bars, and instead of speeding down the highway with Dean Moriarty in a Cadillac, I was being chauffeured by my parents in a rented Toyota. Then again, the goal was the same. Enlightenment. Or reality. Or whatever you want to call it.

We drove from San Francisco down to Santa Barbara, stopping in little coastal towns along the way. My mother sat up front with my dad, while I slouched comfortably in the backseat, watching the world go by. I watched the lush green hills and the mountains rising in the distance, the palm trees, the endless sky, the ocean stretching out toward the horizon. And I wanted to be amazed. I wanted to marvel, awestruck, at the beauty of nature and the majesty of the Earth. Isn't that what you're supposed to feel at the sight of these things? Awe? Only I didn't. I didn't find it marvelous or majestic, not compared to the ideas dancing around in my head. For years we had been trying to unravel the nature of the reality outside the window, when all I really wanted to unravel was the world inside my mind. Then again, was there a difference? *For all intents and purposes, there is no outside.*

Maybe the problem was that I knew that the things outside my window—the trees, the sky, the mountains, the ocean—were just the tip of the cosmic iceberg, negligible in the scheme of things. That none of them was ultimately real. Nabokov once wrote that "'reality' [is] one of the few words that means nothing without quotes" around it, and I was beginning to understand exactly what he meant. I traced my finger across the glass window, drawing quotation marks around a mountain: "mountain." But as I looked at it, all I saw were my own brown eyes staring back: "me."

The golden age of cosmology had passed awfully quickly, and it still wasn't clear what was going to take its place. I understood the draw of Susskind's FRW/CFT, but the decay of de Sitter space that's required to get there is governed by eternal inflation. You can't talk about eternal inflation without describing the universe beyond our cosmic horizon as part and parcel of the same cohesive reality. With its inherent God's-eye view, it seemed to me that eternal inflation was unphysical and incoherent from the start—so why invoke it at all?

Of course, invoking it brought invariance back to the S-matrix and made string theory viable again. Or did it? A quick Internet search on my cell phone led me to a Bousso paper entitled "Cosmology and the S-Matrix." Reading it in the backseat, I grew unconvinced that having observers like the Census Taker out at the infinite boundary of a flat universe was enough to give meaning (substance, reality) to the S-matrix in the first place. Because the point about the S-matrix wasn't simply that you have to stand at the farthest possible corner of the universe, looking back. It was that you have to stand *outside* the system.

As Hawking had said, the S-matrix works like a charm for describing laboratory experiments because, as observers, we can stand completely outside the system being observed. From behind Wheeler's plate-glass window, we can see what goes in and what comes out, our own existence entirely irrelevant. But when it comes to cosmology, there's no window. When the system is the universe, there's no outside. "The difference between cosmology and the S-matrix," Bousso said, "is that in the S-matrix you're outside looking in, and in cosmology we're inside looking out." It was Russell's barber paradox meets horizon complementarity: when you try to take the view from outside the brackets and include it in your description of the inside, things go horribly awry. You can't be inside and outside simultaneously. You can be inside the universe *or* you can have an S-matrix. Really, there's not much of a choice.

As for the Census Taker, I thought, you can push him out to arbitrarily large distances, but at any given time he's still inside the universe. And while he can access a near-infinite amount of information in his causal patch, he can't access all of it, thanks to the simple but unavoidable fact that he can't measure himself.

In the paper, Bousso pointed out that the impossibility of self-measurement was not only a poignant problem for de Sitter space but one that would ultimately plague *any* space, including FRW.

"This is just a particularly bad version of a more general problem that arises whenever one part of a closed system measures another part," he wrote. "This includes any measurement of the global state of the universe, independently of causal restrictions. Obviously the apparatus must have at least as many degrees of freedom as the system whose quantum state it attempts to establish (in practice it usually has orders of magnitude more)." He concluded, "No realistic cosmology permits the global observations associated with an S-matrix."

Even the Census Taker will hit an impassable limit. His light cone might grow big enough to engulf the whole universe, but it will never engulf him, too. He can never be both subject and object in a single frame. As long as he attempts to describe the physics of a universe that contains him, his description will be thwarted by pathological self-reference, the kind of Gödelian uncertainty that Wheeler saw as containing the key to ultimate reality. *Always have this uncertainty . . . can't decide true or false values from the inside.*

If self-reference undermined invariance even in an FRW universe, I thought, there didn't seem to be much hope for reality anywhere at all. The confusing coauthorship structure would be equally confusing in any universe. Even if my father and I waited for billions of years, surviving the apocalyptic decay of our vacuum into a lower-energy universe, which in turn would plummet after billions of years more, and still we waited, big bang after big bang, until finally we hit upon solid ground where our light cones would come their very closest to consensus, even then Brockman and Matson would have every right to reject our proposal. Coauthorship was as much an illusion as everything else.

At least it was a good illusion, I thought as we cruised down the highway, chatting excitedly about physics and the encounters we'd just shared. When we weren't discussing the nature of ultimate reality my dad would blast music—Radiohead, Beck, Bob Dylan, the Roots—and my mother would dance in her seat, snapping her fingers, bobbing her shoulders, and making up her own inexplicable lyrics as she sang along. I supposed that someone outside the car looking in might have found it an odd family vacation: the three of us driving down the California coast, debating the status of string theory or the meaning of the holographic principle and meeting with various physicists along the way. But for me, on the inside, it was just family.

At the Kavli Institute in Santa Barbara, Joe Polchinski's office was small compared to David Gross's captain's quarters, but it was still pretty sweet.

"Not a bad view," my father joked, pointing to the Pacific Ocean directly outside the window.

Polchinski laughed. "Sometimes when I'm working I see dolphins swim by."

"I knew it!" I muttered.

He turned to me. "Have you been here before?"

I nodded. "A few years ago I arranged a debate between David Gross and Lenny Susskind."

Polchinski's face lit up. "That was you? I heard about that!"

He took a seat in his desk chair while my father and I settled into the couch that faced a blackboard covered in equations. Polchinski seemed reserved, but thoughtful and kind.

Bousso had said that in certain settings, you could think of D-branes as the fundamental ingredients of ultimate reality, and Polchinski was the one who had discovered D-branes in the first place. If we wanted to know more about them, we had come to the right oceanfront office.

"Can you tell us what D-branes are?" I asked.

To understand D-branes, Polchinski said, you have to start with strings. By the 1990s, physicists had discovered not one but five consistent string theories in ten dimensions, whimsically named Type I, Type IIA, Type IIB, SO(32), and $E_8 \times E_8$. When it comes to theories of everything, no one wants five. After all, if there's just one right answer and you've found it, you're done. If there are five possible answers, you've still got a lot of work left to do in order to figure out which one is right.

Strings, Polchinski reminded us, can be open, like tiny shoelaces, or closed, like little rubber bands. Of those five string theories, some had only closed loops; others had both open and closed. In fact, if a theory has open strings in it, it *must* have closed strings, too, since two open strings can always join together to form a loop, even though the reverse doesn't hold true. It's a good rule, considering that gravitons are closed strings. If you had a theory with only open strings, you wouldn't have gravity—which, of course, was the whole point of the thing.

"In the early days, most of the focus was on closed strings," Polchinski said, "because they seemed to give a complete description of what you need in a unified theory." And one of the most remarkable things to come out of closed strings, he explained, was T-duality.

The idea behind T-duality was this. Closed strings get their energy in two ways: from their vibrations and from their wind-

With Joe Polchinski at the Kavli Institute for Theoretical Physics   W. Gefter

ing number. The winding number comes into play because the strings can wind themselves around a tiny, curled-up, compact dimension of space, their energy growing as they stretch with every loop around the circle. The winding number is a kind of potential energy, the taut spring of a rat trap waiting to snap. Vibrational energy is kinetic.

As you vary the size of the compact dimension, there's a trade-off between the string's vibrational energy and its winding energy. The larger the radius of the dimension, the more stretched the string and the greater its winding energy; the smaller the dimension, the more localized the string's position, which, by quantum uncertainty, means the more erratic its momentum and the greater its vibrational energy. Physics, however, doesn't care about the difference between the two forms of energy—the only observable value is the total energy of the string. There's no experiment, even in principle, that could tell the difference between a string with high vibrational energy and low winding energy and a string with low vibrational energy and high winding energy so long as they sum to the same amount. That means that there's no experiment, even in principle, that could tell the difference between a space of radius R and a space of radius 1/R. Which, when you think about it, is fucking crazy.

Size is not invariant! I wrote in my notebook. *What's big from one perspective looks small from another.*

It was a pretty mind-blowing idea. Given the radical differences between the physics of big things, like planets, and small things, like subatomic particles, you'd think—with apologies to the girl from my philosophy class—that size matters. Turns out it doesn't.

Think of what that means for the big bang, I scrawled. *Shrink the radius of the universe small enough and eventually it starts to look bigger again. Bounce, not bang.*

"Strings have a natural vibrational size, and you can imagine taking some space and you start making it smaller and smaller," Polchinski said. "The question is, what happens when the space gets smaller than a string? T-duality shows that if you put a string in a box and you make the box smaller and smaller, what you find rather remarkably is that when the box gets smaller than the string, there's kind of another way to look at the system in which the box is getting larger again. A new spacetime emerges. That's a slogan we have: emergent spacetime. Spacetime is not fundamental. The spacetime in the final picture is not the one you began with—it somehow emerges from the stringiness of space. But it's indistinguishable from the original, so neither one is more fundamental than the other. They are both emergent."

"So instead of thinking of them as two different spacetimes, you can think of them as one spacetime looked at in two different ways?" I asked.

"Exactly." He nodded. "Two ways of looking at it. It's a duality."

It's also a fundamental difference between a world built of point particles and a world built of strings. Particles don't have winding energy, because dimensionless points have nothing to wind. According to particles, big is big and small is small. Strings, however, see geometry differently. The idea that point particles are really one-dimensional strings doesn't just change the nature of matter—it changes the nature of spacetime, too.

Physicists, Polchinski continued, had studied T-duality in theories with closed strings, but no one had bothered with open strings. "I had grad students," he said, "and grad students need problems to do. So I said, why not try the same thing with open strings and see what happens? As with all good problems for students, they couldn't figure it out on their own, but we figured it out together. What we found was that the same sort of thing happens: the box gets smaller and smaller, and then past a certain point it grows bigger and bigger. But the funny thing is, when the box gets bigger, it's no longer empty. There's something in it. A submanifold. A brane."

Unlike closed strings, open strings don't have a winding number. Even if they wrap themselves around a compact dimension, their ends can still move around freely, so they don't create the same kind of tension. In fact, their ends *have* to move around freely. In order to preserve the Poincaré invariance of spacetime—in order to avoid choosing a preferred frame and violating relativity—the strings' endpoints have to be allowed to end at any point in spacetime, a democratic principle known as a Neumann boundary condition.

Without winding numbers, energy for open strings comes only in the form of vibrations—which means that if you shrink the size of one compact dimension down to zero, it never turns around and grows bigger again. It simply blinks out of existence, leaving the open strings in a spacetime with one less dimension than they started with.

That doesn't sound like a problem, until you remember that any theory with open strings inevitably contains closed strings, too. And with closed strings in the picture, things get a little weird.

Shrink the size of the compact dimension down to zero, and as far as the closed strings are concerned it grows big again, maintaining the full dimensionality of the original spacetime. Meanwhile, from the perspective of the open strings, spacetime loses a dimension.

How could there be a single spacetime that appears to have nine spatial dimensions to the closed strings but only eight to the open ones? Actually, the problem was even more subtle than that, because the majority of an open string is physically identical to a closed string. Only its endpoints are different. So the question, really, was how there could be a single spacetime that appears to have nine spatial dimensions according to both the closed strings and the open strings *except* for the open strings' endpoints, which only see eight.

Remarkably, Polchinski and his grad students solved the puzzle: when the shrinking compact dimension begins growing larger again, the freely moving open strings suddenly find their endpoints stuck to an eight-dimensional submanifold within the full nine-dimensional space. That way, the endpoints experience eight dimensions, while the rest of the open strings and the closed strings all enjoy the full nine. In other words, when the new spacetime emerges from the shrinking dimension, the open strings' boundary conditions change. A Neumann boundary condition is swapped for a Dirichlet one; instead of being free to end at any spacetime point, the strings are nailed to fixed points.

Only that couldn't be the end of the story—because the whole reason for the Neumann boundary condition was to preserve the Poincaré symmetry of spacetime. Dirichlet boundary conditions commit the forbidden crime against relativity: they treat certain reference frames differently than others, choosing a preferred surface in space. You'd think the whole exercise was shot. There was no point in trying to preserve the integrity of spacetime for open strings if you were just going to throw out relativity in the process.

But, again, Polchinski saw the solution: treat the spatial submanifold to which the strings' endpoints are nailed as a dynamic object. An object that can *move*.

If the spatial surface can move freely throughout the full nine-dimensional space, carrying the endpoints of the open strings with it, then the democracy of reference frames remains intact, Poincaré symmetry is restored, T-duality holds for open strings, and open and closed strings can happily coexist in the same universe.

It was a pretty amazing creative leap. What looked from one perspective like empty *space* looked from another like an *object*. It was like that image where you see two faces in profile and then suddenly you see that what looked like empty space between them is actually an object, a vase. Polchinski saw that what appeared to be the background space between strings could also be a vase. Given that the entire goal of quantum gravity was to unify spacetime with the objects it contains, it was a pretty big deal. And because it was a kind of membrane born of Dirichlet boundary conditions, he called it a D-brane.

"The D-brane was an object in its own right," Polchinski told us. "It could move, oscillate, break. That was totally unexpected." And it didn't have to be eight-dimensional. It could have any number of dimensions.

You can also stack several D-branes together. "When you put a lot of D-branes on top of each other," Polchinski said, "they'll start to warp space and eventually they'll form a black brane—a black hole that's extended in more dimensions."

In fact, it was by comparing the background-space perspective and the vase perspective at a black brane that Maldacena first discovered AdS/CFT: the idea that took down any last shred of invariance among strings, particles, and dimensions, the idea that demonstrated exactly how the holographic principle works.

"Maldacena's duality showed that gauge theory, which we thought we understood really well, and string theory, which we didn't understand very well at all, are really the same theory," Polchinski said. "That's tremendously striking. It's the deepest thing we know about gravity."

Deepest thing we know about gravity, I scribbled. *That it emerges as a holographic projection. That it's an illusion. That it's not real.*

"Does that mean that quantum physics and spacetime are two ways of looking at the same thing?" I asked.

"Well, there's a tension between the two, and holography means that quantum mechanics wins," he said. "The quantum framework survives unmodified. What changes is the nature of spacetime."

"Because it's no longer invariant?"

"That's right," Polchinski said. "Spacetime is no longer fundamental."

Polchinski's discovery of D-branes not only led to AdS/CFT, it also fueled the so-called second superstring revolution. The first revolution occurred back when Schwarz and Green realized that string theory was a theory of quantum gravity. It was a major leap forward—until they found themselves burdened with five theories of quantum gravity, waiting for another revolution.

The second revolution kicked off in 1995 when Ed Witten suggested that all five string theories might be different aspects of a single theory, which he dubbed M-theory. But it wasn't until Polchinski discovered D-branes that he could prove it.

With D-branes in hand, Witten was able to show that all five string theories were just different ways of looking at a single theory, M-theory. Actually, to make that work, he had to include one more theory in the mix, only it wasn't a string theory at all. It was a theory known as supergravity.

Supergravity is the theory you get when you make supersymmetry into a local, rather than global, symmetry. I had already seen how that worked with the gauge symmetries: view a wavefunction from a slightly different perspective and you shift its phase, but because the speed of light is finite, you can't shift it throughout the entire wavefunction at once; you can shift it only in one local region. That creates a misalignment of the phase—a misalignment we call a force.

Supersymmetry is the symmetry that allows you to shift your perspective in a way that swaps fermions and bosons, but once again the speed of light prohibits you from doing so everywhere throughout space at once. That means that while fermions and bosons are swapped within your reference frame, they might not be in another—a misalignment that once again requires a force to patch up the differences.

What force is capable of repairing supersymmetry? Intriguingly, gravity. Only in this case, the graviton requires its own supersymmetric partner, the gravitino. Together, the graviton and gravitino constitute a new force: supergravity.

Just as gravity accounts for the mismatches between accelerated and inertial observers—*to turn a curve into a line you have to bend the paper*—supergravity accounts for the fact that what looks like a fermion to one observer looks like a boson to another.

With Polchinski's D-branes, Witten could finally show that all five of the ten-dimensional string theories plus eleven-dimensional supergravity were different aspects of the unified M-theory, related to one another by dualities such as T-duality and S-duality, which mapped high energies in one string theory to low energies in another. That meant that you could start in any one of the six theories and, using dualities, find your way to any other one. If you start with eleven-dimensional supergravity and compactify one of the dimensions to a circle, you get Type IIA string theory; compactify that dimension to a line segment and you get $SO(32)$. Turn up the energy of $SO(32)$ and you'll find yourself in the low-energy regime of Type I string theory; apply T-duality to $SO(32)$ and you're in $E_8 \times E_8$. Turn up the energy of $E_8 \times E_8$ and an extra dimension of spacetime emerges, landing you right back in eleven-dimensional supergravity.

It was clear that D-branes were powerful objects. But were they ultimately real? I had my suspicions. D-branes were *made* of spacetime, and Polchinski had said that spacetime wasn't fundamental.

"Are D-branes fundamental?" I asked.

"D-branes are still not the final answer," he said, "but in some ways they are closer to the final answer than strings themselves. Strings were the wrong starting point. The holographic principle is much closer to the right starting point. String theory . . ." He paused. "I don't want to say it has withered away, but . . ."

I laughed. "But it's withered away?"

"You have black holes and you have a quark-gluon plasma, and you can use either one to understand the other," Polchinski said. "And you never have to mention strings. Strings aren't on either side of that duality, but they provide the logical connection that turns one into the

other. The theory is no longer string theory. Strings are just one classical limit. We're looking for the whole thing."

"M-theory?" I asked.

"That's right."

"So it wouldn't be correct to think that the world is 'made of
strings.'"

"Not at all."

In fact, I realized, it was starting to look like it wouldn't be correct
to say that the world was made of *anything*. In each of the five string
theories, the elementary strings, which are supposed to be the most
fundamental ingredients of reality, all give rise to different sets of particles. You'd think that would be enough to show that the five theories
weren't really the same after all—but you'd be missing half the picture.
If for each string theory you list all the elementary particles *and* all the
composite particles—particles made out of multiple strings—the lists
suddenly appear identical. What's elementary in one theory is composite in another—and yet, thanks to all those dualities, all five theories
are equally true.

It was a powerful blow to the whole reductionist enterprise, the
one that for centuries had scientists thinking that if they could just
find the smallest objects out of which everything else was made, they'd
understand the workings of the world. It was also a powerful blow to
anyone who happened to be driving down the California coast with her
father in search of ultimate reality. String theory was stating it pretty
clearly: there are no basic ingredients. They only appear basic depending on your perspective.

"In Maldacena's duality, the strings emerge as composite states of
gluons," Polchinski said. "And in matrix theory, which is another approach to M-theory, you have enormous matrices that reconstruct an
eleven-dimensional spacetime and there are no strings anywhere. String
theory was a way of getting to the holographic principle. That's really the
fundamental thing."

"But M-theory must have some ontology, right?" I asked. "Can you
guess what it might be? Are there strings, branes, particles? Some totally
new object? Is there space? Time?"

"I can't even guess," Polchinski said. "It's remarkable to know so much about many limits and yet have no good idea of what they are limits of! Holography is clearly part of the answer. The fundamental variables are probably very nonlocal, with local objects emerging dynamically."

Is it possible that there is no ontology at all? I wondered. Is it a theory made of nothing? Then again, if it's really a theory of everything, wouldn't it have to be?

"So string theory got us to the holographic principle, to AdS/CFT," I said. "How can we now apply it to our de Sitter universe?"

"Anti–de Sitter space is a pretty special space," he said. "It's like putting gravity in a box. But the problem that many of us keep coming back to is that we don't live in a box. There are no walls to our universe. So that's the question that needs to be answered."

My dad tags a chalkboard at the Kavli Institute for Theoretical Physics
A. Gefter

We don't live in a box, I thought, but we do live inside a cone. I still didn't understand why we didn't just surrender to the incoherency of eternal inflation and stick a cosmic hologram on each observer's boundary in a de Sitter space that would no longer decay. Sure, we'd lose every last invariance, but who were we kidding? Weren't we headed there already?

"What about the observer-dependent de Sitter horizon?" I ventured.

Polchinski thought for a moment. "Have you talked to Tom Banks? He has a rather different take, which I think emphasizes more the point you just made. That somehow each individual observer should have their own version of holography."

As it happened, Tom Banks was en route to Santa Barbara.

13

Smashing the Glass

The blaring ring of the telephone woke me from a sound sleep.

"Dad's having a kidney stone attack." It was my mother calling from the room next door.

"Shit!" I said, snapping awake. "Is he okay?"

"I don't know. We might end up going to the hospital. We need to keep the car here in case. You should take a cab to the campus."

I threw on sneakers and ran to their room. My father was lying on the bed, doubled over in pain, moaning. This wasn't the first time this had happened. I knew it would eventually pass, that he'd be fine. Still, he was the doctor. He was the one we all turned to when *we* were sick. Seeing *him* sick was like seeing the world turned inside out.

"I don't have to go," I told him. "I'll cancel it and stay here."

He cringed, clutching his stomach. "You . . . should . . . go," he stammered, pushing the words through the pain. "Kick . . . physics . . . ass."

"Go," my mother reassured me. "There's nothing you can do. I'll call you if anything changes."

I wasn't sure how long it would take for the cab to get there, so I called one right away. It came immediately, and I found myself arriving at the Kavli Institute forty-five minutes early to meet Banks. I headed into the common room to wait.

The place was empty except for one man who was pouring himself a cup of coffee. Was it Banks? I had glanced at a photo of him online a few weeks back, but I'd never had a good memory for faces. I offered up a big smile and said hi, figuring that if it was him, he'd react accordingly. The man looked at me, gave a slight nod, and returned to his coffee. Surely not the reaction of someone awaiting a blind interview. I sat down on the couch to wait.

The man took his coffee and headed outside, sitting down at a table in the courtyard just behind me, perhaps ten steps away. I waited, patiently biding my time.

I was excited to talk to Banks. I didn't know much about his work, but I did know that he and his collaborator Willy Fischler had a theory of holographic spacetime—one that, as Polchinski suggested, was even more observer-dependent than the other ideas we'd encountered, one that embraced our current de Sitter situation rather than holding out for billions of years in the hopes of transitioning to flat FRW space. Then again, Susskind had said that formulating physics in a single observer's patch in de Sitter space "probably can't be right because the de Sitter space can decay." I was curious to know what Banks would make of that.

The more time passed, the more I began to suspect that the man with the coffee was in fact Banks—only now it was way too late to say anything without making the situation even more awkward.

So I just sat there. And he just sat there. *For thirty minutes*.

Finally I decided to email him to say that I had arrived early and that I was in the common room whenever he was ready. He emailed me back. I could hear him typing: *Hi, Amanda. I'm in the courtyard*.

With an awkward smile, I joined Banks in the courtyard. I felt like an idiot. I sat down at his table, shaking his hand, apologizing for the confusion and explaining why my father couldn't join us. Banks nodded. He was quiet and reserved, but as soon as we launched into a physics discussion his demeanor transformed and he became friendly and animated.

I figured I'd dive right in and ask what I now knew was the million-dollar question: "How can we apply the holographic principle to our de Sitter universe?"

"The thing about de Sitter space," Banks began, "is that no matter how long you live, you can only access a finite region. Your horizon will always have a finite area. In the conventional picture of spacetime from general relativity, you'd say that there's more universe out beyond what we can see, beyond our horizon. But the holographic principle tells us that, no, there's a complete description of everything beyond the horizon right here in stuff that's causally connected to us. Each observer's universe, their causal diamond, is finite but complete."

Susskind had been hesitant to apply complementarity to the de Sitter horizon, but Banks wasn't hesitating at all. "This idea just follows black hole complementarity to its logical conclusion," he said. Information never leaves the observer's light cone; it just piles up at the horizon, scrambled and burnt by the Hawking radiation.

A causal diamond is the diamond formed by the intersection of an observer's past and future light cones, the total region of spacetime with which that observer could ever interact. A complete but finite universe.

I already knew the problem with a finite universe: there's not enough room for invariance. Invariant definitions of an S-matrix, of particles and strings, required an infinite boundary, infinitely far away. Finite boundaries a finite distance away couldn't cut it. As I had learned a long time ago, particles—and, in turn, strings—were irreducible representations of the Poincaré symmetry group. But horizons break Poincaré symmetry. It was precisely that fact that led to Hawking radiation, to the holographic principle, and to horizon complementarity. In a world with horizons, observers can't agree on what's a particle and what's empty space. What's more, neither one of them is more right than the other. In a de Sitter universe like ours, even the most stable building blocks of reality are rendered observer-dependent.

"What happens to the S-matrix?" I asked. "Don't you need an infinite region to get any kind of invariance?"

"You're right," Banks said. "If the causal diamond can ever become infinite, then all observers agree and there are gauge-invariant observables like the S-matrix in asymptotically flat space. But in de Sitter space this never happens. . . . Susskind et al. want to define some kind of observables in zero cosmological constant, approximately supersymmetric FRW. They want de Sitter to be unstable to decay into FRW."

"You don't think it will?"

"No," Banks said. "That idea is based on eternal inflation and the string landscape, and those are ideas that I think are just wrong." One reason they're wrong, he explained, is that they rely on the notion that spacetime can undergo quantum fluctuations.

"It can't?" I asked, shocked.

"Not in the holographic picture of spacetime," he said.

Banks explained that thanks to the holographic principle, it was now possible to encode all the properties of spacetime into the language of quantum mechanics. That, of course, was the holy grail. Quantum gravity.

The properties of spacetime fell into two categories: causal structure and scale. The causal structure tells you which points can communicate with each other and which can't—that is, the positions of light cones. Scale tells you how big things are.

I was surprised to hear that causal structure could be encoded into quantum language. Given the conceptual gulf between relativity and quantum mechanics, you'd think that light cones would have nothing to do with anything remotely quantum.

But the key, Banks explained, was commutativity.

I already knew a little something about commutativity. I knew, for instance, that certain pairs of measurements—certain "operators"—can't both be specified to arbitrary precision. One of these pairs was position and momentum, and another was time and energy; these were conjugate pairs linked by uncertainty. Uncertainty tells you that the order in which you perform the measurements matters. Measure position first and you obliterate information about momentum; measure momentum first and you smear out the position. The order matters, which is to say, the operators don't commute.

"If you have two spacetime regions that can't communicate—that is, points in spacetime so far apart that light couldn't possibly have traveled between them—then the quantum operators associated with those regions commute with each other," Banks said.

That made sense. After all, it wasn't quite accurate to simply say that position and momentum operators don't commute—the true statement was that position and momentum operators don't commute *within a single reference frame.* Within a single light cone. If you came across position and momentum operators that *do* commute, you'd know you were talking about causally separated events—events that lie outside each other's light cones.

"That commutation expresses the lack of causal connection between them," Banks said. "When operators *don't* commute, they interfere with each other. If you didn't have quantum interference, you could send signals faster than light. So the causal structure of spacetime tells you which quantum operators commute and which ones don't. But you can also read that backward. You can start with the algebra of quantum operators, which tells you what commutes and what doesn't, and from that you can deduce the causal structure of spacetime."

Voilà—quantum spacetime. Well, almost. In addition to causal structure, you still needed scale. Quantum commutation relations can tell you that two points are too far apart to communicate, but they can't tell you how far apart they are.

The holographic principle, however, can. "The holographic principle tells you that the number of quantum states—the entropy—measures an area," Banks said. "The area of the region's boundary. So if the holographic principle is true, we now have a way to state completely all the properties of spacetime in the language of quantum algebra."

"And that tells you that spacetime can't fluctuate?" I asked.

"That's right. Fluctuating spacetime is an old idea, but it's wrong. The holographic principle tells you that the properties of spacetime are encoded into which quantum operators commute with each other and how big the space of states is, the Hilbert space, the entropy. Those are things that don't fluctuate. It's the values of variables that fluctuate in quantum mechanics, not the size of the Hilbert space or the commutation relations. When you talk about the uncertainty principle for position and momentum, what fluctuates is how much you know about either one. But the commutation relation between them is just there.

It's exact; it doesn't fluctuate. So holographic spacetime tells us that geometry doesn't fluctuate. And that has a very, very profound influence on how we should think about string theory."

That was putting it lightly. If geometry—that is, spacetime—didn't fluctuate, that meant no eternal inflation and no string landscape. No decay to FRW. No infinite, flat space. No invariance. No reality. Just us, sitting here in de Sitter space. Finite. Stuck.

"Okay," I said. "So you have an observer sitting in a causal diamond in de Sitter space, surrounded by a finite horizon. But the horizon is observer-dependent. It's not like in AdS/CFT, where you have one boundary for the whole universe. Here there's a cosmic hologram for each observer?"

"AdS/CFT is a special case," Banks said. "The areas of the causal diamonds are taken to infinity in a special way. Observers are related by a symmetry transformation of the space, so they're all equivalent to each other. That's not true in de Sitter space. As a result, it's much more observer-dependent. People trying to squeeze everything into the AdS/CFT paradigm remind me of movie producers who are trying to make the fifth sequel to *Saw*."

"What happens when you have more than one observer?"

"There's a beautiful galaxy called the Sombrero galaxy, which is not gravitationally bound to our local group. If we're really in de Sitter space, we will eventually see the Sombrero galaxy's light redshift farther and farther away. It will appear to approach our horizon, and as it does, it redshifts more and more until it disappears from sight. The only trace of it left will be the uniform background of radiation coming from the horizon. But if you were sitting on the Sombrero galaxy, you would see *us* approach the horizon and become absorbed into that temperature while you're just sitting there having a cup of coffee and thinking you're perfectly okay. Those are two completely equivalent descriptions of the same physical system, but they use degrees of freedom that can't be measured by a single observer."

I nodded. "They only contradict each other if you try to take a God's-eye view."

"That's right," Banks said. "You can't say that our story or the Sombrerinos' story is the right one. Neither is more real than the other. In quantum mechanics I can talk about a particle's position or its momentum but I can't talk about both simultaneously, even though either description is equally valid. Likewise, I can't be both the accelerated and the inertial observer."

You can't be inside and outside a horizon. Safe and Screwed.

"Holographic spacetime is constructed observer by observer," Banks continued. "If I think about two observers, like you and me, there's a large region of spacetime which we can both explore for long times. So you have some description of that spacetime and I have some description of that spacetime. Those descriptions are individually complete."

"So a second observer is . . . what? A copy?"

"Relativity tells us that no observers are special. There has to be a gauge equivalence between causal diamonds, so everything outside my horizon is a gauge copy of the physics I can observe right here. So if you think of every possible causal diamond, you have an infinitely redundant description of the same quantum system seen by different observers."

Banks was taking the holographic principle to its logical conclusion, but the conclusion was pretty astounding. Susskind had taught me that no observer can see both elephants—or, more accurately, that there's only one elephant. There only *seems* to be two when you mistake a copy for the real thing. Not that there was a "real thing." There were only copies of copies, each as illusory as the next. But if redundant elephants seemed fucked up, what exactly were we to make of redundant *universes*?

"So you have infinitely redundant descriptions of the same quantum system, and spacetime emerges when you put all these descriptions together," Banks continued.

"Spacetime is emergent?" That made sense, given everything Polchinski had told us about M-theory.

"Well, time isn't emergent, but it is observer-dependent. Space arises from the quantum mechanical relations between different observers. The key is that where two observers' causal diamonds overlap, there has to be consistency between what they both see. And that's actually a very, very strong constraint."

"What would satisfy that constraint?" I asked. What could possibly remain invariant from causal diamond to casual diamond? If all observers had different views of the same underlying reality—what's the reality?

"It doesn't have many solutions," Banks said. "The one Willy Fischler and I found is what we call a black hole fluid. It's a state that doesn't look like anything we're used to. There are no particles. There's not even spacetime. There's just a quantum system in which all the degrees of freedom are interacting constantly. It's a homogeneous and isotropic state with maximal entropy."

No particles, no spacetime? Perfectly homogeneous? It sounded an awful lot like my father's nothing to me.

"We believe that the black hole fluid existed at the very beginning of the big bang," Banks said. "Inflationary cosmology was created to explain why the universe started off so homogeneous and isotropic, which is such an improbable or low-entropy state. But the black hole fluid is homogeneous and isotropic, but it's also at maximum entropy. So then the question becomes, why doesn't the world look like the black hole fluid now? And the answer is, you can't have life or biology or any kind of complexity in that state. But if the black hole fluid only has a finite number of states, you could imagine that it eventually finds itself in an improbable low-entropy state. That idea goes back to Boltzmann. He said, if a system has finite entropy and it's in equilibrium, so it's cycling through every possible state, if you wait long enough it will happen to find itself in a low-entropy state where complexity can emerge. Willy and I are working on that idea—how can we start with the black hole fluid and yet explain how you get low enough entropy for something complex and interesting, like life, to happen? That's our goal, but we haven't gotten there yet."

"I was reading Luboš Motl's blog and he said that you are 'building physics from scratch,'" I said. "Do you feel that way?"

"Luboš is partly right. However, none of the ideas that went into holographic spacetime could have existed without the work of all the people who contributed to our understanding of string theory. So, in the unlikely event that it turns out I'm right, then 'if I've seen further than others, it's because I stood on the shoulders of giants.' I assume you know that Newton said that as a concealed dig at Hooke, who was a dwarf."

I had heard that, but it still made me laugh. Everyone thinks Newton was being eloquently modest, when really he was just being a dick.

"There seems to be a kind of paradigm shift under way in thinking about cosmology—from a God's-eye point of view to saying that you have to think about it from the point of view of a single observer," I said.

"Yes, but I don't think it has permeated the community yet. The people at Stanford think that way, but they're thinking differently from me. They're trying to put everything that was done in the God's-eye point of view, like eternal inflation and the string landscape, into this new framework, whereas I think those things were just misconceptions. They want to take the observer's view of the world and try to squeeze the God's-eye view into it. I think they're wrong, but at least it's an idea for an idea," Banks said, quoting Wheeler.

"I've been trying to find out what the most fundamental ingredients of the universe might be," I said. "It's becoming clear that strings aren't the answer."

"That's right," Banks said. "We have these models, and in some extreme range of parameters where everything becomes calculable, they look like they are describing strings. But the real theory doesn't have strings in it in any fundamental way. I think a lot of people would agree that we don't really understand what's fundamental. My view is that the fundamental principle is the holographic principle, this relationship between geometry and quantum mechanics."

When our conversation ended, I called the cab company for a ride back to the hotel. Fifteen minutes later the same cab driver who had dropped me off pulled up in front of the institute.

"Are you a scientist?" he asked me as we headed back down the coast.

"No," I said. "A writer. I was just doing some research."

"What kind of research?" he asked.

"Physics," I said. "The nature of reality."

"I was watching a show on TV the other day and they were talking about the smallest things that exist. The *smallest* things! Some kind of little atoms. You know something about that?"

"Quarks?" I ventured. I didn't bother explaining that "small" didn't mean anything anymore. That what looked like the smallest thing in one reference frame could look like the biggest thing in another. I figured it wasn't a good idea to blow someone's mind while he was driving.

"That's it, quarks!" he said, satisfied. "Interesting stuff."

"They might be holographic encryptions of gravity in ten dimensions," I mumbled.

"What's that now?" he asked.

"I said they're interesting," I yelled up to the front seat. "So *small!*"

Back at the hotel I was grateful to find my father sitting up in a chair reading. With a particularly bad bout of pain-induced vomiting, the stone had lurched farther down its route, where it now sat comfortably, his pain temporarily subsided. My mother, looking exhausted and relieved, was on the bed, knitting.

"How was Banks?" my father asked.

"Completely fascinating."

I recounted what Banks had told me about holographic cosmology. It wasn't like Markopoulou's universe, I explained, in which we were all looking at finite chunks of the same larger universe. No, according to Banks, every observer's finite chunk was *the* universe. The whole thing. Self-consistent and complete. "Your universe is all there is," I told my father.

He looked thoughtful. "Does that mean that you are just a copy of me?"

"Well, we already knew *that*," my mother said.

"I'm pretty sure this is my universe and you are all copies of me," I said.

"Please," my mother said, rolling her eyes. "We were here way before you."

I laughed, but the question of exactly whose universe this was, was a legit one. Banks had taken Smolin's slogan to a whole new level. The first principle of holographic cosmology must be that there is nothing outside *my* universe.

That evening, while my father rested in bed, I took a walk with my mother. We wandered aimlessly through the quaint Santa Barbara streets and eventually made our way to the harbor. The air was warm, the sky a soft pink; the boats rocked gently at their moorings.

"I can't stand seeing Dad in pain," I said as we looked out over the marina.

"Tell me about it," she said. "You should have seen him in the middle of the night. It was awful."

"But he'll be okay?"

She offered me a reassuring smile. "Of course he will."

"I keep thinking about the book," I said. "And about Brockman and Matson. They don't think coauthorship can work. Neither does physics. Banks says we each have our own universe and you can't talk about more than one at a time. I guess part of me wants to try to write a book on my own, and another part of me can't handle the thought of it. Do you think he would be upset? Do you think he would hate me?"

I wasn't sure what had brought on my sudden confession. Something about seeing him lying in bed in agony—I felt guilty. Like his kidney stone was a direct manifestation of my betrayal. Like my secret consideration of sole authorship had calcified and lodged itself in his urinary tract. Like if I didn't get it out in the open, here and now, it would lodge itself somewhere inside me. Like it would keep on amassing until it formed a black hole in my center and sucked me in. Like I was Screwed.

My mother laughed. "He could never hate you! Working on physics with you and being your partner in crime mean the world to him. I think he would be terribly disappointed to not have that anymore. But I don't think it's about authorship. The writing part was always for you. He's a thinker. You're the writer."

"But that's just it. I'm not. I'm *pretending* to be a writer."

"Maybe it's time to stop pretending," she said.

I sighed. "I guess I'm afraid that I won't know how to do that. Matson said I need to write with my own voice. I'm just not sure where to start. I'm not even sure I *want* to start. Not without Dad."

Every way I looked at it, it looked wrong. How could I write a physics book without him? If it hadn't been for him, I would never have learned a word about physics in the first place. I would never have done any of it—not only because he roped me in with his question about nothing, but also because without him there to share my excitement at every turn, none of it would have mattered anyway. I guess that was the thing about invariance. I guess that was the thing about other people. Even though they have minds we can never verify, let alone know, even though they sit on the other side of an unbridgeable abyss, even though their irreducible otherness is exactly what makes us so fucking lonely, it's also the only thing that makes us feel like we are real. Like life is real. Not even the answer to the universe would have made it all seem worthwhile without my father there, echoing it back to me.

I looked at her pleadingly. "Tell me what to do."

"I can't," she said.

I pouted. "Since when?"

"He'll be proud of you no matter what," she said. "And so will I. I'm sure you'll make the right decision."

I gazed out across the harbor, my eyes following the calm water until it met the mountains, which looked small off in the distance under the now fuchsia sky. I knew that coauthoring a book committed the mortal sin against holography, that cutting across horizons was bound to produce something voiceless and untrue. Then again, I wasn't entirely sure that my father and I counted as two separate observers, like Safe and Screwed. We always knew what the other was thinking. We spoke in half-uttered thoughts, we easily finished each other's sentences, and there had been more than one occasion on which I had called him to share some new revelation and reached his voicemail only to find out moments later that it was because he had been calling *me* at precisely the same moment to share *precisely the same revelation*. If I hadn't been one of the telepathic parties in question, I never would have believed it. Such preternatural events occurred with enough frequency that those who knew us best, like my mother, were convinced that we were two halves of a single brain.

On the other hand, we had our differences. Where he was eternally patient, I was impetuous. Where he was unflappably kind, I was moody and cynical. He was laid back and intuitive; I was anxious and logic-bound. He refused to cross the street against a Don't Walk sign, even if the road was empty for miles, while I felt more comfortable breaking rules than following them. And when he had a brilliant idea, like the H-state, he preferred to sit on it awhile, as if it were an egg in need of incubation, while I preferred to grab hold of it and strap it to our chests like dynamite, then send us running for the highway into oncoming traffic.

I wanted to listen to Matson. I wanted to find my voice, to venture outside the justified line and head for the margin. I needed to find a way to write from inside my own reference frame, but here in my frame I didn't know what it meant to talk about physics without talking about my father. For me, from the beginning, growing up and discovering the nature of the universe had always been one and the same, though I suppose that's true for everyone. My world had always been a strange hybrid of life and physics, and if reality was my Snark, I thought, perhaps my book would have to be a strange hybrid, too.

"Holy shit," I muttered.

"What?" my mother asked.

"I think I know what to do."

It was all hitting me now. Horizon complementarity demanded more than sole authorship. It demanded *first-person narration*.

"I'm going to write the whole thing," I said. "The whole story. About Dad and the H-state and Princeton and Wheeler . . . about all of it."

My father and I might not live in the same universe, I thought, but he appeared in mine as convincingly as I did in his. His name didn't have to be on the cover—he could be *in* the book. He would be my sidekick, or maybe I'd be his. We'd be like Don Quixote and . . . Don Quixote's equally delusional father. I would write the book in the first person according to the mandate of Brockman and Matson and the laws of physics, and, most important, we would still be in it together, partners in crime. Our book had started off as an idea, a symbol, the answer to the universe. But I realized now that growing up meant recognizing the book for what it really was: a story. No, my story. *Smash the glass, reach in.*

"So it'll be like a memoir?" my mother asked.

I smiled, exuberant. "Exactly."

It was the only logical possibility: a book that contained its author on the inside, a maze so convoluted that what appears from one standpoint to be author is from another point of view character. A top-down, self-excited, first-person, Gödelian gonzo cosmology. A fucking memoir.

My mother was eyeing me warily. "Am I going to be in it?"

"If you're lucky," I said.

She shot me a stern look. "You better not make me look bad."

When we returned to the hotel, I nervously told my father about my idea to write a physics book that was also a memoir, and I watched as the logical necessity of the thing washed over his face.

"That's it," he said fervently, grinning. "*That's* the real book."

He said it as if that were the book we had been talking about the whole time, only we didn't know it. Or as if *he* had known it and was just sitting around waiting for me to see the light. I was seized by a momentary paranoia. Had he planned this all along? Was he teaching me some kind of *lesson*? Maybe it was the passing of the kidney stone, but he even seemed a little relieved, as if he was more comfortable in the role of character than author. I was more than a little relieved myself, knowing that he was happy and on board and that I had found a way to try to be that other kind of writer, the one he had once worried I hadn't become.

Thirty thousand feet above the Earth, heading back toward the East Coast, I began to sense the enormousness of what we had encountered on the road. Cosmology was truly on the cusp of a radical paradigm shift. Although it had yet to fully permeate the physics community, let alone the wider public, it was on its way and there was no stopping it now. As Bousso said, it was inevitable.

Paradigm shifts in cosmology are few and far between. There was Copernicus's *De Revolutionibus,* published in 1543, which sparked the so-called scientific revolution and sent the Aristotelian worldview, which had reigned for more than a millennium, into crisis. Aristotle had argued that the universe was a finite set of nested crystal spheres that carried the Moon, Sun, planets, and stars in perfect circles around the Earth, which sat at the center of it all, unmoving. Objects, Aristotle said, moved up or down according to their relative abundances of nature's four elements—air, fire, earth, water—then settled into absolute rest. Copernicus, however, suggested that the Sun sat at the center of the solar system and that the Earth itself was moving, a claim that pushed what we now call science beyond the bounds of common sense toward something more like ultimate reality. It also blurred the division between celestial and terrestrial, introducing a relativity of motion that could no longer support Aristotle's physics of absolute up and absolute down. By attributing the apparent motion of the stars to the motion of the Earth, Copernicus freed the stars from their fixed sphere and scattered them at various distances throughout space, opening up the possibility of an infinite universe. Tycho Brahe's observations of the paths of comets shortly after obliterated the existence of the spheres altogether, leaving the planets and stars to wander on their own, untethered, through vast stretches of unnervingly empty space.

These developments raised difficult questions. For example, if the Earth is moving, how can we ever say that an object is at rest? What makes the planets move? If there's nothing but space where the crystal spheres used to be, what mysterious force keeps the planets in their orbits? Galileo answered the first question: Aristotle's defunct distinction between invariant motion and rest was replaced by Galilean relativity, which showed that there is no difference between rest and inertial motion. Newton answered the next two: the same force that had an apple fall toward the Earth had the Earth fall toward the Sun. Gravity. But Newton's theory raised a question of its own: what the hell is gravity? That question hung in the air for three hundred years, weightless.

Einstein's general relativity unveiled the true nature of gravity and sparked the possibility of a dynamic spacetime, a changing cosmos, one that could expand and contract and begin in a big bang. But the big bang wasn't complete until 1980, when inflation filled in the missing pieces of the puzzle. Unexpectedly, inflation had brought a paradigm shift of its own. It went eternal, swapping our single universe for an endless multiverse. At the same time, it wrought a new crisis, the measure problem, which undermined the basic predictability of science.

Meanwhile, Einstein's theory of gravity had come into deep conflict with quantum mechanics, the theory of matter. When Hawking pitted the two against each other, the result was the black hole information-loss paradox, which looked to be a lose-lose situation until Susskind's epiphany—*restrict to a single observer's point of view*—provided an escape. That, in turn, brought with it yet another shift, one that would further shatter our notions of invariance and ultimate reality, a shift too young to have a name but would likely sound something like "holographic," one that swaps the multiverse back for a single universe.

Gazing out the window and down into the clouds, it dawned on me how amazingly unlikely our own story was turning out to be. For fifteen years my father and I had been running around trying to answer our own questions about the universe, living in the small, surreal world we had created for ourselves, as if it were all just our own private little game. We had come to California to hang out with Bousso and Susskind and Banks thanks to a conversation we once had in a Chinese restaurant, and somehow we had been wrenched out of our little world and into something so much bigger. It was as if we had bumped into Galileo just as he turned his telescope to the sky or brought Hubble his coffee as he calculated the distances and redshifts of the galaxies. As we stupidly carried out our personal mission, we had stumbled, Forrest Gump style, into the trenches of history.

If Max Born thought that the notion of reality had become problematic by the 1950s, I thought, he'd be freaking out now. M-theory's dualities had undermined the reality of size, dimensionality, geometry, topology, particles, and strings. What looks like spacetime in one view looks like an object in another. High energy becomes low energy. Elementary becomes composite. Big becomes small.

Personally, I was relieved to be an ontic structural realist. An ordinary realist would lose her shit. Nearly every remaining shred of ontology was vanishing before our eyes. But dualities preserve structure. All of those drastically different physical pictures are descriptions of the same mathematics. Structure is safe in the face of ontological underdetermination, and the universe had never seemed so underdetermined as it did now. Jesus, as far as the world's leading physicists are concerned, there is *no ontology*. Which is to say, the world isn't made of anything at all.

Now I understood what Susskind and Gross had meant when they said that they didn't know what string theory was. It wasn't because there were some holes in the mathematics or a lack of plausible experiments or even a missing principle or two. It was because they literally didn't know what the theory was describing. Particle physics is the theory of particles. Quantum field theory is a theory of quantum fields. General relativity is a theory of gravity and spacetime. String theory is a theory of . . . *what*? Not strings. Not particles, either, or even branes. So many physicists had been working so hard constructing the mathematical edifice of the theory, but they had no idea what it was supposed to be describing. "I can't even guess," Polchinski had said. If strings looked like branes at low energy, and branes looked like strings in extra dimensions, and particles looked like strings in different geometries, then none of them could be invariant. None of them could be real.

Adding to Born's anxiety, Banks had taken any remaining hope for invariance and sent it packing across the horizon. Taking the lessons of the holographic principle and horizon complementarity to their logical conclusions, Banks argued that information couldn't be lost across de Sitter horizons. Each observer's universe, though finite and bounded, was the whole show. If no observer was missing information—if all of reality was contained completely within an observer's light cone—then anything beyond the observer's horizon couldn't be new information. It had to be a redundant copy of information the observer already had. Another description of the same elephant. A new it from the same bit. An isomorphic element of the same structure. A gauge copy. Unreal.

To Susskind's deepest question in cosmology, Banks answered firmly in the negative. Objects on the other side of the cosmic horizon are not real. Which, ultimately, meant that *nothing* was real. After all, horizons are observer-dependent. My de Sitter horizon was not the same as my father's, which meant that an object outside my universe could be inside his. If objects outside horizons aren't real, then something that wasn't real according to me could be real according to my father, and vice versa. Reality is no longer observer-independent.

Thanks to Banks, we could cross the universe off the ultimate reality list. Individual, observer-dependent universes were all that remained, along with the cardinal rule of the new cosmology: you can't talk about more than one universe at a time. It was the kind of solipsism that Wheeler had tried so hard to avoid. Still, there had to be some kind of consistency conditions to relate what different observers see. If each observer's universe is another complementary, redundant description, what exactly is it a description *of*? Banks said it was a homogeneous black hole fluid. My father said it was nothing.

My father, of course, was thrilled with these developments. All of them seemed to support his intuition that everything had to be nothing, the H-state, which meant that nothing other than nothing itself could be invariant or real. I was a bit thrilled myself, because I knew that a universe with any ontology at all would be impossible to explain. A universe made of nothing, however, could potentially explain itself. Like a self-excited circuit.

But we weren't down to nothing just yet. One ingredient on the IHOP napkin persevered: *the speed of light*. It made sense that it would be the last one standing. Everything had been rendered observer-dependent, but observers themselves are *defined* by light cones. If all you need to turn nothing into something is a boundary, all you need to have a boundary is a finite and invariant speed of light. I couldn't imagine how we would ever cross that one off the napkin. It wasn't a new question. In fact, it now dawned on me with horror, it was exactly the same question we had started with when we confronted Wheeler in Princeton: if observers create reality, where do the observers come from?

14

Incompleteness

I took hold of my laptop, subscribed to the first step taken, and began to type.

I was working in a magazine office when the lie was born. That was the idea, anyway—"working" in an "office." In reality I was stuffing envelopes in the dusty one-bedroom apartment of a guy named Rick. The idea was that I worked for Manhattan *magazine. The reality was that I worked for* Manhattan Bride.

I wrote about faking some press credentials to crash the Science and Ultimate Reality symposium with my father. I wrote about our cryptic conversation with Wheeler, and about loitering on Einstein's lawn. I wrote about my secret plan to become a journalist and about scoring the *Scientific American* piece. I wrote about sneaking into the photograph at the Davis conference and escaping the awkward dinner by fleeing in Timothy Ferris's car. I wrote about my father's ideas about nothing, how it was an infinite, unbounded, homogeneous state, and I wrote about my ideas about something, how it was defined by invariance and how it slips away every time you reach for it. I wrote it not as a Journalist or an Academic or an Author. I wrote it as me. Twenty pages poured out of me in what felt like an instant. I titled the proposal

"Crashing the Ultimate Reality Party," stuck it in an email addressed to Matson and Brockman, and hit "Send."

I arrived at the Harvard Science Center and made my way to the fifth floor, but when I got there a flood of students and professors came pouring down the hallway. "Is the Witten talk in there?" I asked a student emerging from the room where the lecture was scheduled to take place.

"Too many people," he said. "We're changing rooms."

In California, my father and I had learned so much about M-theory—its instantiation of the holographic principle in AdS/CFT, its radical revision of the nature of spacetime, and its suspiciously vacant ontology, all of which seemed to turn the universe inside out. But there was one thing about M-theory we still didn't know: what the hell did the M stand for?

Everywhere I read about it, I was told, "Nobody knows what the M in M-theory stands for." Perhaps it's *magic*, they said, or *mother*. Steven Weinberg guessed that it stood for *matrix*. Sheldon Glashow wondered if it was an upside-down W, for *Witten*. Even Stephen Hawking wrote, "No one seems to know what the M stands for, but it may be 'master,' 'miracle,' or 'mystery.'"

How can nobody know? I thought. Ed Witten made it up and the guy's still alive. Why doesn't someone just ask him?

When I heard he'd be speaking at Harvard, I figured I'd try.

I followed the crowd to the new classroom and tried to snag a seat, but it quickly became apparent that the bigger room wasn't big enough. Whoever had organized this thing had forgotten that Witten was a full-blown superstar. Soon we were switching rooms again, this time heading to another building on campus, one with a larger auditorium. I rushed to keep up with the ever-growing crowd as they made their way outside, pushing and shoving one another out of the way in a race to score a seat. I had never seen physicists get so rowdy. As a group, they aren't exactly the most aggressive or athletic bunch, but with a Witten lecture on the line, these guys were ready to sprint, tackle, and pound anyone who got in their way. Running alongside them, I noticed a blind guy in the midst of the stampede, waving his walking stick and trying to keep his footing as the other physicists trampled past. I momentarily slowed down and considered helping him maneuver through the crowd. But he'd probably rather do it on his own, I told myself, then charged, elbows out and full speed ahead, toward the auditorium door. *M* is for *mayhem*.

I body-checked my way into a seat, and Witten's lecture finally began. He was a strange character—unnervingly tall and broad-shouldered, with a large rectangular head that had inspired my father to refer to him as "Ed the Head." But despite his looming presence and intellect, Witten spoke in a soft, whispery, high-pitched voice, an incongruity that lent him an air of otherworldliness. (*M* is for *mother ship?*) He was widely considered the smartest man alive.

Actually, this was the second time I had seen Ed Witten. The first was many years earlier, about a month after the Davis conference in 2003, when the American Physical Society held a meeting in Philadelphia. I had scored myself a press pass and my father had snuck in to watch with me as Wheeler was awarded the Einstein Prize. Afterward, we were taking an escalator down to the ground floor when we noticed Witten standing a few steps ahead of us. We had watched as he stepped off the escalator and into a revolving door. He attempted to push the door clockwise, and when it didn't budge, he inexplicably continued pushing it for several seconds before deciding to turn around and try pushing it the other way. My father and I had looked at each other, trying not to laugh out loud. We were thinking the same thing: that is the smartest man alive.

Now, at Harvard, I watched as he scribbled incomprehensible equations on the blackboard, his pants covered in chalk. I had no idea what he was talking about. It wasn't just that the subtleties of his argument were lost on me, or that the math threw me off—I couldn't even figure out what the topic was. Everyone else, however, seemed thoroughly engaged. Even the blind guy was following along. I had grown accustomed to the feeling that I could see through the equations to the concepts beneath, but today they were an opaque reminder that, no matter what Smolin said, I *was* on the outside of this universe, circling its periphery like someone locked out of her house, sizing up the interior, pathetically trying to find a way in. *M* is for *mediocre*.

When the talk ended, I approached Witten, introduced myself, and asked if he might have some time to talk with me while he was in town. He paused. What seemed like minutes passed. He paused for so long that I began to wonder if I should just walk away, and I was about to when he said, "I can do it at my hotel, the Inn at Harvard, tomorrow after I've finished my breakfast at eight-thirty A.M."

Eight-thirty A.M.? There were very few people on the planet for whom I'd be willing to be anywhere at that hour. Johnny Depp, probably. Fiona Apple. A resurrected Albert Einstein. And, apparently, Ed Witten.

I nodded, and before I could say another word, he turned around to talk with someone else.

The following morning, at the crack of dawn, I made my way into the Inn's dining room, which sat beneath a vaulted glass atrium, bathed in the painfully early morning glow. I spotted Witten and awkwardly joined him at his table, hoping to God he remembered our ten-second conversation the day before and didn't think I was some weirdo off the street hoping to polish off his yogurt.

It was more than a little intimidating, talking to Ed the Head in person. It was obvious that he had no interest in small talk and that I should jump right in with questions. "What the hell were you talking about yesterday?" seemed like a bad way to start. But what was a good way? What exactly does one say to the smartest man alive?

M is for *mute*.

"What are you working on these days?" It was the best I could come up with.

"I'm working on an application of physics ideas to math, particularly to understanding knots better," he said in that surreal whisper.

Knots? Was that what yesterday's lecture had been about?

We discussed knots for a bit before I transitioned into dualities and their erosion of ultimate reality. I was eager to get his perspective on M-theory's elusive ontology. (*M* is for *missing*.)

"Initially, when people talked about string theory, they said okay, point particles are really just strings," I said. "Then with the second revolution, with M-theory, we found there weren't just strings but branes of every dimension. And now with dualities we see that strings are equivalent to particles again in certain situations. Is there a fundamental entity that everything is made of?"

"The dualities would contradict that idea, because in each description a different aspect of the theory is fundamental and other things are derived," Witten said. "There are fundamental ideas rather than fundamental physical objects."

Fundamental ideas rather than fundamental physical objects. It was like structural realism meets Berkeley's *esse est percipi.* (M is for *mind-dependence?*)

"You ushered in the second string revolution," I said. "Do you foresee a third one coming?"

"My crystal ball is cloudier than when I was younger. By definition a revolution is hard to foresee. But in the mid-eighties and mid-nineties, before the second revolution happened, there were kind of hints that something was going to happen—I didn't know what, of course. I don't have that same feeling now, but perhaps other people do. . . . If I had my druthers I'd like to go deeper into what's behind the dualities, but that's really hard. Maybe that's the third string revolution. Or maybe it's something that won't be understood for a long time."

My mind flashed back to the revolving door. "When you are doing everyday things—say, going to the grocery store or the dry cleaner's—are you thinking in eleven dimensions?" I asked. It seemed like a plausible explanation.

"Sometimes I'm able to continue thinking that way in my daily life and come up with key ideas while I'm doing an errand or something. Two of my key ideas came while I was riding on an airplane."

I smiled. Then I asked *the* question.

"What does the *M* in M-theory stand for?"

"I didn't mean to confuse everyone with that," he said. "The way I described it at the time was that the *M* stood for *magic, mystery,* or *membrane,* according to taste. But I thought my colleagues would understand that it was really for *membrane.* Unfortunately, it got people confused."

So there it was. *M* is for *membrane.* Magical miracle mystery solved. All that confusion just because nobody got the joke. To be fair, it was likely the first and only time Ed Witten had ever cracked a joke. But still.

> From: Katinka Matson
> To: Amanda Gefter
> Subject: RE: Proposal
>
> Hi Amanda,
>
> This is terrific—fresh and fun. Let's talk about next steps.
>
> Best,
> KM

The next steps involved polishing up the proposal, chatting with potential editors, signing a book deal, and, after six years more wonderful than I could have imagined or schemed, leaving my job as an editor at *New Scientist* to find out what would happen when I stopped pretending, sat down, and began to write.

Driving home from the *New Scientist* office on my last day of work, my mind flashed back to the family car trips we used to take when I was a kid. A few times a year we would drive from Philadelphia to Stamford, Connecticut, to visit my father's parents. My father would blast Bob Dylan, my mother would point out turns and exits as his mind inevitably wandered, my brother would put on headphones and sleep, and while the world passed by my window, I would curl up with a book and read.

My grandparents' home was a kind of autocracy that made our house look a hippie commune by contrast. My grandfather, a retired physician, was dauntingly strict and serious, with a piercing intellect and an insatiable appetite for knowledge. As a kid, I would stare with wonder at their looming library, at the hundreds upon hundreds of books on every subject imaginable: architecture, politics, art, ethics, religion, philosophy, science. My grandfather would find me standing there, staring, and challenge me to a game of chess. Seated at the card table in the midst of the books, we would play, and with each move I made he would ask, "Are you sure you want to do that?" I would reconsider and change strategy and he would ask again until finally I found the proper move to satisfy him. While I studied the board for potential lines of attack, he would debate with me on matters of moral philosophy, speaking in a wildly grandiose vocabulary. Each time one of his words was met by a blank stare, he would send me over to the massive dictionary to look it up.

In his spare time, my grandfather wrote a treatise on biomedical ethics and composed crossword puzzles in Latin. In my spare time, I would wait until he left the room, move one of his countless trinkets not half an inch from its place, and then watch with perverse delight as he returned and, within seconds, adjusted it back to its initial position.

When it came to my grandfather, there were some things you just couldn't touch. One of them was Einstein. When I later wrote my article on Fotini Markopoulou, I showed it to my grandfather. As the piece was about loop quantum gravity and its attempt to reconcile general relativity with quantum mechanics, the *Scientific American* editors had titled the article "Throwing Einstein for a Loop." My grandfather, glancing at the headline, declared it an insult to Einstein and dropped it on the coffee table, unread.

The pedestal on which my grandfather placed Einstein wasn't metaphorical. It sat next to a window, holding a large bust, Einstein's bronze eyes keeping watch over the living room. He wasn't the only one. Hanging over the television across the room was a sculpture of Homer in relief. As a kid, I would sit on the couch between them, looking at one and then the other, my eyes sweeping back and forth as if I were watching a tennis match, surveying those twin pillars of the world: words and ideas, story and science.

Just when everything had fallen into place, the ground gave way beneath me.

It started one afternoon when I was browsing the physics papers on the arXiv and spotted a new one: "Black Holes: Complementarity or Firewalls?"

Firewalls? I was intrigued. The paper was written by Joe Polchinski along with Ahmed Almheiri, Donald Marolf, and James Sully. I grabbed a cup of coffee and sat down to read.

In the paper, the authors once again sent Screwed plunging into a black hole, then compared his view of reality with Safe's. But rather than worrying about the illegal cloning of quantum bits, this time they were worried about entanglement.

To begin their thought experiment, Polchinski and crew waited until the black hole had evaporated away to less than half its original size. That was important, I remembered, because it's not until the half-way point that Safe can extract even a single bit of information from the Hawking radiation. Then they sent Screwed hurtling toward the horizon as Safe watched from afar.

Now consider, they said, a bit of information—call it B—just out-side the black hole's event horizon. In Screwed's reference frame, B is part of the vacuum. Screwed, after all, is in an inertial frame, one de-void of boundaries, allowing all the positive and negative frequency modes of the vacuum—all those pairs of virtual particles and antipar-ticles, those uncertainty-born fluctuations of zero-point energy—to cancel out to a perfect zilch.

Guaranteeing their cancellation is entanglement, a form of quan-tum superposition in which two particles—such as a virtual particle and antiparticle pair—are described by a single wavefunction, a whole that's greater than the sum of its parts. Because the two particles form a single quantum state, their properties will remain correlated no mat-ter how far apart they are. If, upon measurement, one is found to have a positive frequency, the other is guaranteed to have a negative fre-quency, a correlation that ensures everything adds up to zero and the vacuum remains a vacuum. The bit B, according to Screwed, is en-tangled with its opposite bit, A, deep in the black hole's interior.

Safe, however, disagrees. According to Safe, B is not a virtual vac-uum mode but a real particle, a bit of Hawking radiation. That, I had learned back in London, was because the horizon restructures the vacuum, separating what was once a virtual particle from its antiparti-

cle mate, severing their entanglement, preventing them from canceling each other out and leaving instead a net positive gain, promoting B
from virtual to real, transforming, bit by bit, what was once an empty
vacuum into what is now a seething swarm.

No, Safe insists, B is entangled not with its counterpart inside the
horizon but with another bit of Hawking radiation—call it R—that
emerged earlier in the evaporation. That's required to prevent information loss. As Susskind had insisted—and Hawking eventually
conceded—Safe will never see information vanish. Instead of disappearing into the black hole, Safe sees it burn up at the horizon, scrambled beyond recognition and then radiated back out. Because it's
scrambled, the information no longer resides within a single Hawking
particle, but in the entangled correlations among the radiation.

Thus, the paper's authors said, we have a paradox. Screwed says B
is entangled with A. Safe says B is entangled with R. And quantum
mechanics says that one of them has to be wrong. A bit can't be fully
entangled with more than one other bit. Entanglement is monogamous.

As I sipped my coffee, I wasn't worried. The contradiction sounded
exactly like the kind that Susskind's complementarity was designed to
resolve. After all, complementarity allows you to describe either what's
on one side of an event horizon or what's on the other—but never both.
Just restrict your description of B to a single frame, Safe's or Screwed's,
and entanglement will always appear monogamous, I thought. No
problemo.

But as I kept on reading, problemo. Unlike the original cloning
paradox, this one doesn't disappear upon restriction to a single observer's frame—because the measurement on B takes place outside
the horizon for *both observers*. They are still in causal contact; they can
communicate. Screwed can measure B, find that it's entangled with A,
then turn around and tell Safe, who insists that, au contraire, B is entangled with R. The contradiction lives in a region where the two observers' light cones overlap. Complementarity prevented quantum
cloning because by the time the duplication appeared, Safe and
Screwed could no longer compare notes. In this case, they can. "Complementarity," the authors concluded, "isn't enough."

Someone, it seemed, had to be wrong. If it's Safe—if B is entangled with A and not with R—then the correlations strewn across the cloud of Hawking radiation are severed and information is lost. But who would be willing to allow information to be lost after physicists had just spent decades finding it? Hawking would have to change his mind, again. Elephants would disappear from the universe. The brilliant moral of AdS/CFT—that black holes are dual to information-preserving quark-gluon plasmas—would be wrong. The Schrödinger equation would fail. Quantum mechanics would become nonsensical. Nearly all progress in fundamental physics over the last thirty years would spiral down the drain.

No, information loss was not an option—which meant that Screwed had to be wrong. B was entangled with R, not with A.

Unfortunately, that wasn't any better. Severing the entanglement between B and A is equivalent to inserting a horizon—after all, that's what a horizon does. The vacuum no longer cancels itself out. Instead of emptiness, there are particles. *Hot* particles. Planck-temperature, molten, scalding particles. A firewall.

"Perhaps the most conservative resolution is that the infalling observer burns up at the horizon," the paper concluded. Screwed is even more screwed than we thought.

Now I was starting to sweat. If Polchinski and his crew were right—if B is entangled with R and not with A—we'd have to throw out not only horizon complementarity but *general relativity* as well. After all, the very fact that Screwed should find himself in a vacuum, with nothing out of the ordinary occurring at the horizon, is a consequence of the equivalence principle. Relativity had always said that Safe would see Screwed get burned to a crisp at the horizon. But that was only from Safe's perspective. In Screwed's own frame, nothing bad was ever supposed to happen. He wasn't supposed to feel any heat, just like the guy falling off the roof doesn't feel any gravity. Now this paper was suggesting that Safe's perspective is the *right* perspective. That inertial and accelerated frames aren't equivalent, however happy Einstein's thoughts. And if that's the case, all bets are off.

I didn't know what to think. There was no way the firewall argument could be right, and yet I couldn't see how it was wrong. But it was okay, I told myself—Susskind would. Surely he'd read the paper, spot the flaw, and everything would be back on track in no time. There was no need to panic.

Sure enough, Susskind posted a paper on the arXiv called "Complementarity and Firewalls." I breathed a sigh of relief—the madness was over.

Or so I thought. Weeks later, Susskind retracted the paper, the notice on the arXiv reading merely, "Withdrawn because the author no longer thinks it is correct." Just like that, we were back to firewalls.

Over the next few weeks I kept a close eye on the arXiv, waiting for a solution to appear. Bousso posted a paper: "Observer Complementarity Upholds the Equivalence Principle." Then he withdrew it. Daniel Harlow posted one: "Complementarity, Not Firewalls." Then he withdrew it.

What the hell was going on? I wondered. It seemed that the arXiv had turned into a complete clusterfuck. Papers appearing one day, disappearing the next? Everyone was quick to shoot down the AMPS paradox—which had been named for its authors' last initials—because the firewall scenario seemed so obviously wrong. Yet with every attempt it became increasingly clear that those damn firewalls were not about to be extinguished. The physics community was nearing a panic. I could feel it.

"Just ask them what's going on," my dad said on the phone. "I'm sure they don't really think firewalls are real. They can't. Ask Susskind. Hell, ask Polchinski."

I emailed Susskind first. "Is complementarity in trouble?"

"I don't think the concept of complementarity is in danger," Susskind replied, "although the AMPS paper is pointing to a much deeper understanding of it. I think it's less of a question of 'Complementarity or firewalls?' as it is 'When complementarity and when firewalls?'"

Complementarity, he said, would have to hold up until the black hole evaporation's halfway point in order to prevent cloning. After that, the firewall would take over.

I wanted to feel halfway better, but I had always been more of a black-hole-is-half-empty kind of girl. So I emailed Bousso, hoping for better news.

"I originally thought that complementarity, properly implemented, would get rid of firewalls," Bousso told me. "But I now believe that complementarity is not enough. The answer may be firewalls, but I'm hoping that by asking what it takes to avoid firewalls, we'll learn something deeper, hopefully something about the fundamental description of the infalling observer, and thus of cosmology."

You want a description of Screwed? I thought. Just look around.

Stressed, I emailed one of the AMPS responsible for the mess, namely, P.

"Do you really think that horizon complementarity is incorrect?" I asked.

"I am very puzzled," Polchinski replied. "I do not see any outcome that is fully satisfying. I expected that complementarity would survive in some transmuted form, but as I look at the follow-up papers, I do not see anyone doing any better with this than we did."

I texted my father: *Okay. It's officially time to freak out.*

I sat outside on my balcony, sipping a glass of wine, trying to make sense of things. The Moon loomed large over the city. The air was thick and warm, bathed in the flaxen glow of Jupiter, a steady light in the western sky perched high above the skyline, the river's glassy, dark water catching flecks of lightfall below, the whole city awash in stillness. Everything holding its breath.

I felt like the world we'd spent all these years building was about to crumble—or, more to the point, burn. I had finally gotten the book deal. I was finally ready to write the book my father had invented for us, the one in which we were supposed to solve the riddle of the universe, and now everything we had learned about the universe was evaporating right in front of me.

Could the firewall argument possibly be right? Could horizon complementarity possibly be wrong? It would undo everything. It would mean that spacetime was invariant again, that we'd have to add it back to the IHOP napkin, along with *particles/fields/vacuum*. It would mean that the multiverse could actually exist, that reality was not an approximate concept, that the second observer was not a copy, that the Aleph was not so hypothetical or secret after all. It would mean that coauthorship was legit, my decision to go solo unjustified, my dissertation retroactively bogus. It would mean that Einstein's happiest thought had turned seriously depressing, that heavier vaginas do fall faster than light ones, that the laws of physics aren't the same for every observer, and that there *is* a preferred frame, namely, the one in which you don't get incinerated out of nowhere by a fucking firewall. That there was a real, observer-independent world out there, one that was not nothing but something, something fundamentally and devastatingly inexplicable. It would mean that after seventeen years, I was just as clueless as when this whole thing had started.

From: Leonard Susskind
To: Amanda Gefter
Subject: Firewalls

Dear Amanda,

We are having a meeting at Stanford on firewalls. It was called together on very short notice, but all the "players" will be there. The meeting is small and by invitation only, but I'd be happy to have you attend as an observer.

From: Amanda Gefter
To: Leonard Susskind
Subject: RE: Firewalls

Dear Lenny,

That sounds amazing—I would love to attend as an observer. Though if there's a choice, I'd prefer to be the accelerated observer.

* * *

A few weeks later I packed a bag and headed west again, to Palo Alto.

When I arrived at the conference I saw a slew of familiar faces among the crowd. Susskind was there, of course, as was Bousso. Banks wasn't around but his collaborator, Willy Fischler, was. I spotted Juan Maldacena, Don Page, John Preskill . . . so many brilliant thinkers whose contributions had, over the last few decades, erected an unbelievable theoretical edifice that was now on the verge of collapse. They looked nervous. Polchinski, the unlikely troublemaker, was there, as were A, M, and S. A bust of Einstein watched over the lounge. He looked nervous, too.

When it was time to head into the conference room, I made my way toward the back. "The room is divided into three sections," Susskind had explained to me. "The comfortable chairs up front are for participating physicists. The slightly less comfortable chairs behind them are for physicists who aren't speaking. And the uncomfortable chairs in the back are for observers."

As I sat down and readied myself for the first speaker, I couldn't help but smile. Everything I knew might be wrong and I had been relegated to the uncomfortable chairs—but I had been *invited*. They say there's a first time for everything. I looked around. I wasn't trespassing anymore.

Polchinski opened the proceedings by walking everyone through the AMPS argument. "We thought Lenny would set us straight," he said. "I'm glad to see you're all as confused as we were."

Soon Lenny got up to speak. "I am pretty much deaf and I can't hear you when you ask questions, so don't bother," he began. Everyone laughed. "What?" he asked. "Did you say something?"

But as he spoke, Susskind grew more serious, wondering aloud whether firewalls might be telling us that reality is even more observer-dependent than it seems.

"When I was a young person," he said, "I think what I was thinking was that A equals R. I thought it was redundant to describe both. But I became scared of that after you guys pointed out how crazy that would be. And I'm still scared of it."

That A equals R—that the vacuum mode inside the black hole and the early Hawking particle outside the black hole are two radically different descriptions of the very same bit—made perfect sense to me, fitting, as it did, every intuition I had developed over the course of our journey. As far as I was concerned, that *was* the deep insight of Susskind's horizon complementarity and always had been. It was certainly the point Banks had driven home with his and Fischler's holographic spacetime. But the problem now, the one that AMPS had pointed out, was that there's a time during which Screwed can see *both* descriptions, A *and* R, simultaneously, violating entanglement's monogamy, creating the deadly firewall.

Okay, let's assume for a moment that firewalls are real, said Douglas Stanford, a young physicist, when he stood up to speak. What would that mean? Perhaps what's happening is that the singularity is moving from the center of the black hole out to the horizon. "The black hole has no interior," he said. "The singularity is the edge of space."

Well, sure, that was true for Safe, I thought. For all intents and purposes, there is no other side. The problem is, it *shouldn't* be true for Screwed. Einstein says Screwed is in an inertial frame—there's no edge of space for him. I shifted in my uncomfortable seat. Why was everyone so calm? If the black hole has no interior for Screwed, every profound advance in theoretical physics in recent history would unravel. Why wasn't everyone freaking out?

"At least that way all the weird physics is happening at the singularity, and we don't know what equations apply there anyway." Stanford shrugged.

It was at this point that Andy Strominger, the Harvard string theorist, lost his shit. "You have flat space and a singularity appearing of out of nowhere?" he shouted from his seat, incredulous.

"That's exactly the problem with firewalls!" Bousso shouted in reply.

"Well, I'm just glad you're saying it in such a transparently absurd way!" Strominger yelled, his voice dripping with sarcasm. "I think it's wonderful that someone drew a singularity coming out of flat space with a straight face."

I felt vindicated by Strominger's outrage. This was no time to be polite.

But the mood in the room shifted when a young McGill University physicist named Patrick Hayden gave an inspired talk. The AMPS paradox, he said, is based on the assumption that Screwed can make a measurement on B that will reveal its entanglement with R before he falls into the black hole, where he will find that B is also entangled with A—if it weren't for the firewall, that is. But we have to ask, Hayden says, what would it really take for Screwed to perform that measurement? What would it take to decode the scrambled Hawking radiation and extract the information from the correlation between B and R? In practice, picking out a correlation in Hawking radiation is even harder than, say, looking up a word in a dictionary after it's been burned in a fire. The Hawking radiation is *seriously* scrambled. Decoding its information would require the most powerful computer imaginable. Namely, a quantum computer.

Quantum computers exploit the power of quantum superpositions to quickly perform calculations that an ordinary computer couldn't pull off even given billions of years. Whereas an ordinary bit of information— the kind manipulated by an ordinary computer—is either a 0 or a 1, a quantum bit, or qubit, can be a 0, a 1, or a superposition of 0 and 1 at the same time. As you add more qubits, the number of simultaneous states a quantum computer can occupy grows rapidly. Ten qubits can be in 1,024 states simultaneously. Twenty qubits can be in well over a million. Three hundred qubits can be in more states simultaneously than there are particles in the universe. That a quantum computer can perform so many calculations simultaneously means that it can, in principle, factor large numbers into primes, search vast databases in an instant, and, quite possibly, decode a cloud of Hawking radiation. Who cares that the largest quantum computers built to date had only a handful of qubits? The question, Hayden said, is about what is measurable *in principle*. Quantum computation is as good as computation gets. If a quantum computer can't compute something, it's not computable. Period.

"The argument in the AMPS paper is that hypothetically we can decode the radiation and then jump into the black hole," Hayden said. "But in the spirit of complementarity we have to be as operational as possible. . . . Can you decode it on a quantum computer? And if so, on what time scale?"

Standing at the blackboard, Hayden took us through a number of calculations, showing what it would mean to put the Hawking radiation through a series of universal 2-qubit gates in a given amount of time. His conclusion? "Decoding the radiation would take exponential time." That is, for every additional bit of information, the amount of time Screwed needs to decode the radiation increases exponentially. Given a black hole of any size, by the time Screwed and his quantum computer are finished decoding—by the time he can figure out that B is entangled with R—the black hole has long since evaporated and the threat of a firewall is long gone.

The next morning, before the meeting resumed, I spotted Hayden and Harlow sitting on a couch in front of an equation-clad chalkboard. I was eager to ask Hayden about his talk, which I hadn't been able to get out of my mind all night.

"Just because you can't decode the information, does that really mean it's not even there?" I asked. "I mean, just because we can't measure that B is entangled with R, does that automatically mean B isn't entangled with R?"

"Think about it like quantum complementarity," Hayden said. "There you want to say, just because you can't measure position and momentum simultaneously doesn't mean the particle doesn't really have a position and momentum simultaneously. But that's exactly what it means. It could be the same kind of thing here."

It was a good point, I thought as I headed back into the conference room. Still, one would be inclined to wonder why the hell what is operationally possible should have *any* connection to what is ontologically existent. It did—quantum mechanics made that perfectly clear. But why? If reality came in the Einsteinian brand—sitting out there, independent of observation—then the connection would be inexplicable. There was only one way to explain why what we can *know* determines what can *exist*: reality is radically observer-dependent. And if reality is radically observer-dependent, I thought, then firewalls are going down.

Inside, Harlow took the floor and expressed his agreement with Hayden. "It seems there is a fairly robust conspiracy preventing [Screwed] from measuring R. . . . Even though they are in the same light cone, they are not computationally accessible to a single observer."

Echoing Susskind's speculation that A equals R, Harlow wondered if we ought to adhere to "strong complementarity." "Strong complementarity would say, the way to think about this is that [Safe] has some quantum mechanical theory and [Screwed] has some quantum mechanical theory and there's some criteria about how much they have to agree. But they only have to agree on things they can both measure."

I was grinning in my uncomfortable chair.

Ordinary complementarity says that when there's an event horizon around, you have to restrict to a single observer's reference frame, rather than taking an unphysical God's-eye view. It was the bold claim that rendered spacetime observer-dependent. *Strong* complementarity took things to a whole new level. It says that you have to restrict to a single observer's reference frame *regardless of whether there's an event horizon*. After all, in the AMPS scenario, the discrepancy in Safe's and Screwed's descriptions occurs in a region where the two observers are not yet separated by a horizon. Strong complementarity doesn't just make spacetime observer-dependent—it makes *everything* observer-dependent.

On the other hand, Harlow said, his face darkening, "that doesn't seem to agree with AdS/CFT, because that suggests there should be one quantum mechanical description that you can put in one Hilbert space."

But we don't live in AdS! I silently protested. Banks was right: they were trying to make the fifth sequel to *Saw*. We live in de Sitter space, which, unlike AdS, has observer-dependent horizons. If we want to understand cosmology, we have to stop tailoring everything to AdS and deal with *this* universe.

"Somehow these two things, [A and R], have to be the same operator," Harlow said, "but I have a love/hate relationship with that idea."

I could see Susskind nodding. "We all do."

From: Leonard Susskind
To: Amanda Gefter
Subject: Watch the arXiv

Amanda,

Watch the arXiv early next week for papers by Harlow-Hayden and by me. Something is happening.
Lenny

Something is happening?

What exactly does one do when one gets an email from one of the greatest living physicists reading only, "Something is happening"? Apparently one runs laps around one's living room, then jumps up and down in front of an annoyed cat, shouting, "Something is happening! Something is *happening!*"

"What do you think is happening?" my dad asked on the phone. Six weeks had passed since the Stanford meeting.

"What do I think? I think he resolved the firewall paradox, and I think he did it by making things even more observer-dependent. I think he gave in to A equals R and strong complementarity, and he knows he can do it because of Hayden and Harlow's claim about quantum computation."

Come Monday, I was perched in front of my computer hitting the refresh button on my arXiv search, waiting for something new to pop up under Susskind's name. At 9:30 P.M. it did: "Black Hole Complementarity and the Harlow-Hayden Conjecture."

I read it as quickly as I could. I could barely see past my own smile.

"Bousso and Harlow have [advocated] a strong form of complementarity that can be described by saying each causal patch has its own quantum description," Susskind wrote. "In [Screwed's] quantum mechanics, B is entangled with A and not with the outgoing radiation. In [Safe's] description, B is entangled with R. . . . It's obviously premature to declare the paradox resolved, but the validity of the HH conjecture would allow the strong complementarity of Bousso and Harlow to be consistent, without the need for firewalls. For these reasons I believe that black hole complementarity, as originally envisioned by Preskill, 't Hooft, and Susskind-Thorlacius-Uglum, is still alive and kicking."

Susskind said it was premature to declare victory, but as far as I was concerned, Susskind, Bousso, Harlow, and Hayden had put out the fire. And with firewalls out of the way, my father's and my mission was back on track.

At the same time, I realized there was a lesson to be learned from the whole fiasco. The firewall paradox wasn't trying to tell us something about black holes. It was trying to tell us something about quantum mechanics.

If each observer has his or her own quantum description, as strong complementarity demanded and as Banks had been claiming all along, we were going to need a new understanding of quantum physics. In ordinary quantum theory, there's one Hilbert space and entanglement is absolute. In a firewall-free, holographic world, there's one Hilbert space *per observer* and entanglement is relative to a given reference frame. Things were going to have to change.

Luckily, I was pretty sure it was exactly the kind of change we needed to finally get to the bottom of ultimate reality and the origin of existence. In our quest, my father and I had found that invariant after invariant gave way to observer-dependence. But every clue we uncovered was rooted in the assumption that reality is governed by quantum mechanics. Hawking radiation, the holographic principle, top-down cosmology, M-theory, holographic spacetime, strong complementarity—every last one of them assumed quantum mechanics from the start. If they withered down the ontology of the universe, it was because quantum mechanics withered down the ontology of the universe. I could see clearly now that if we wanted answers, they were going to come from one question alone, the same one that Wheeler had asked that day in Princeton, the one he had decided to focus on when he knew time was ticking down: *How come the quantum?*

15

Into the Margin

"If different observers give different accounts of the same sequence of events, then each quantum mechanical description has to be understood as relative to a particular observer. Thus, a quantum mechanical description of a certain system (state and/or values of physical quantities) cannot be taken as an 'absolute' (observer-independent) description of reality, but rather as a formalization, or codification, of properties of a system *relative* to a given observer. . . . In quantum mechanics, 'state' as well as 'value of a variable'—or 'outcome of a measurement'—are relational notions."

As I read Carlo Rovelli's paper, a gospel choir sang "Hallelujah" in my head.

How had I never heard of this before? It was so simple. It was so brilliant. It was exactly what we needed.

As Wheeler had emphasized in his journals, the central problem of quantum mechanics was coauthorship—the problem of the second observer. Or, as Wheeler put it, "What happens when several observers are 'working on' the same universe?" It was precisely this problem that Rovelli had set out to solve in his 1997 paper "Relational Quantum Mechanics," which I had stumbled upon amid a desperate search of the physics literature for some new insight into the quantum mystery.

Rovelli began by comparing the problem of the second observer to the problem of Lorentz transformations in special relativity. To account for the fact, first observed in the 1887 Michelson-Morley experiment, that all observers measure light to be moving at the same speed regardless of their own state of motion, Hendrik Lorentz proposed that objects will physically contract or stretch in exactly the right proportion to cancel out the effect of an observer's motion and leave the speed of light constant. That was in 1892, more than a decade before Einstein published his special theory of relativity. These Lorentz transformations successfully accounted for the constancy of light's speed, but you only had to think about it for a second to realize that it was completely fucking insane. If I'm measuring how long it takes a light beam to travel the length of a road, and I'm running down the road as I measure it, how the hell would the road *know* to shorten itself by the exact right amount to cancel out the effects of my speed relative to the light's speed as it conspires to fool me into thinking that light always travels at 186,000 miles per second? Not to mention the question of what physical process the road would employ to shorten itself at will. Lorentz transformations produced the right answers, but, like quantum mechanics, they seemed bat-shit crazy.

If the equations of special relativity had already been written down by Lorentz in 1892, Rovelli asked, "What was Einstein's contribution? It was to understand the physical meaning of the Lorentz transformations." Lorentz had the right structure but the wrong story. It was, Rovelli said, "quite an unattractive interpretation, remarkably similar to certain interpretations of the wavefunction collapse presently investigated. Einstein's 1905 paper suddenly clarified the matter by pointing out the reason for the unease in taking Lorentz transformations seriously: the implicit use of a concept (observer-independent time) inappropriate to describe reality."

In other words, it wasn't that the lengths of objects were magically changing to fool observers. It was that space and time were observer-dependent concepts. Give up their invariance, and suddenly everything starts to make a lot more sense.

Could a similar reinterpretation of quantum phenomena make sense out of all the bizarre, ad hoc explanations for wavefunction col-

lapse and the paradox of the second observer? To Rovelli, previous solutions, such as Bohr's ontological divide between observer and
observed or Wigner's privileging of consciousness as some metaphysical force, "look very much like Lorentz's attempt to postulate a mysterious interaction that Lorentz-contracts physical bodies."

"My effort here is not to modify quantum mechanics to make it
consistent with my view of the world," Rovelli wrote, "but to modify my
view of the world to make it consistent with quantum mechanics."

So what was it that had to be modified? "The notion rejected here
is the notion of an absolute, or observer-independent, state of a system; equivalently, the notion of observer-independent values of physical quantities." In other words, Rovelli wrote, "a universal observer-
independent description of the state of affairs of the world does not
exist."

That day back in IHOP when my father and I made our list of possible ingredients of ultimate reality, we didn't bother listing reality itself. "It would be like listing cake as an ingredient in cake," my father
had said. But now it was looking as though we should have, if only to
have the mind-melting pleasure of crossing it off the list. According to
Rovelli, reality itself was observer-dependent. Which meant, however
insane it sounded, that reality itself *wasn't real*.

Once you abandon the notion of observer-independent quantum
states, the paradox of the second observer disappears. After all, the
paradox comes from the contradictory descriptions that Wigner and
his friend give of the same events. But the descriptions are contradictory only if we assume that there's a single reality they are both describing. Wigner's friend says the atom's wavefunction collapsed; Wigner
says it hasn't and that the atom and his friend are now in a superposition state. Which is it *really*? According to Rovelli, there's no *really*. It
collapsed relative to Wigner's friend. It didn't collapse relative to
Wigner. End of story.

"Bohr and Heisenberg's key idea that 'no phenomenon is a phenomenon until it is an observed phenomenon' must therefore apply to
each observer independently," Rovelli wrote. "This description of physical reality, though fundamentally fragmented . . . is complete."

On one hand, I wasn't surprised. Or at least I shouldn't have been. Everything I had learned up until now had fully prepared me for this moment. Is the elephant dead and crispy outside the horizon or alive and terrified inside? It depends who you ask. There's no God's-eye view that contains the "truth" of the matter. The "truth" of the matter is observer-dependent. On the other hand, Rovelli did seem to be taking the observer-dependent thing to a whole new level. He was making *everything* observer-dependent, and reinventing quantum mechanics in the process.

Fundamental physics proceeds by paradox. It always has. It was a paradox that led Einstein to relativity: the laws of physics had to be the same for everyone and, given the relational motion of light, the laws of physics couldn't be the same for everyone. A paradox led Polchinski to D-branes: open strings had to obey T-duality and, given their boundary conditions, open strings couldn't obey T-duality. Another paradox led Susskind to horizon complementarity: information had to escape a black hole and, given relativity, information couldn't escape a black hole. And yet another led the entire physics community to wonder whether each observer has his or her own quantum description of the world: entanglement had to be monogamous and, given the equivalence principle, entanglement couldn't be monogamous.

There's only one way to resolve a paradox—you have to abandon some basic assumption, the faulty one that created the paradox in the first place. For Einstein, it was absolute space and time. For Polchinski, it was the immovability of the submanifold to which the open strings attached. For Susskind, it was the invariance of spacetime locality. For everyone involved in the firewall mess, it was the idea that quantum entanglement is observer-independent.

Quantum mechanics short-circuits our neurons because it presents yet another paradox: cats have to be alive and dead at the same time, and, given our experience, cats can't be alive and dead at the same time. Rovelli resolved the paradox by spotting the inherently flawed assumption: that there is a single reality that all observers share. That you can talk about the world from more than one perspective simultaneously. That there's some invariant way the universe "really is."

I called my father and we discussed Rovelli's paper for hours, debating its implications for the meaning of ultimate reality until the Sun came up. Or until the Sun came up *relative to me.*

Rovelli's relational quantum mechanics, which rendered wavefunction collapse observer-dependent, allowed him to tackle the thought experiment that Einstein himself had waged against quantum mechanics, hoping it would rip the theory apart at the seams: EPR.

A staunch realist, Einstein resented quantum theory's claim that somehow a particle doesn't have properties until it's measured. If quantum mechanics is probabilistic, he said, the probabilities reflect our subjective ignorance, not some objective uncertainty in reality itself.

Einstein (E), along with Boris Podolsky (P) and Nathan Rosen (R), proposed a thought experiment to prove it. The gist of the EPR experiment was simple: you have two quantum-entangled particles, say an electron and a positron. Because they're entangled, the two particles are described by a single wavefunction, which might have zero total spin. Consequently, if the electron's spin is measured up, the positron's must be down, and vice versa. Their spins have to be anti-correlated so that they always add to zero.

An entangled electron-positron pair is created somewhere in the middle of Connecticut. The particles part ways. One arrives at my door here in Boston; the other makes its way down toward Philadelphia. I decide to determine the electron's spin. Spin can be measured along any spatial direction: *x, y,* or *z.* I choose *x:* my electron's spin is up. Meanwhile, in Philadelphia, my father is about to measure the positron's spin along the *x*-axis. But the outcome of his measurement is already determined: it has to be down. He makes the measurement a mere fraction of a second later. Sure enough, it's down.

Einstein had a serious problem with this. How did my father's particle "know" about my measurement across several state lines? Any signal that my particle could have sent to my father's would have had to travel faster than light to get to Philly in time for the measurement. But of all people, Einstein wasn't about to allow for superluminal speeds, or what he called "spooky action-at-a-distance." The only reasonable explanation, according to EPR, was that the particles each had well-determined spins all along: the electron's spin was up before I measured it and the positron's spin was down from the start. There was some way that things *really were,* independent of either of our observations. After all, we could have chosen to measure our spins along the *y* or *z* axis—so the particles had to have set spin values along each axis from the get-go. These determined outcomes, or "hidden variables," aren't reflected in the formalism of quantum mechanics. Ergo, said EPR, quantum mechanics is incomplete. Probabilities represent uncertainty in our knowledge, and not in reality itself.

Unfortunately for Einstein, John Stewart Bell shattered EPR's reality. He calculated that any hidden-variables theory would produce the wrong probabilities for the outcomes of multiple measurements—unless, that is, the hidden variables operated via spooky action-at-a distance. The only way to save a reality out there, independent of observers, was to violate the locality at the heart of relativity.

Bell's theorem was tested over and over in different ways in labs around the world, all of which came up with the same result: Einstein was wrong. After a particularly damning lab test in 2007, *Physics World* ran an article headlined "Quantum Physics Says Goodbye to Reality."

Even though Einstein had set the whole quantum revolution in motion, he couldn't accept what the theory was telling him about reality. You'd think he would have learned to ignore his philosophical prejudices after the whole expanding-universe fiasco. But no. He wanted to retreat back behind that thick plate-glass window where he could passively observe a reality that couldn't care less that he was observing it. But it was too late. Quantum mechanics had already smashed the glass, and there was Bell, stomping on the shards.

Bell showed that my electron has no defined spin—it's in a superposition of spin up and spin down—until it randomly chooses a value upon measurement. By randomly choosing a value, it also sets the value for my father's positron. *Instantaneously.* Faster than light. Since there's no way anyone could use this superluminal effect to transmit information, it wasn't a flagrant violation of relativity. But it was walking a pretty fine line.

Ever since Bell, physicists have surrendered to spooky action-at-a-distance. That's just the weird way things are, they said. But it never seemed quite right. And now, reading another paper by Rovelli, I understood exactly why.

"Einstein's reasoning requires the existence of a hypothetical superobserver that can instantaneously measure the state of [Amanda] and [her father]," the paper stated. "It is the hypothetical existence of such a nonlocal superbeing, and not quantum mechanics, that violates locality."

The point was this: the EPR problem arises because it seems like by collapsing the wavefunction of my electron, I mysteriously collapse the positron's wavefunction several states away. But relational quantum mechanics tells a different story. When I measure my electron, its wavefunction collapses *relative to me*. As far as my dad is concerned, the electron's wavefunction hasn't collapsed at all, and I'm in a superposition of "having measured spin up" and "having measured spin down." Nothing superluminal is going on. He can take a trip up to Boston and collapse that superposition to find out that the electron's spin is anti-correlated with his positron's, but that's a totally legit, local quantum interaction. From my point of view, nothing happens faster than light. From my father's point of view, nothing happens faster than light. The only way to see anything happen faster than light is to be a third observer who can see what's happening in Boston and Philly simultaneously, which is impossible. As long as you restrict to what individual observers can see, no laws of physics are violated.

Of course, this sounded familiar. What was the solution to the black hole information-loss paradox? Realizing that no single observer can see both sides of a horizon. The solution to the apparent backward causation in top-down cosmology? Realizing that no single observer can be outside the universe to see causality violations. The solution to the firewall paradox? Realizing that no single observer can witness polygamous entanglement. Now the solution to the EPR paradox? Realizing that no single observer can see both measurements simultaneously. I couldn't be sure, but I was sensing a pattern here.

* * *

I called Rovelli at his home in Marseilles, in the south of France.

"Would it be fair to sum up the situation by saying that the interpretational difficulties in quantum mechanics come from trying to describe the world from an impossible view from nowhere?" I asked.

"Yes," Rovelli agreed enthusiastically. "If we can renounce the view from nowhere, the view from the outside, and accept the idea of referring to observers, then the difficulties all go away. That's what I think."

"Okay, so you always have to restrict to one frame of reference at a time, and the quantum weirdness arises when you compare your measurements to someone else's. But what about interference, like in the double-slit experiment? How is it that within a single reference frame you can see an interference pattern?"

"I like this question," Rovelli said. "It forces thinking! When we say *interference*, what do we mean? In order to have interference, we need two of something to interfere. In the double-slit experiment, the interference is between the component of the electron passed through one slit and the component that passed through the other slit. But there is no electron passing through any slit in nature. This is just our language to mean 'If I was at the slit and was measuring, I would see the electron here or there.' *Interference* is not a term denoting what is actually happening. It is a term referring to a comparison between what is observed by one observer and what would be observed by another observer. In your language, it is a term denoting the comparison between two different reference frames."

That made sense. Interference was the interference of phases, and phases were just reference frames. Points of view. The strangeness of superposition wasn't the strangeness of many *worlds*. It was the strangeness of many *frames*.

"As I've been looking at a number of different developments across physics, I'm coming to the realization that you always run into trouble when you try to describe physics from an impossible God's-eye view," I said, "and that for anything in physics to make sense you have to define it in terms of a single observer's reference frame. It's starting to seem like there is one universe per observer. Like there's no way to talk about *the* universe."

"I understand what you mean," Rovelli said. "This is the challenging aspect of modern physics. It forces us to renounce the previous image of a clear, objective, well-defined, perfectly describable state of affairs of the world. Quantum mechanics has asked us to give this up. This does come with a heavy metaphysical burden. Are we ready to rethink the universe in these terms? I think we have to take our physics seriously."

"It's interesting that a lot of the progress physicists have made in understanding reality has come from accepting that more and more things are relative," I mused, thinking of all the invariants that had slipped right off the IHOP napkin.

"Exactly!" Rovelli replied excitedly. "Exactly. When people heard that the Earth was round, it was very complicated conceptually to accept it. How could the people in Sydney be walking upside down? Then eventually people understood, there's no real up and down; they are relative. And they got used to it. Then it was hard to understand that motion was relative. Then it was hard to understand that simultaneity was relative. And I think quantum mechanics is a step in the same direction. It's telling us that the world is even more relative than expected. If one observer sees spin up, that doesn't necessarily mean that it's fixed forever for everyone else."

Not fixed for everyone else. I couldn't help but think of Wheeler. *How preposterous to think that each has to invent the universe afresh.* "Did John Wheeler ever come across your work on relational quantum mechanics?" I asked. "I just spent time reading through his journals, and for years he seemed to be struggling with the question of multiple observers in quantum mechanics. He would ask things like, 'Why does my measurement of the electron's spin fix it for every other observer?'"

"Absolutely, yes," Rovelli said. "I had a very warm relationship with John Wheeler. He got interested first in my work on quantum gravity. He sent me a very enthusiastic letter in his typical flowered style, which is today hanging on the wall of my office, inviting me to Princeton to deliver one of the first talks on loop quantum gravity—which, needless to say, raised unhappy criticisms from Ed Witten and David Gross, who were sitting uncomfortably in the audience."

Criticism? From David Gross? I couldn't imagine it.

"Later, when I wrote the relational quantum mechanics paper, he reacted again enthusiastically and sent me a beautiful letter and a package of everything he had written on the argument, all tied into a handmade folder with an orange cover and a picture of his preferred 'it from bit' image on the cover. My relational paper owes a lot to Wheeler's intuition, obviously. It was Wheeler who understood that information had to be the right concept. But he was always thinking about *the* observer, observing the world and making it happen and being part of it. He was focused on the circular path that this structure generated, and did not find a way to make sense of the relation between observers. I think he did appreciate my opening 'the observer' into a multiplicity of observers, with any physical system being one, and especially the analysis of the coherence that quantum theory gives to the information of different observers. But he was aged when my paper came out, and I don't think he wrote anything on it.

"I met him again later," Rovelli continued. "He was extraordinarily warm and friendly with me, and I can't forget his eyes. We went for a long walk. He was talking a lot, but with a very soft voice, which I could barely hear. He was continuously stopping and looking at me. I felt a lot of affection from him. The first day I arrived in Princeton, he came early in the morning to pick me up at the bed-and-breakfast where I was staying—I was still sleeping when he rang me up from the reception—to have breakfast with me. He then took me for a walk toward the Institute. We started walking in silence and then he said to me, 'Carlo, I did this another time.' I said, 'What?' He said, 'Coming to pick up somebody early for breakfast and walk him for the first time toward the Institute.' I looked at him, and he said, 'It was Albert Einstein, when he escaped the Nazis and arrived here. I received him, like I receive you today.' He was like that, always emphatic, but always capable of touching you deeply. He showed me the rooms where he first discussed the atomic bomb and where he talked to Einstein about writing to Roosevelt. . . . Sorry for getting lost in these memories. John has always been my hero, and you can imagine how I felt, as a young man, when my hero wrote to me a letter full of compliments. I keep all his sweet cards. The last one of these is dated '95, which is when I wrote the first relational paper. It says: 'Dear Carlo, continuing plea-sure that you exist in this puzzling world of ours. Warm good wishes, John.'"

My eyes welled up as Rovelli recounted his times with Wheeler. I worried that if we talked much more about him I'd start sobbing into the phone like a weirdo, so I changed the topic.

"It seems like in relational quantum mechanics it must be impos-sible for observers to measure themselves," I said.

"It's a fascinating point," Rovelli replied. "It has enormously raised my curiosity, and I've spoken with philosophers who were interested in that. I've never come to total clarity about it. Indeed, the whole rela-tional view is somehow related to the impossibility of total self-measurement. The entire structure of quantum mechanics tells us that our information is always limited. The quantum mechanical world is intrinsically probabilistic; we only have partial information about things. Formally, if an observer could measure himself completely he could violate quantum mechanics. But I can't articulate that. It's some-thing I find fascinating, though."

"There seem to be hints of a relationship to Gödel's incomplete-ness," I suggested.

"Yes," he said. "Absolutely. I just haven't been able to work it out."

Wheeler hadn't been able to work it out, either, the whole Gödel busi-ness. He had sensed a profound connection between propositional logic, self-reference, and quantum mechanics. In his mind, determin-ing the truth-value of a proposition, such as *Snow is white* or *My pants are on fire,* was akin to collapsing a quantum wavefunction. In his vi-sion, if all observers—all who had ever lived, were living, would live—collectively assigned values to enough Boolean yes/no propositions, together we could build the universe. But the flaw in Wheeler's plan was becoming ever clearer: there is no collective universe. My pants may not be on fire relative to me, but they could well be on fire relative to some other point of view. As Rovelli had so elegantly demonstrated, wavefunction collapse, along with the truth-value of a proposition, was observer-dependent. Wheeler wanted co-authorship, but reality wasn't having it. When he drew his U-diagram, Wheeler had assumed that the giant eye looking back at itself was a stand-in for a multitude of eyes, for countless observers all gazing at one and the same universe. A giant eye for a God's eye, one that encompassed every possible reference frame simultaneously. But if I had learned anything from black hole physics, horizon complementarity, top-down cosmology, the firewall paradox, and now relational quantum mechanics, it was this: there's just one eye. One eye per universe, with the cardinal rule that you can't talk about more than one at a time. Either you were in Safe's reference frame or you were in Screwed's. Mine or my father's. Observers could never see across more than one reference frame. If they could, physics would break down. Wheeler knew that this was a participatory universe—he just thought there could be more than one participant at a time. For a man who was on a mission to go everywhere, talk to everyone, and ask anything, nothing was worse than the specter of solipsism: one man (one worm, one rock), one universe.

"The Einstein-Rosen-Podolsky 'experiment' is valuable among other reasons because it shows *two* observers participating in the making of reality," Wheeler had written in one of his journals. "We are not concerned, as some are, with turning back from quantum mechanics, but with going on, to two or more observers, two and more 'systems,' two and more observations, ultimately to see how the iron pillars and papier mâché are combined to make reality. What does the EPR experiment show?"

Rovelli had the answer: the EPR experiment showed that there's no single reality shared by all observers. Everyone's stuck having to papier-mâché a world for himself.

Wheeler knew that Gödel's incompleteness theorem was hiding something, some clue to understanding quantum mechanics and the universe, but he was looking in the wrong place: on the outside. Even Gödel himself had made the same mistake. He wasn't particularly distraught over incompleteness, because he was pretty sure that we could decide the undecidable from here, outside mathematics. Statements like "This statement can't be proven by this mathematical system" were undecidable from within the system, but looking down on it from the outside, we could still claim it to be true. Only that claim couldn't be a mathematical claim—you can't be doing mathematics outside mathematics. It had to be something else. Something flimsier. Something like "intuition." Intuition, Gödel said, was a valid enough decider of truth and falsity. Like Wheeler, he was hopeful that we could always impose truth-values from the outside; he believed in the power of the human mind to compensate for the deficits of our mathematical systems. Then again, he also starved himself to death because he was convinced his food was being poisoned, so he wasn't exactly the epitome of optimism.

Everything I had learned about physics led me to my own intuition: there is no outside. You can't step outside mathematics, the universe, or reality. They are one-sided coins.

Wheeler knew that the universe was a one-sided coin, but he wanted to flip it all the same. The tension he saw between the individual and the collective, the inside and the outside, the self-excited circuit and the Gödelian observer knocking at the door was *the* tension, the *ultimate* tension, which sat deep in reality's core, the buried heart of physics, a dense, gnarled pit that bore the universe's impossible form. Quantum mechanics needed external observers to collapse wavefunctions, but general relativity did away with external observers. Quantum gravity had to resolve that paradox while obeying Smolin's slogan: *The first principle of cosmology must be "There is nothing outside the universe."*

How else could an impossible object come to exist but from within its very own architecture? Where else but inside could the seed of creation lie for an object with no outside? The universe *had* to be a self-excited circuit, the spark of creation igniting in its belly, existence that hoists itself into being by its bootstraps. If observership was a prerequisite for existence, then the universe had no choice but to observe itself.

I thought back to what my father had said all those years ago: *There's something about reality you need to know. You think there's you, and then there's everything else outside you, but it's all just one thing.* We like to think of ourselves as standing apart from the world, from nature, interlopers who mysteriously awoke one day to find ourselves in a universe that is something other than us. But we are pieces of the universe, fleeting patterns that the universe momentarily indulges, then dissipates. We are, as Wheeler envisioned, the universe looking at itself. How are we to hold a mirror to ourselves when we ourselves are the mirror?

We are stuck inside the universe—which means we can't give a consistent description of the universe without also describing ourselves. But Gödel's theorem showed that self-referential statements can't be proven from within the system that's stating them. What, then, of cosmology's self-referential statements? "Within" is all we have. They simply can't be proven. In physics, "proven" means "measured," and measurements are about gathering information. The Gödelian incompleteness of the universe seemed to place fundamental limits on the amount of information we can access. If self-referential statements can't be proven by physical measurements, then observers can't measure themselves.

As Rovelli had confirmed for me, "the whole relational view [of quantum mechanics] is somehow related to the impossibility of total self-measurement. The entire structure of quantum mechanics tells us that our information is always limited." He wasn't the only one. Bousso had said as much when describing the failure of the S-matrix to describe cosmology: he called it part and parcel of "the more general problem that arises whenever one part of a closed system measures another part. . . . Obviously the apparatus must have at least as many degrees of freedom as the system whose quantum state it attempts to establish." Indeed, the philosopher of science Thomas Breuer used a Gödelian argument to prove that "no observer can obtain or store information sufficient to distinguish all states of a system in which he is contained."

If an elephant could measure itself, collapsing its own wavefunction, then it needn't exist relative to anything outside itself—in other words, it would simply, inherently exist. It wouldn't be observer-dependent. It would just *be*. In an act of self-affirmation or quite possibly suicide, Schrödinger's cat would collapse its own wavefunction before anyone opened the box. But quantum mechanics—through the uncertainty relations, complementarity, EPR—has already proven that if we assume that elephants inherently exist in some objective, observer-independent way, *we get the wrong answers*.

By relativizing everything, Rovelli had rejected any kind of ontological distinction between observer and observed, leveling the playing
field to a quantum monism where every perspective is a possible reference frame, none of them any better than the next. That did away with
the seeming paradox that an observer can't be a subject and object simultaneously, and yet somehow the observer is a subject and object
simultaneously. I am the subject *relative to me*. I am an object *relative
to my father*. There's no God's-eye view from which both would appear
true at the same time. But again, that hinged on the impossibility of
self-measurement. If I could measure myself, I'd be both subject and
object and quantum physics would fall apart. The prohibition on self-
measurement upheld Wittgenstein's intuition that "the subject does
not belong to the world: rather, it is a limit of the world."

Rovelli had shown that quantum mechanics seems bat-shit crazy
as long as we assume the existence of a single reality shared by multiple observers. Give up that notion and all the quantum weirdness begins to make perfect, non-spooky sense. We can dissolve the problem
of the second observer by embracing the cosmic solipsism that physics
demands. It's not the brand of solipsism that Everett or Wigner momentarily considered, in which there is just one absolute observer. The
solipsism that radical observer-dependence implies is itself observer-
dependent—as Rovelli emphasized, observer in one frame is observed
in another.

But that's only true if observers can't measure themselves. If they
could, quantum states would be absolute, global logic would be Boolean, interference patterns would disappear, quantum monism would
splinter into a dangerous dualism, Ladyman's realism would give way
to Einstein's, the moon would hold steady in an invariant sky, and my
father and I would hang our heads in defeat, because we would in fact
be working on the same universe, one that is something, not nothing,
something whose existence would forever go unexplained. Thank God
for Gödel.

Everyone had always taken Gödel's theorem to be a deeply pessimistic statement about the limits of knowledge. But in a universe that
is nothing, limits are exactly what we need.

I had already learned the implications of limited viewpoints when it came to horizons—horizons mark the edges of an observer's reference frame, and the area of the horizon is a measure of how much information that observer can ever access. Now I saw that the self-referentiality inherent in a universe that contains its observers on the inside also limited an observer's information—it was a kind of logical horizon. Was our positive cosmological constant—our de Sitter horizon—some kind of physical manifestation of a Gödelian incompleteness?

How interesting, I thought, that the shift from invariance to observer-dependence always seemed to be sparked by the discovery that some feature of nature long thought to be infinite, or perhaps zero, was actually finite. In relativity, the finite speed of light, long believed to be infinite, rendered space and time observer-dependent. In quantum theory, Planck's finite constant was long held to be zero; it rendered all physical features connected by uncertainty relations observer-dependent. Most recently, the discovery that the entropy of a region of spacetime, which everyone had assumed to be infinite, was actually finite, made spacetime itself observer-dependent. The speed of light, Planck's constant, entropy—they all represented nature's most fundamental limits. The limits were the clues. If we could find the limits, we could find reality. Or the lack thereof.

Wheeler believed that information, binary bits related by the logical rules of the calculus of propositions, were the atoms that constituted reality. "Logic as building material," he had scrawled. But logic had turned out to be observer-dependent—"yes" in one reference frame looked like "no" in another. Back in the lounge at the Tribeca Grand, Fotini Markopoulou had told me that we needed to use non-Boolean logic—observer-dependent logic—to account for the fact that each observer has only partial information. Boolean logic was just the ordinary logic we usually assume holds true, with its basic rules, like if p is true then not-p is false, or if p implies q and p is true then q is true. There was also the crucial law of the excluded middle: a proposition, p, is either true or false; there's no third option. Non-Boolean logic— *quantum* logic—openly defies the law of the excluded middle. A proposition p can be true *and* false, depending on who you ask.

But I now saw that logic becomes non-Boolean only when you

compare the perspectives of two or more observers. According to any one observer, p is either true or false. We only violate the law of the excluded middle when we try to view p from more than one reference frame at the same time. Classical logic tells us that the particle passed through one slit or the other. Non-Boolean logic offers a third option: it went through both. But the point is, there's no observer who can see it go through both. That would require an impossible God's-eye view, akin to seeing inside and outside a black hole's horizon. *No observer sees both elephants.* Look at the slits and you'll see that it only goes through one. *The phrase "both elephants" is totally misleading.* Statements such as "The photon travels two paths simultaneously" are wrong. They assume there's some singular reality, a way things "actually are." There isn't. Nature has shown us otherwise. Here is what we know: when we compare two possible perspectives of the photon's path, mistakenly assuming a singular reality that both perspectives share, it *looks* as though the photon travels two paths simultaneously. It *looks* as though logic is non-Boolean.

We never see simultaneously-alive-and-dead cats because superpositions represent a multiplicity of points of view, and, by definition, a given observer has only one. Superpositions reek of a God's-eye view. We see *evidence* of superpositions in interference patterns, but interference patterns, as Rovelli said, are comparisons of multiple reference frames. And the key thing, I now realized, was that if reality weren't observer-dependent, we wouldn't see interference. Every perspective would be equivalent; you could map one onto the next, each proposition value lining up perfectly with the others, true on true, false on false, in the same way that perspectives line up in Susskind's FRW universe, straight line on straight line, an alignment that signals invariance and ultimate reality. If, however, reality is radically observer-dependent—if, that is, the universe *is* nothing, then we *need* interference to literally cancel out the disagreements between our perspectives. Interference—the physical manifestation of non-Boolean logic—exists because nothing is real. Or because reality is nothing. With this "many frames" interpretation of quantum mechanics, that bat-shit experiment was actually starting to make sense.

Logic had to be non-Boolean for the same reason that gravity ex-

ists. It was like a logical gauge force. In general relativity, every observer's local patch of spacetime is flat, but when you try to stitch
together many local patches, they don't always line up properly and
you end up with a warped spacetime that gives rise to the force of gravity. Likewise, when it came to quantum measurements, each individual observer's local logic was Boolean—logic became non-Boolean
only when you tried to piece one reference frame together with other
points of view to form a single reality. The truth-values of propositions
don't match up. Just as gravity exists to account for why an inertial
observer appears to be accelerating from another point of view, quantum interference exists to account for why a true proposition appears
false from another point of view. Stitched together, the local logics create a warped logical space. Non-Boolean logic is a fictitious logic.

Proposition: *Snow is white.* Truth-value according to me: *yes.*
Truth-value according to some other guy: *no.* In the old-school Boolean view of the world, our mismatched information spelled catastro-
phe. But physics is now telling us that we can't talk about both truth
assignments at once. They are noncommuting gauge copies, two
quantum-mechanics-violating clones of the same elephant, two different its from the same bits. As Susskind had said, it all came down to
the misuse of the word *and.* Not yes and no—yes *or* no. "Quantum
gravity may not admit a single, objective, and complete description of
the universe," Bousso had written. "Rather, its laws may have to be
formulated with reference to an observer—no more than one at a
time." Pick a reference frame. Pick a local Boolean algebra. Pick an
eye.

And really, wasn't that what quantum mechanics was trying to tell
us all along? Like the uncertainty principle. We can't assign precise
values to both position and momentum or to both time and energy simultaneously. And what did "simultaneously" mean? *Within a single
reference frame.*

Bohr and Heisenberg knew all that. They knew that the values of
complementary features were relative to the measuring apparatus.
Where they went wrong was in thinking that once a feature was measured and its wavefunction was collapsed, its value was then fixed for
all observers everywhere. The only way to make that work would be to

consider the observer some special thing standing outside the laws of physics. That's exactly what Bohr did, and Wheeler tried to follow in his footsteps. But nagging in the back of his mind, Wheeler knew it wouldn't hold up: *Elementary phenomena are impossible without the distinction between observing equipment and observed system, but the line of distinction can run like a maze, so convoluted that what appears from one standpoint to be on one's side and to be identified as observing apparatus, from another point of view has to be looked at as observed system.* It was Rovelli who finally found the way through the maze. Indeed, all observers are, from some other reference frame, the observed. Reality is radically observer-dependent.

Einstein's spooky action-at-a-distance was spooky precisely because it was derived from a view from nowhere. And Einstein, of all people, should have known better. He thought that entanglement undermined locality, light's ultimate speed limit. But what *really* undermined locality was the reference frame he was using—one that simultaneously encompassed two light cones. Of course, it probably didn't occur to him to worry about describing physics across light cones because, even though he himself had discovered that space and time change from one frame to the next, he still believed that some basic features of reality were invariant. If he had been right—if an electron's spin was always up for all possible observers—then his God's-eye view wouldn't have caused any trouble. But it did cause trouble. *Spooky* trouble. The problem wasn't with locality. It was with reality. Despite Einstein's weirdly uncharacteristic commitment to old-school realism, the basic features of reality were *not* invariant. They were observer-dependent. If you tried to describe them from a God's-eye view, you'd get the wrong answer.

I knew that Einstein had talked a big realist game, but seriously? It hadn't occurred to him—to *Einstein*—that some seemingly invariant things might turn out to be relative? *Relativity?* Really? That hadn't rung a bell?

I needed to gather my thoughts, piece together these disparate strands, so I hopped on a train bound for Philadelphia.

When I rang the bell at my parents' house, I was greeted by a thick silence. No barking, no whimpering, no thump of a tail. At eleven years old, Cassidy had developed a tumor in her leg, which had swelled to the size of a grapefruit. Like a trouper she had held that leg in the air for nearly a year, until the pain became too much. The vet informed us that amputation would be difficult and wouldn't buy much time. My mother was the only observer in the room when the universe relative to Cassidy came to an end. My father placed her bowl and her chain on his desk. I cried into the phone when they told me the news. The knowledge that everything was an illusion was hardly a comfort. She was the sweetest illusion I had ever known.

Inside, I was eager to look back to the beginning, to the H-state, to try to make sense of things. "Didn't you keep some notebooks back when we first started working on this?" I asked my father.

"I think they're in one of the cabinets in the library," my father said. "You're welcome to try to find them."

In the library, countless stacks of books barricaded the cabinet doors. I rolled up my sleeves and began moving them, piling them on the couch and on rare patches of bookless floor. The stacks encoded a chronology of my father's intellectual interests, like rings in a tree. In front were more recent acquisitions: books on cosmology and quantum gravity. Behind them were relativity and quantum mechanics, followed by astrophysics and astronomy. When I finally reached the last stacks, the books were more diverse: a biography of Einstein, Schrödinger's *What Is Life?*, a compilation of Bob Dylan lyrics. There were a few books by the philosopher Alan Watts, including one entitled *The Way of Zen*.

I flipped through the yellowed pages. My father had once told me that when he was a teenager on summer vacation, he was lying on a hammock in the backyard of his family home, not two miles from the home in which he and my mother would later raise me, reading *The Way of Zen*. "The book was talking about the illusion of the ego," my father had told me, "and the duality of subject and object. I was totally blown away by this idea, which was so simple and yet so profound. It had such an effect on me that I became hyperaware of everything around me. I was so in the moment. And then a bee landed on the page and pooped on it and then flew away. So I circled the stain on the page and wrote in the margin, 'A bee pooped here.'"

When he told me that story, I found myself wondering what I would have done if a bee had shat on my teenage reading, which was pretty much the opposite of everything Zen. Most likely I would have circled the stain in my Sartre and written, "Figures." But the funny thing is, unbeknownst to my father, when I had snuck out of this house to get my first tattoo at the age of fourteen, I had, amidst my existentialism and angst, gotten a tattoo of the Chinese character for *Zen*, which looked like a little Hawaiian man carrying a tiki torch on my hip, because even though I was rebelling, I really just wanted to be like my father, to have the kind of subterranean wisdom I saw lurking behind his eyes, which were large and brown and sloped downward at the edges so that he appeared perpetually sleepy or stoned, eyes I had inherited from him and regarded not as mere genetic facsimile but as a secret handshake. It was his Zen-like thinking that led my father to his epiphany, the H-state, a way of thinking about nothing that made it the ontological equal of everything, and it was his H-state that led me to tell a little lie, and to dream up a life, and a book, and a universe.

I placed the book down gently, careful not to disturb any sentimental excrement. I moved some Buddhist texts, a philosophy of space and time, a collection of poems by William Carlos Williams. Williams, I remembered, was my father's favorite poet, maybe because of the surreal and Zen-like quality of his work—*So much depends on a piece of bee shit*—or maybe because he, too, had been a doctor at the University of Pennsylvania and because he, too, had gone home at night to live another life, one man writing poetry, the other tracking down the secret of the universe, which I suppose was really the same thing.

When I finally managed to open the cabinet doors, I spotted the stack of hardbound notebooks. I carried them into my old bedroom and sprawled out on the bed to read.

My father had filled perhaps a quarter of each of the notebooks, leaving the remainder blank and then inexplicably starting a new one. Every entry expounded on the meaning of the H-state and agonized over the same, infuriating question: why would it ever change?

"Everything and nothing are simply dualities which, in their extreme, become one and the same. . . . Both everything and nothing are co-embedded in the H-state. So for everything to 'appear' out of nothing is not a big conceptual leap. But how did it change from a featureless 'void' to an inhomogeneous feature-full universe of 'things'?

"Everything, including time, space, energy, and matter, is simply an expression of what appear to be changing faces of the H-state, but which ultimately never changes," he wrote elsewhere. "How can it? It is by definition perfect homogeneity and thus without change."

Eventually my father pieced together an answer. "We can turn to three different well-established laws of nature to explain this. In fact, it so well follows as a consequence of each of these laws that for the h-state not to have expressed itself in forms would have been a violation of our basic scientific principles."

The first was the second law of thermodynamics. The H-state, he wrote, having only one unique configuration, has zero entropy, which, by the second law, is compelled to increase. At the same time, being maximally homogeneous, the H-state has infinite entropy. "The H-state is both ultimate order and disorder, and neither! It is a merging of the two. Thus ultimate order and disorder are the same! The universe must derive from the H-state and ultimately 'return' to it."

The second arbiter of change was symmetry breaking. "The H-state, in its perfect homogeneity, has by definition perfect symmetry. A perfectly symmetric state is perfectly unstable. . . . Physicists have come to understand that the features of our world are manifestations of broken symmetries. In fact, all of our conservation laws, such as the conservation of energy, angular rotation, etc., are themselves expressions of underlying symmetries. But if everything is derived from broken symmetry, all must originate from perfect symmetry: the H-state."

The third piece of the puzzle was quantum mechanics, my father wrote. According to the laws of quantum mechanics, "nothing in the universe can have a perfectly defined energy level. So with the H-state. Uncertainty demands that the H-state give up its homogeneity." Quantum fluctuations, he said, "create intrinsic oscillations that form the basis of 'thingness,' like a wheel must rotate around a motionless hub.

"Since all is perfectly the same," he noted, "one cannot specify a location or time within the H-state. There are no locations; all locations are the same. There is no time; all instants are the same. Nothing changes, neither spatially nor temporally nor in any other dimension. But all of our fundamental laws of nature dictate that such a state cannot persist. It is unstable. The laws of thermodynamics, symmetry breaking, and quantum mechanics prescribe the transformation of the H-state, from nothing to something. Since there are no places or times within the H-state, the change will happen at all places and all times. You could say that the origin of the universe comes from a point, but it is infinite in size. . . . Homogeneity is ultimate reality. Patterns are conventional reality. . . . Nothingness cannot exist. It is unstable."

My father's reasoning was impressively similar to the kind of reasoning I had encountered from physicists over the years. Wilczek, for instance, had written, "The most symmetric phase of the universe generally turns out to be unstable. One can speculate that the universe began in the most symmetric state possible and that in such a state no matter existed. . . . Eventually a patch of the less symmetric phase will appear—arising, if for no other reason, as a quantum fluctuation. . . . This event might be identified with the big bang. . . . Our answer to Leibniz's great question 'Why is there something rather than nothing?' then becomes 'Nothing is unstable.'"

Of course, we had since learned the problem with the whole "nothing is unstable and quantum fluctuations will change it to something" idea. It was a global story told by an omniscient narrator, a narrator with an impossible God's-eye view, a reference frame situated outside the H-state, relative to which the H-state, which by definition has no outside, would change. What's more, it presupposed quantum mechanics, leaving Wheeler's question dangling, unanswered: *Why the quantum?*

But now, thanks to Rovelli, I had the beginnings of an answer. "Why the quantum?" was the same as "Why non-Boolean logic?" But non-Boolean logic, I now knew, was a fictitious logic, the logic that crops up when you cut across horizons, when you try to describe reality from multiple points of view simultaneously. Quantum logic is non-Boolean because reality is radically observer-dependent. Because there's no singular way things "really are." There was my "really" and my father's "really," but never both.

The H-state couldn't change because it has no outside. But from here on the inside, it could *appear* to change, as if something is just what nothing looks like from the inside. Here on the inside, with a finite speed of light, observers can't see the whole thing. Their perspective is bounded. But when you put a boundary on the H-state, it's no longer the H-state. It's no longer nothing. It's something.

Of course, the problem with *that* idea was that it required a finite speed of light to define the light cones that in turn define an observer's perspective. The stubborn speed of light was still hanging on, untouched on the IHOP napkin, leaving one last ingredient irreducible and inexplicable. *If observers create reality, where do the observers come from?*

I flipped the pages of my father's notebooks, frustrated. With a star drawn next to it, my father quoted Lao Tzu: "To find the most precious of pearls doesn't compare to discovering the source of all things."

A few days later, I awoke to find a paper lying on the floor outside my bedroom door. My parents had both left and gone to work. Groggy, I picked it up. There was a note attached: *A clue? L, D.*

I sat back down on my bed to read. It was a typed transcript of a talk given by a French astrophysicist named Laurent Nottale. Strangely, the conference from which the talk came was not about physics but about Buddhism. I laughed. Physics and Buddhism? Throw in Bob Dylan and some oatmeal-raisin cookies and you've got my father's paradise.

Relativity, Nottale explained in the talk, is about emptiness, the emptiness of motion and of space and time. Einstein's happiest thought was that a man in free fall can't feel his own weight. "Thanks to this," Nottale said, "he has realized that gravitation that looks so solid and so universal has no intrinsic existence." It's observer-dependent. It's not ultimately real.

"Form is emptiness, emptiness is form," Nottale says, quoting the Heart Sutra. "This is what relativity tells us."

When I got to page ten, I discovered that my father had highlighted a section of text. "Form is emptiness because it is always possible to find a reference system in which the thing disappears. At this stage, it can really help us to understand in which reference system the thing disappears. The answer is that it is in the proper reference system . . . the self reference system, in itself. . . . It is true for any property we may consider. This property can disappear in the proper reference system. Consider whatever you want, like a color, a form, an object, a mass, a particle, and put yourself into it, in the interior of the thing, then the thing disappears. In the color, there is no color. . . . What makes the color is the wavelength. If you are smaller than the wavelength, the concept of color does not even exist. It disappears completely. If you are *in* light, participating in its motion, light and time disappear (this is what Einstein understood at the age of fifteen and what led him to build ten years later his first theory of relativity). Therefore, in motion, there is no motion, in position, no position, in particle, no particle."

If you are in light, light disappears.

I dropped the paper in my lap. *Holy shit.*

That was it.

That was the answer.

The teenage Einstein had asked what light would look like if you were running at the same speed alongside it. But what happens when you invert the question and ask, what does the universe look like to light? What does a photon see?

Light, by definition, uses up its entire spacetime quotient on space, leaving none for time. In other words, it sees all of space in no time. From my point of view, the light leaving a star 5 million light-years away takes 5 million years to reach my eye. But from the light's point of view, its journey is instantaneous. From the light's point of view, the speed of light is not the speed of light. It has no speed. It is everywhere at once in a single instant. A photon doesn't see the universe. A photon sees a singularity.

It sees *the H-state*.

It was all dawning on me now. So much depended on the speed of light: the existence of horizons, light cones, information bounds, reference frames, observers. As long as it was invariant, so were they.

Wheeler had worried about light's invariance, too. On August 27,

1985, he had written in his journal, "My picture (U-diagram) shows a reflexive system, but one with at least one primitive element, that dashed line." That dashed line was the finite and invariant speed of light. Wheeler's self-excited circuit could explain away everything— except that.

But now I saw that the speed of light *isn't* invariant. It isn't real. *Here's the reality test. If you can find one frame of reference in which the thing disappears, then it's not invariant, it's observer-dependent.* Nottale had pointed out the one frame in which the speed of light disappears: the frame of the light.

Horizons—the last remaining ingredients of reality, the last remnants of a sandcastle fading back into the boundless sameness of the beach, that dashed line reaching back through cosmic history, the final bulwark that stood between something and nothing—were built of light, light frozen in place by acceleration and by gravity.

But horizons don't have horizons.

Boundaries don't have boundaries.

The boundary of a boundary is zero.

That evening, I invited my father to have dinner with me at our Chinese restaurant, the one where he had first asked me about nothing.

It was a little corny, I knew, forcing that kind of cinematic symmetry on my own life. But it felt right. It reminded me of how far we'd come, and at the same time of how so little had changed. Besides, I knew how much he loved the cashew chicken.

We arrived at the restaurant and sat down at what we swore was our original table, though I suspected that was some kind of mutual false memory. After we ordered our food, I pulled out my notebook. "Okay," I said. "I've made a list of the key clues." I read them off one at a time.

One: Nothing is defined as an infinite, unbounded, homogeneous state. Which means that "something" is defined as a finite, bounded state. To turn nothing into something, you need a boundary.

Two: There are no nonzero conserved quantities. Everything is in some sense nothing.

Three: All of physics seems to be defined on boundaries. On horizons.

Four: The laws of physics make sense only within the reference frame of a single observer, a single light cone.

Five: Given a single reference frame, an entire observer-dependent cosmic history will unfurl, thanks to top-down cosmology and Wheeler's delayed choice.

Six: Horizon complementarity and holographic spacetime suggest that nothing beyond my horizon is real. As if the region carved out by my light cone is the be-all and end-all of reality.

Seven: The positive value of our cosmological constant ensures that for any given reference frame there exists an inescapable and observer-dependent boundary. The universe is fundamentally fragmented. Wait forever and you'll never see the whole thing.

Eight: The low quadrupole in the cosmic microwave background seems to suggest that the entire universe is the size of the observable universe.

Nine: The relational nature of quantum mechanics and the inescapable limitations of Gödelian self-reference ensure that a subject can never be an object in its own frame, and that in turn the world is always broken into pieces.

Ten: M-theory, our best description of the physical world to date, appears to have no ontology.

Eleven: Reality is radically observer-dependent. Every possible ingredient of ultimate reality, every item on the IHOP napkin, has been crossed off. Nothing is invariant. Nothing is ultimately real.

"Those clues paint a pretty striking picture," my father said.

That was an understatement. The whole thing was kind of uncanny, how it all fit together. But into what exactly? Into Nothing?

"What do you think it all means?" I asked.

"I'm thinking what you're thinking," he said. "That everything is nothing, the H-state, and that it only looks like something when you have a limited, internal perspective. And you *have* to have a limited, internal perspective because no external perspective exists. There's no outside. But there's still some transformation you can make, by going to the point of view of the light, of the horizon, to get you back to the nothing that's always there."

That *was* what I was thinking. My father's definition of nothing as an infinite, unbounded homogeneous state carried two implications: nothing has no outside, and nothing will never change. At first that had seemed like a nonstarter—if it can't change, how could the universe ever be born? But Smolin's first principle of cosmology held the answer: the origin has to come from *inside the nothing*. Given some internal reference frame with a boundary, a universe is born, its history unfurling from present to past. A top-down universe that exists only relative to its reference frame. Beyond the bounds of the frame, there's nothing.

The failure of the God's-eye view signaled the nonexistence of any reality beyond a single observer's point of view. After all, why would all of physics be defined in terms of a single reference frame unless the single reference frame *defined* the universe?

"You know, if you started from the premise that internal reference frames create the universe by transforming a boundless nothing into a bounded something, then you'd actually *expect* physics to make sense only within a single frame at a time, with nonsensical redundancies cropping up any time you mistakenly talk across horizons," I said. "And if a single reference frame marks the edge of reality, you'd think you might see evidence that there's nothing beyond the cosmic horizon."

"Evidence like the low quadrupole in the CMB?" my father asked, grinning.

I grinned back. It was a tantalizing prospect.

One thing was clear: the key to existence was a boundary. From the beginning I had worried that light cones alone wouldn't be enough. Given infinite time, any given light cone would engulf the entire H-state, turning the something back to nothing again. It seemed you needed something more permanent—something like the kind of perpetual boundary dark energy provides. Then again, maybe it was enough to say that no observer can measure himself. Maybe Gödelian incompleteness and the impossibility of self-measurement keep the nothing at bay, the world always carved in half, observer and observed.

To turn nothing into something, you needed information bounds, a finite amount of information to get it from bit. "Something of an information-theoretic character is at the bottom of physics, spacetime, existence itself," Wheeler had written on his way to the hospital. "This is a quick sentence if anyone asks me for a last word before I leave this Earth."

I had wondered why, of all the profound thoughts rolling around in his head, Wheeler had chosen that one to be his final word on the nature of reality. Why not the self-excited circuit? Or the boundary of a boundary? Why information?

Now I was beginning to understand what information really was: asymmetry. To register a bit of information, you need two distinguishable states: black or white, spin up or spin down, 0 or 1. You need twoness. After all, entropy was a measure of missing information, and with entropy comes symmetry. A smoothly distributed gas, the epitome of high entropy, looks pretty much the same everywhere—it's highly symmetric. And what's symmetry? It's redundancy of description, a redundancy of information. If you need to describe a five-pointed snowflake, the only information you require is the information describing one of the points along with the fact that there are five of them. You don't need individual information for each of the points because it's just the same information repeated over and over again. A five-pointed snowflake is symmetric in that its information repeats five times. The more symmetry something has, the less information it contains.

My father's H-state was a state with no differentiation whatsoever. A state of perfect symmetry. That meant it had zero information, which made sense, considering it was nothing. So how do you get information from the H-state, turning nothing into something? You put a boundary on it. The boundary breaks the symmetry, creating information. But the boundary is observer-dependent, and so is the information it creates.

"In classical physics you can find out the state of a system, and then someone else can come along and find out the state of the same system, and you'll both agree," Zurek had told us. "In quantum mechanics, that's generally impossible."

I understood now why it was impossible. Making a quantum measurement amounted to choosing a reference frame. In the non-Boolean superposition of every possible point of view, there's no differentiation, no information. Quantum interference makes sure of that. When you make a measurement by selecting a single point of view, a Boolean logic, yes or no, you break the symmetry of the superposition and create, at random, a bit of information. "That's the participatory business," Zurek said. "That's a hint of how this universe is built."

"If the world wasn't quantum mechanical," I said to my father, "then it couldn't possibly have arisen from nothing. I don't mean that in the usual way people mean it, like quantum mechanics takes some state that you call 'nothing' and then uses the uncertainty principle to transform it into something. That's trivial. That's assuming quantum mechanics from the start. What I mean is that if the world weren't described by quantum mechanics, then logic would be Boolean, and reality would be invariant. All observers would agree on the truth-value of propositions. They'd agree on what's real. There would be no interference between their perspectives; physics would be classical. But when you have invariance, you have *something*. Something you can't explain away. There would be a reality that's ontologically distinct from nothing, and then you'd be stuck with this unbridgeable divide that logic can't fix. It doesn't make sense to say that the universe came from nothing but is now something—there's no way to get from one to the other. It does make sense if the universe *is* nothing. But if the universe is nothing, then only nothing is ultimately real. Nothing is invariant. And that lack of invariance shows up, to us, as quantum mechanics."

"So a universe that had some kind of real existence, a universe that was *something*, wouldn't be described by quantum mechanics?"

"That's my hunch," I said. "Wheeler always knew that the quantum was a clue. I think it's a clue to the observer-dependence of reality. To the fact that everything, at bottom, is nothing."

"You know the story of Plato's cave?" my father asked. "All the prisoners are chained up in the cave and they can't see the real world outside, only the shadows on the wall? That's supposed to be a negative thing, like they'll never know reality. But the truth is, you have to be stuck inside a limited reference frame for there to be any reality at all! If you weren't chained to your light cone, you'd see nothing. The H-state."

I nodded. "You'd have no information. You need the broken symmetry, the shadow, to have information and information gives rise to the world. It from bit."

I couldn't help but grin with excitement. The message was clear: having a finite frame of reference creates the illusion of a world, but even the reference frame itself is an illusion. Observers create reality, but observers aren't real. There is nothing ontologically distinct about an observer, because you can always find a frame in which that observer disappears: the frame of the frame itself, the boundary of the boundary.

"If physicists discover an invariant someday, the game will be up," my father mused. "That would rule out the hypothesis that the universe is really nothing."

That was true. But so far, at least, every last invariant had gone the way of space and time, rendered relative and observer-dependent. Spacetime, gravity, electromagnetism, the nuclear forces, mass, energy, momentum, angular momentum, charge, dimensions, particles, fields, the vacuum, strings, the universe, the multiverse, the speed of light—one by one they had been downgraded to illusion. As the surface appearance of reality fell away, only one thing remained. Nothing.

In my mind, these conclusions were greeted by a sudden fanfare. Lights flashed and balloons and confetti came pouring down from the ceiling, like we were the hundredth customer in the grocery store checkout line. Throngs of cheering people filled the restaurant and crowded around us, applauding. Looking out into the crowd, I noticed some familiar faces. There, in a flowing skirt, was Fotini Markopoulou, and next to her I saw Carlo Rovelli and Lee Smolin. Alan Guth was there with his giant yellow backpack, as was James Ladyman, dreadlocks spilling down his back. Timothy Ferris stood dangling his car keys from his finger, and Andy Albrecht was laughing and waving his hand as if to say, *Don't worry about it!* Off to one side I spotted the Panama hat: Brockman and Matson were there. Behind them I saw Phil, from *Scientific American,* and for a minute I swore I caught a glimpse of Rick from *Manhattan Bride*. Through the commotion I heard Susskind's Bronx inflection, and I looked over to see him standing with Raphael Bousso and Tom Banks. Joe Polchinski was there, too, as was Ed the Head, and next to him, in his wheelchair, was Stephen Hawking. I spotted Kip Thorne in the back donning a Star Trek uniform, and I saw a crop of bushy hair that I presumed to be Zurek. Scurrying happily around everyone's feet were seven rats tittering and shouting, "Gotcha!"—one of them sporting a large X-shaped bandage where the base of its tail would have been. Suddenly the crowd grew quiet and parted, creating a path for an older man who slowly made his way toward our table. As he neared I saw that it was Wheeler. He shook my father's hand, then mine. He smiled, a perceptible gleam in his eye. "I told you persistence would be rewarded."

In real life, the restaurant was quiet, the world still littered with unanswered questions. In real life, we clinked our glasses together, smiled, and drove home.

Back in my old bedroom, I curled up in bed with my notebook, my eyes retracing those clues.

My father's definition of nothing had made it possible to cross that ontological divide between nothing and something, and the radical observer-dependence of every ingredient of reality down to reality itself made it possible to cross back. We had found the universe's secret: physics isn't the machinery behind the workings of the world; physics is the machinery behind the illusion that there *is* a world.

Still, so many questions remained. It wasn't clear yet what the new paradigm in cosmology would bring—Hawking and Hertog's top-down cosmology, Banks's holographic spacetime, or something else altogether. It wasn't clear what new ingredients might still be buried among the dualities of M-theory. A positive cosmological constant seemed to be required to ensure that nothing would look like something, but would an ultimate theory ever uniquely determine its value, or was its value irrelevant in the same way that the particular values of the speed of light or Planck's constant were? How would the mystery of dark matter be resolved? Would some new twist in the data appear in the tunnels of the Large Hadron Collider or the patterns mapped by the Planck satellite that would send everything we know spinning on its head?

Personally, I was glad for all the unanswered questions—they meant my father and I weren't done yet, that we were still in it together. For me, chasing the mysteries of the universe and growing up had always been one and the same, and I wasn't ready to be done growing up just yet.

We are each the author of our own universe, I wrote in my notebook, *but it's comforting to know that there is some other reference frame in which my father and I are both characters, side by side, partners in crime.* It was my last sentence on Earth, just in case I needed it.

I thought back to that day in Princeton when we first crashed the party. I thought about Wheeler's four questions and how we might answer them after everything we'd learned. It from bit? Yes, but each observer creates different its from the same bits, bits that are themselves observer-dependent and arise from the asymmetry wrought by a finite reference frame. A participatory universe? Participatory, yes; a universe, no. It was one participatory universe per reference frame, and you can only talk about one at a time. Why the quantum? Because reality is radically observer-dependent. Because observers are creating bits of information out of nothingness. Because there's no way things "really are," and you can't employ descriptions that cross horizons. How come existence? Because existence is what nothing looks like from the inside.

It was time to start writing my book. From a justified line into the margin. I took a deep breath.

It's hard to know where to begin. What even counts as a beginning? I could say my story begins in a Chinese restaurant, circa 1995, when my father asked me a question about nothing. More likely it begins circa 14 billion years ago, when the so-called universe was allegedly born, broiling and thick with existence. Then again, I've come to suspect that that story is only beginning right now. I realize how weird that must sound. Trust me, it gets weirder.

Acknowledgments

I cannot express the depth of my appreciation to the many physicists who have patiently and generously offered me their time and wisdom over the years, and who have no idea what a profound influence they've had on my life. I'd especially like to thank Lenny Susskind, Raphael Bousso, Fotini Markopoulou, Joe Polchinski, Alan Guth, Tom Banks, Carlo Rovelli, Wojciech Zurek, Kip Thorne, Lee Smolin, and James Ladyman.

This book would not exist without the help of my kickass agents, Katinka Matson, John Brockman, and Max Brockman, who now (embarrassingly) know just how long I've dreamt of working with them. I want to thank them for taking a chance on me and for helping me to find my voice.

It has been a joy and privilege to work with the editorial team at Random House, and particularly with my editor Mark Tavani, who took on this project when others fled and who always remained in my corner. This book would have been a mess if it weren't for the genius copyeditor, Sue Warga, and the ever-helpful production editor, Loren Noveck. Thank you also to Luke Dempsey—wherever you are—for believing in this book from the start.

My extreme gratitude goes to the library staff of the American Philosophical Society, and particularly to Charles Greifenstein, for affording us access to the John Wheeler journals and for preserving a priceless piece of intellectual history.

I owe a special thanks to the brilliant, kind, and hilarious Maggie McKee, who graciously read the entire manuscript and offered invaluable guidance, and to Hester Kaplan, for encouraging me to tell stories.

Thank you to Dan Falk for reading a bulk of the book and for ten years of friendship and physics.

A huge thank you to Philip Yam at *Scientific American* for giving me my start as a journalist, and to all the talented editors and science journalists I've had the pleasure of working with since. I am forever indebted to the entire *New Scientist* family, and especially to Michael Brooks, Michael Bond, and Valerie Jamieson for fighting for me and supporting me over the years.

Thank you to Samantha Murphy and Rebecca Rodriguez, my 24/7 lifelines and the best friends a girl could have. To Winston Loach for believing in me, and to Joe Kitsch for the inspiration. To Christina Shock Weiss, Stephanie Dresner, Kevin Kerrigan, Natasha Wehrli, Katherine Tomkinson, and all the friends who have lived with and encouraged my insanity over the years.

My family, and especially my grandmother Winnie Gefter, mean the world to me. I'd like to express my enduring admiration for William Gefter, who never underestimated the power of ideas, and my inexpressible love to Harry and Marion Bergelson, who are no longer with me but who helped me grow in ways I'm still discovering today.

This journey would never have gotten off the ground without the unfailing support of my mother, Marlene Gefter. She and my brother, Brian Gefter, have been my inspirations and my best friends—and for enduring countless hours of physics conversations at the dinner table they deserve some kind of medal.

Finally, a shout-out to Hunan Restaurant in Ardmore, PA, where my cosmic adventure first began. I'll never forget the cashew chicken.

Glossary

Accelerated observer An observer whose speed or direction of motion is changing. Goes by the name Safe. Can be found outside black holes.

Acceleration The rate of change in velocity with respect to time.

AdS/CFT Juan Maldacena's breakthrough 1997 discovery that string theory (with gravity) in an anti–de Sitter space with five large and five tiny dimensions is exactly equivalent to a conformal field theory (without gravity) on its four-dimensional boundary. What look like strings on one side of the equivalence look like particles on the other; what look like five large dimensions on one side look like four on the other. Because neither description is more real, this introduced ambiguities in the nature of reality, and was the first convincing example of the holographic principle in action.

Anthropic principle The seemingly tautological statement that features of our universe must be compatible with our biological existence. Why? Perhaps we live in a multiverse in which features vary from one universe to the next, and we find ourselves, unsurprisingly, living in the one we can live in. Or perhaps, as John Wheeler suggests, observers play a role in creating the universe that created them.

Anti–de Sitter (AdS) space A space in which a negative cosmological constant bends space like a saddle at every point, distorting geometry in such a way that light can travel to spatial infinity and back in a finite amount of time. In AdS space, every observer's light cone overlaps, so everyone sees the same universe.

Antiparticle A particle with the same mass but opposite charge as its ordinary counterpart. Or, equivalently, an ordinary particle moving backward in time.

Arrow of time The notion that time only moves in one direction, namely, forward.

Baryon Any particle made of three quarks, including the neutrons and protons in the heart of every atom.

Big bang The claim that the early universe was hot and dense, then expanded and cooled. That's really all there is to it.

Black hole A region of spacetime where gravity is so strong not even light can escape.

Black hole information-loss paradox When black holes evaporate via Hawking radiation and disappear from existence, what happens to the stuff that fell inside? If it escapes, Einstein's theories of relativity are wrong. If it doesn't escape, quantum mechanics is wrong. Neither relativity nor quantum mechanics is wrong.

Boson A force-carrying particle whose quantum spin is an integer; examples are a photon, which carries the electromagnetic force and has spin 1, and a graviton, which carries the gravitational force and has spin 2.

Causal diamond The entire region of universe that a given observer could have accessed in the past and can access in the future. Because it is formed by the intersection of an observer's past and future light cones, it is shaped like a diamond.

Census Taker Leonard Susskind's hypothetical observer who lives in FRW space, the end product of any series of vacuum decays in eternal inflation. The Census Taker's light cone will continue to grow forever, allowing him to measure, in principle, any region of the universe—except himself.

Commutation If $A \times B = B \times A$, then A and B commute. If $A \times B \neq B \times A$, they don't. Commutation relations tell you whether order matters. Quantum uncertainty, for instance, tells you that, within a single reference frame, it matters whether you measure a particle's position first and then its momentum or its momentum first and then its position, because the more accurately you measure the first, the less accurately you can measure the second.

Conserved quantity A quantity whose invariance is ensured by the laws of physics. All experiments suggest there are no nonzero conserved quantities in the universe.

Cosmic microwave background (CMB) Remnant radiation from the big bang, composed of photons whose frequencies have been stretched to the microwave region by the expansion of the universe and now heat empty space to 2.7 Kelvin.

Cosmological constant Originally a term in Einstein's equations of general relativity, it is now believed to be the energy inherent to the vacuum itself. If its value is positive, it exerts a negative pressure on the vacuum, causing space to expand at an accelerated rate. If its value is negative, it exerts a positive pressure, causing space to fold in on itself at every point.

D-brane A membrane that from one point of view looks like a region of space where open strings can end, and from another point of view looks like an object that can move or form a black hole.

Dark energy The unidentified force causing the expansion of the universe to accelerate. Most likely suspect: Einstein's cosmological constant.

Dark matter A hypothetical form of matter that does not interact via the electromagnetic or strong nuclear forces, but whose gravity is thought to hold stars in place within galaxies.

Decoherence The process by which quantum superposition states (and the interference patterns they produce) are quickly destroyed by interactions with their environment, which explains why we hardly ever see any simultaneously dead and alive cats.

Delayed-choice experiment Wheeler's version of the double-slit experiment in which the observer's decision to either measure the interference pattern produced when a particle travels two paths or to measure which single path the particle takes, thereby destroying interference, occurs after the particle has presumably traveled either both paths or just one. In other words, the observer's choice of measurement determines the history of the universe *after* it has already occurred.

De Sitter horizon In de Sitter space, the universe expands at an accelerated rate. Light can only travel a finite distance even given infinite time, because as the light crosses any given distance, the distance itself grows. For any inertial observer, there is a region of the universe from which light will never be able to reach him. The event horizon separating the accessible portion of the universe from the dark portion is known as a de Sitter horizon. It is observer-dependent, as it is relative to the observer's location.

De Sitter space A space formed when a positive cosmological constant exerts an outward pressure, causing the expansion rate of the universe to accelerate, diluting the density of matter until the universe is empty except for the cosmological constant itself. In de Sitter space, every observer is surrounded by a unique horizon, so that no two observers in de Sitter space ever see the same universe.

Diffeomorphism transformation A method of translating between misaligned points of view by introducing a force, like gravity. *To make a curve match up with a line, bend the paper.* This is the key tool of general relativity and an example of a gauge transformation.

Double-slit experiment The classic, bat-shit crazy quantum experiment in which particles are fired at a screen that has two tiny slits in it. On the other side of the screen is a photographic plate to record where the particles land. When both slits are kept open, light fired at the screen creates an interference pattern of dark and light stripes on the plate, suggesting that light exhibits wave behavior. When single photons are fired, they will, one by one, build up the same interference pattern, a bizarre fact that seems to tell us the individual photons are traveling through both slits simultaneously. If a detector is placed at one of the slits to measure the photon's path, the photon will only travel through a single slit and the interference pattern disappears.

dS/CFT A hypothesized analogue to AdS/CFT, in which the physics of our de Sitter universe is dual to a conformal quantum field theory living on the lower-dimensional boundary of the universe. Such a formulation, however, would describe a global universe that no observer can ever access, since any observer in de Sitter space is stuck in a finite region surrounded by an event horizon.

Duality A one-to-one mathematical equivalence between two radically different physical descriptions.

Entanglement A form of quantum superposition in which two systems are described by a single wavefunction, so that information about the system resides in neither system individually but in the correlations between them, which persist in spite of spatial separation.

Entropy A measure of the amount of information it would take to characterize every detail of a physical system.

Epistemology The study of what is possible to know and how we may or may not know it.

EPR A thought experiment conceived by Einstein, Boris Podolsky, and Nathan Rosen in an attempt to argue that quantum theory could not be a complete description of reality. If features, say spins, of two particles are correlated, then measuring the value of one seems to instantaneously decide the value of the other, no matter how far apart the particles are. EPR concluded that since relativity forbids instantaneous action-at-a-distance, there must be so-called hidden quantum variables that allow the particles to have well-defined values at all times, even prior to measurement. Bell's theorem has since shown the local hidden variables interpretation to be flawed, while Carlo Rovelli's relational interpretation of quantum mechanics resolves the EPR paradox by revealing the outcome of quantum measurements to be observer-dependent.

Equivalence principle Einstein's happiest thought: there's no difference between an accelerated frame without gravity and an inertial frame with gravity. In other words, gravity is ultimately not real—it's a gauge force that accounts for the mismatch between reality as seen from an inertial frame and reality as seen from an accelerated frame.

Eternal inflation According to the theory of inflation, the universe begins in a false vacuum that eventually decays, but, thanks to the uncertainty principle, it doesn't decay everywhere at the same time. When one region decays, it briefly expands faster than light, forming a bubble universe causally disconnected from the original vacuum. Remaining portions of the vacuum decay, too, forming other bubbles. The false vacuum grows faster than it decays, so there is always more left over and it never stops blowing bubbles. Any feasible theory of inflation is bound to go eternal and create an infinite multiverse.

Event horizon A surface in spacetime that light cannot cross. Horizons divide spacetime into causally disconnected regions.

False vacuum A state that is temporarily stable but is not the lowest possible energy of the system. Given enough time it will decay, dropping down to its lowest energy state.

Fermion A matter particle whose quantum spin comes in half integers, such as an electron, which has spin ½.

Fictitious force A force that arises as a consequence of an observer's perspective. Because they are gauge forces, the four fundamental forces of nature—gravity, electromagnetism, and the strong and weak nuclear forces—are fictitious.

Friedmann-Robertson-Walker (FRW) space A simple, homogeneous, expanding or contracting universe.

Gauge A phase or reference frame.

Gauge boson A particle that carries a gauge force.

Gauge force A force that exists in order to account for mismatched descriptions between two reference frames. The electromagnetic force, for instance, exists so we don't confuse two different descriptions of the same electron—whose phase changes from one reference frame to another—for different electrons.

Gauge symmetry All gauges—or reference frames—are created equal. None offers a truer version of reality than another.

General covariance Einstein's key principle that there's no preferred way to slice spacetime into space and time—slice it any way you want and the fundamental laws of physics remain unchanged. In a world with both accelerated and inertial reference frames, getting general covariance to hold requires diffeomorphism transformations.

General relativity Einstein's masterpiece, which puts inertial and accelerated observers on equal footing by introducing gravity, which bends spacetime in precisely the right way to align mismatched reference frames and ensure that we don't mistake different descriptions of reality for different realities.

Global A large-scale description that encompasses many light cones, larger than any single observer could ever see.

Gödel's incompleteness theorem If a sufficiently complex mathematical system—one that is capable of making statements about itself—is consistent, it can't be complete. That is, it will contain statements that are fundamentally unprovable.

H-state A state of infinite, unbounded homogeneity. Otherwise known as nothing.

Hawking radiation When an event horizon is present, observers will no longer agree on whether space is empty or filled with particles. Those observer-dependent particles are called Hawking radiation.

Higgs boson An excitation of the Higgs field.

Higgs field A pervasive field that interchanges right- and left-handed particles, allowing them to have mass without violating gauge symmetry.

Hilbert space A mathematical vector space for representing quantum states.

Holographic principle All information necessary to reconstruct the physics of a given region of spacetime can be encoded on the region's lower-dimensional boundary. Or, equivalently, the total amount of information that can fit in a given region of spacetime must be less than one-quarter the area of its boundary in Planck units.

Horizon complementarity You can describe the universe in terms of what's on one side of an event horizon or in terms of what's on the other side, but never both at once.

Inertial observer An observer in uniform (as opposed to accelerated) motion. Goes by the name Screwed. Can be found falling into black holes.

Inflation A brief period in which the universe expanded faster than light; occurred in the first trillionth of a second following its birth.

Inflaton A hypothetical scalar field that existed in the first fraction of a second following the big bang. The inflaton was believed to have begun in a false vacuum state; its decay triggered the superluminal expansion of inflation.

Interference pattern The pattern that arises when waves come together, their phases adding in places where they are aligned and canceling where they are misaligned.

Invariance Sameness. A feature is invariant if it doesn't change from one reference frame to another.

Isomorphism A one-to-one correspondence. For example, the scrambled information in a two-dimensional hologram is isomorphic to the three-dimensional image it projects.

It from bit Wheeler's expression for the notion that a physical object, an "it," is at bottom just a configuration of information, or "bits."

Landscape The set of 10^{500} vacua described by string theory, which are formed by the various ways in which one can compactify extra spatial dimensions. Each vacuum corresponds to its own universe, with its own local laws of physics and its own value of the cosmological constant.

Light cone The region of spacetime encompassing everything with which a given observer can have a causal relation. If something is in your past light cone, you can see it. If it's outside your past light cone, you can't—light that far away hasn't had enough time since the origin of the universe to reach you. If something lies in your future light cone, your actions can affect it. If it's outside your future light cone, it is forever out of reach.

Local Within a single light cone, accessible by a single observer.

Loop quantum gravity A theory of quantum gravity in which spacetime is made up of discrete units of area and volume.

Lorentz symmetry The symmetry that ensures equivalence between inertial frames that are moving at different uniform velocities or are rotated relative to one another.

Lorentz transformation A method of translating between two inertial, or uniformly moving, reference frames, usually by trading one reference frame's time for another's space, and vice versa, while keeping the total spacetime interval the same in both frames. This is the key tool of special relativity and is required to ensure that the speed of light remain constant in all reference frames.

Low quadrupole The lack of temperature fluctuations at scales larger than 60 degrees in the cosmic microwave background.

M-theory A candidate for a holy grail theory of quantum gravity, it is the larger theory of which the five versions of string theory and eleven-dimensional supergravity are mere shadows. It describes objects such as particles, strings, and branes, but none is its fundamental ingredient. In fact, it's not clear whether M-theory has any fundamental ingredients at all.

Manifold A space that is locally flat and Euclidean but may be globally curved and warped. The key rule of general relativity is: *To make a curve match up with a line, bend the paper.* The paper is the manifold.

Mathematical structure A set of isomorphic elements or equivalent representations of a number.

Measure problem (eternal inflation) In the infinite multiverse produced by eternal inflation, everything that can happen does happen an infinite number of times. Calculating the probability for observing anything becomes impossible, as all probabilities are infinity divided by infinity.

Measurement problem (quantum mechanics) Prior to measurement, a quantum system is in many states simultaneously, as evidenced by interference. When we measure it, we find it in a single state. What does it mean to make a measurement? Why should our measurements have any effect on reality?

Multiverse A global collection of causally disconnected universes.

No-cloning theorem An unknown quantum state cannot be copied.

Non-Boolean logic A system of logic that defies the Boolean two-valuedness of true and false, usually by revoking the law of the excluded middle and allowing for values to be both true *and* false.

Non-Euclidean geometry A system of geometry that tosses out Euclid's fifth axiom, which says that parallel lines will never meet. The curved spacetime of general relativity is described by non-Euclidean geometry.

Observer A reference frame, or perhaps the origin of a reference frame, which is bounded in space by the finite speed of light.

Observer-dependent Something that changes when viewed from different reference frames.

Occam's razor The philosophical criterion that when one is given a host of empirically equivalent alternatives, the simplest theory is usually true.

Ontology What exists; the furniture of reality.

Particle An irreducible representation of the Poincaré symmetry group—a definition that only means anything in a spacetime with Poincaré symmetry; that is, a flat spacetime without gravity. In the face of gravity, there's no observer-independent definition of a particle. In all cases, however, it is definitely *not* a tiny ball.

Phase How far along a wave is in its cycle relative to a given observer. The phase is not inherent to the wave—it defines the frame of reference from which the wave is being viewed.

Planck scale The incredibly tiny (10^{-33} centimeters) or, equivalently, incredibly high-energy (10^{19} GeV) scale at which quantum effects on spacetime grow extreme. Looking to smaller length scales or higher energies would collapse spacetime into a black hole, so the Planck scale is the border beyond which spacetime loses all meaning.

Poincaré symmetry The symmetry that ensures equivalence between inertial frames that are moving at different uniform velocities, are rotated relative to one another or are at different locations in spacetime. This is the symmetry of Minkowski spacetime, the flat, gravity-free spacetime of Einstein's special relativity. Particles are only invariant in spacetimes that are Poincaré symmetric.

Proposition A declarative sentence that can be deemed true or false, such as "The Earth is round" or "2 + 3 = 7."

QCD Quantum chromodynamics, the theory that describes how gluons bind quarks together via the strong nuclear force.

Quantum cosmology A theory of the universe's origin and evolution that takes into account its fundamentally quantum nature. Tends to suffer from an extreme form of the measurement problem, since the universe, by definition, has no outside, and therefore no one can measure it.

Quantum gravity The holy grail theory of everything that will unite Einstein's theory of gravity, general relativity, with quantum mechanics.

Quark-gluon plasma A hot gas plasma of free-roaming quarks and gluons, which existed in the earliest moments of the universe.

Redshift The stretching of a photon's wavelength and the corresponding decrease in its frequency and energy. A Doppler redshift results when a photon source is moving away relative to an observer, like the galaxies that are moving away from us as the universe expands. Light can also be redshifted by the expansion of space or by gravity as it travels from its source toward an observer.

Relational quantum mechanics Carlo Rovelli's interpretation of quantum mechanics, which emphasizes the observer-dependence of quantum measurement.

Rindler horizon An event horizon that results from an observer's acceleration. So long as the observer continues to accelerate, light from far regions of the universe will never catch up to him, rendering a portion of the universe dark and causally inaccessible, like a black hole.

S-duality The duality that equates the strong coupling regime of one string theory with the weak coupling regime of another, revealing what look like very different string theories to be different descriptions of the same theory, M-theory.

S-matrix A method of calculating the probabilities for various outcomes of particle interactions, which requires the observer to stand outside the system under study.

Singularity A place where spacetime curvature becomes infinite and the laws of general relativity, along with all notions of space and time, lose their meaning.

Solipsism The belief that I am the only conscious creature in the universe and the only observer who will ever read this sentence.

Special relativity Einstein's theory, which puts all inertial reference frames on equal footing by keeping the speed of light invariant in all frames but allowing space and time intervals to change from one frame to the next, so that what one observer views as time another might view as space. All observers will agree on four-dimensional spacetime intervals.

String landscape The vast collection of some 10^{500} vacua described by string theory, each with its own values of physical constants, like the cosmological constant.

String theory A theory of quantum gravity, positing that all the different types of elementary particles are each vibrations of a single entity: a string. Supersymmetric strings vibrate in nine spatial dimensions.

Strong complementarity Physics only makes sense within the reference frame of a single observer. Quantum mechanically, this means that each observer lives in his or her own Hilbert space.

Structural realism (ontic) The philosophical position that the world is made not of things but of mathematical relationships or structure.

Supergravity A theory that combines general relativity with supersymmetry. Because supersymmetry is a local symmetry—what looks like a boson in one reference frame might look like a fermion in another—it requires a gauge force to patch up the misalignments between frames. That gauge force is gravity.

Supernova An exploding star.

Superposition A phenomenon in which a quantum system is in multiple, mutually exclusive quantum states, such as "dead cat" and "alive cat," simultaneously. We can't measure superpositions directly, because upon measurement they conveniently disappear. We can, however, see evidence of them in interference patterns. Superpositions reflect quantum theory's non-Boolean logic.

Supersymmetry The theory that bosons and fermions are just two ways of looking at a single, unified object. That ought to mean that every known boson is a known fermion in disguise, and vice versa, cutting the number of elementary particles in half. It doesn't. Instead, every known boson is paired with an unknown fermion and vice versa, doubling the number of elementary particles, half of which have yet to be discovered. On the other hand, supersymmetry elegantly cuts the number of families of particles down to just one and, if you make it a local symmetry, unifies gravity with the other known forces. Sometimes Occam's razor cuts both ways.

Symmetry Sameness. The symmetries of a system ensure that certain features remain invariant under transformation.

T-duality A string theory duality that equates a space of radius R with a space of radius 1/R, big with small. It results from the strange way in which strings experience geometry.

Top-down cosmology Stephen Hawking and Thomas Hertog's conception of cosmology in which measurements an observer makes today select the history of the universe from a superposition of quantum possibilities. It's Wheeler's delayed choice on the grandest scale: observers in the present create a 13.7 billion-year cosmic history.

Uncertainty principle The more accurately you measure one member of a conjugate pair, such as momentum or energy, the less accurately you can measure another, such as position or time. The uncertainty principle reflects the noncommutativity of quantum operators: the order in which you measure things matters. This suggests that quantum features are radically observer-dependent.

Underdetermination When a physical situation has multiple, equally valid theoretical explanations and we have no way of deciding the true reality that lies beneath. Structural realism resolves underdetermination because the theoretical alternatives usually share the same mathematical structure, leaving only one true reality, and a knowable one at that.

Universe "We must be prepared to question the very term 'universe.'"
—John Archibald Wheeler

Unruh radiation A hot bath of observer-dependent particles, also known as Rindler particles, that exist relative to an accelerated observer as a result of a Rindler horizon.

Virtual particle A particle—which comes complete with an antiparticle pair—that arises from the vacuum. Time and energy are related by quantum uncertainty, so the more precise the time, the less precise the energy. In a very short span of time, a large amount of energy can fluctuate out of the vacuum, which, by $E = mc^2$, is also mass. The mass takes the form of a particle, but because it is living on borrowed energy, it quickly annihilates with its antiparticle and disappears—unless the two are separated by an event horizon, at which point the virtual particle becomes a real particle and is known as Hawking radiation.

Wave-particle duality The notion that every particle is also a wave at the same time. When you measure it, it's always a particle. The wave aspect is seen in quantum interference, which results from phase differences. Only waves have a phase, so the particle must be a wave. Then again, phase is not inherent to the particle—it defines the reference frame from which you're viewing the particle.

Wavefunction A probability distribution for the outcomes of quantum experiments, it encodes everything it is possible to know about a quantum state.

Wigner's friend Eugene Wigner's thought experiment in which his friend measures the state of an atom in a lab, collapsing its quantum wavefunction from a host of probabilities to a single reality. Wigner, however, is standing outside the lab, so from his perspective the atom's wavefunction hasn't collapsed, but has instead become entangled in a superposition with the wavefunction describing Wigner's friend. Who is right? Did the wavefunction collapse or not?

WMAP NASA's Wilkinson Microwave Anisotropy Probe, a space-based telescope that mapped temperature variations across the cosmic microwave background radiation.

World line An observer's trajectory through spacetime.

Notes

1 Crashing the Ultimate Reality Party

4 **"This weekend," the article read** Dennis Overbye, "Peering Through the Gates of Time," *New York Times,* March 12, 2002.

8 **Heidegger said that the question** Martin Heidegger, "The Quest for Being," in *Existentialism from Dostoevsky to Sartre,* ed. Walter Kauffman (New York: Meridian, 1956), 245.

8 **"no one," wrote Henning Genz, "has ever given us an answer to what exactly defines nothing"** Henning Genz, *Nothingness: The Science of Empty Space* (Cambridge, MA: Perseus, 1999), 5.

13 **"Can one only hope some day to understand 'genesis' via"** John Archibald Wheeler, *At Home in the Universe* (Woodbury, NY: AIP Press, 1996), 24–26.

20 **"The universe and all that it"** John Archibald Wheeler, *Geons, Black Holes, and Quantum Foam* (New York: W. W. Norton, 1998), 340–41.

21 **"Wheeler seeks to . . . turn the conventional"** Paul Davies, "John Archibald Wheeler and the Clash of Ideas," in *Science and Ultimate Reality,* eds. John D. Barrow, Paul Davies, and Charles Harper Jr. (Cambridge, UK: Cambridge University Press, 2004), 10.

2 The Perfect Alibi

32 **"It is true that the universe"** Lee Smolin, *Four Roads to Quantum Gravity* (New York: Basic Books, 2001), 17.

42 **"a radical revision of our attitude"** Niels Bohr, "Can Quantum-Mechanical Description of Physical Reality Be Considered Complete?" *Physical Review* 48 (October 15, 1935): 697.

42 **"may be paralleled with the fundamental"** Ibid.

44 **"The 'paradox' is only a conflict"** Richard Feynman, *The Feynman Lectures on Physics* (New York: Basic Books, 1965), 3:18–19.

3 Smile!

55 ***The most detailed and precise map*** Dennis Overbye, "Cosmos Sits for Early Portrait, Gives Up Secrets," *New York Times,* February 12, 2003, A34.

55 **"NASA today released the best"** NASA, "WMAP Results," press release 03-064, February 11, 2003.

68 **"If an anthropic principle, *why*"** John Archibald Wheeler, *At Home in the Universe* (Woodbury, NY: AIP Press, 1996), 38.

77–78 **"Life's a Sim and Then You're"** Michael Brooks, "Life's a Sim and Then You're Deleted," *New Scientist,* July 27, 2002, 48.

4 Delayed Choices

85 **"If we live in a simulated reality"** John Barrow, "Glitch!" *New Scientist,* June 7, 2003, 44.

89 **"something deeply hidden"** Albert Einstein, "Autobiographical Notes," in *Albert Einstein: Philosopher-Scientist,* ed. Paul Arthur Schilpp, *Library of Living Philosophers* 7 (Evanston, IL: Library of Living Philosophers, 1949).

92 **"The most symmetric *f* phase of the"** Frank Wilczek and Betsy Devine, *Longing for the Harmonies* (New York: W. W. Norton, 1987), 275.

99 **"No search has ever disclosed"** John Archibald Wheeler, *At Home in the Universe* (Woodbury, NY: AIP Press, 1996), 24–26.

101 **"Any exploration of the concept"** Ibid., 27.

101 **"Unless the blind dice of"** Ibid., 45.

102 **"Mice and men and all on"** Ibid., 306.

102 **"Spacetime," Wheeler wrote, "often considered to"** Ibid., 282–83.

104 **"we used to think that the world"** John Archibald Wheeler, "Time Today," in *Physical Origins of Time Asymmetry,* eds. J. J. Halliwell, J. Pérez-Mercader, and W. H. Zurek (Cambridge, UK: Cambridge University Press, 1994), 19.

104 **"There is no more remarkable feature"** Wheeler, *At Home in the Universe,* 42.

104 **"Except via those time-leaping quantum phenomena"** Ibid., 309.

105 **"Can we ever expect to understand"** Ibid., 310.

5 Schrödinger's Rats

114–16 **In 1989 Worrall published an article** John Worrall, "Structural Realism: The Best of Both Worlds?" *Dialectica* 43, 1–2 (1989): 99–124.

115–16 **"Equations express relations, and if the"** Henri Poincaré, *Science and Hypothesis* (New York: Dover, 1952), 162.

6 Fictitious Forces

132 **"'The notion of reality in the'"** Max Born, "Physical Reality," *Philosophical Quarterly* 3, 11 (1953): 139.

132 **"'The shadow of the circle will'"** Ibid., 143.

132 **"'The projection (the shadow in our example)'"** Ibid., 144.

132 **"'The main advances in the conceptual'"** Ibid.

132 **"'I think the idea of invariant'"** Ibid., 149.

133 **"'That these were two, in principle different'"** Albert Einstein, "Fundamental Ideas and Methods of the Theory of Relativity, Presented in Their Develop-

ment," 1920, in *Collected Papers of Albert Einstein* (Princeton, NJ: Princeton University Press, 2002), vol. 7, doc. 31.

140 **"Physics is an attempt"** Albert Einstein, "Autobiographical Notes," in *Albert Einstein: Philosopher-Scientist,* ed. Paul Arthur Schilpp, *Library of Living Philosophers* 7 (Evanston, IL: Library of Living Philosophers, 1949).

147 **"I believe that nature"** Albert Einstein letter to Raymond Benenson, January 31, 1946, Albert Einstein Archives, Hebrew University of Jerusalem.

153 **"One therefore suspects it is wrong"** John Archibald Wheeler, *At Home in the Universe* (Woodbury, NY: AIP Press, 1996), 24–26.

7 Carving the World into Pieces

162 **"I always feel like a criminal"** John Archibald Wheeler, *Geons, Black Holes, and Quantum Foam* (New York: W. W. Norton, 1998), 314.

174 **"I take it to be true"** Albert Einstein, "On the Method of Theoretical Physics," the Herbert Spencer lecture delivered at Oxford University, June 10, 1933, trans. Don A. Howard in "Einstein's Philosophy of Science," *The Stanford Encyclopedia of Philosophy* (ed. Edward N. Zalta, 2010), plato.stanford.edu/archives/sum2010/entries/einstein-philscience.

177 **"interpreted their results as an indication"** Raphael Bousso, "Adventures in de Sitter Space," in *The Future of Theoretical Physics and Cosmology: Celebrating Stephen Hawking's 60th Birthday,* eds. G. W. Gibbons, E. P. S. Shellard, and S. J. Rankin (Cambridge, UK: Cambridge University Press, 2003), 545.

8 Making History

186 **"Spacetime itself may be reinterpreted as"** Edward Witten, "Reflections on the Fate of Spacetime," *Physics Today,* April 1996, 24–30.

188 **"a new and stronger relativity principle"** Leonard Susskind, *The Cosmic Landscape* (New York: Little, Brown, 2005), 336.

195–6 **"Just as Darwin and Wallace"** Steven Weinberg, "Living in the Multiverse," in *Universe or Multiverse?* ed. Brandon Carr (Cambridge, UK: Cambridge University Press, 2007), 39.

196 **"Faced with scientific claims like neo-Darwinism"** Christoph Schönborn, "Finding Design in Nature," *New York Times,* July 7, 2005, A23.

196 **"Martin Rees said that he was"** Weinberg, *Living in the Multiverse,* 40.

198 **"that horror of the spectral duplication"** Jorge Luis Borges, "Covered Mirrors," in *Collected Fictions* (New York: Viking, 1998), 297.

199 **"in which the histories of the universe"** Stephen Hawking and Thomas Hertog, "Populating the Landscape: A Top-Down Approach," *Physical Review D* 73 (2006): 123527.

199 **"but cosmology poses questions of a"** Ibid.

202 **"the universe would be completely self-contained"** Stephen Hawking, *A Brief History of Time* (New York: Bantam Books, 1988), 141.

202 **"This might suggest that the so-called"** Ibid., 144.

204 **"The past has no existence"** John Archibald Wheeler, *At Home in the Universe* (Woodbury, NY: AIP Press, 1996), 126.

9 A Hint of How the Universe Is Built

219–20 **"It does little good to second-guess"** John Archibald Wheeler, *Geons, Black Holes, and Quantum Foam* (New York: W. W. Norton, 1998), 20.

230 **"And when he gets that new idea"** Deborah Byrd, "At Home in the Universe," *Alcade,* Jan./Feb. 1978, 30.

10 That Alice-in-Wonderland Shit

239 **"I don't care if you agree"** Leonard Susskind, *The Black Hole War: My Battle with Stephen Hawking to Make the World Safe for Quantum Mechanics* (New York: Back Bay Books, 2008), 254.

239 **"To be? To be?"** Niels Bohr quoted by John Archibald Wheeler, "Quantum Theory Poses Reality's Deepest Mystery," *Science News,* May 12, 2008. www .sciencenews.org/view/generic/id/32008/description/John_Wheeler_1911-2008.

249 **"'What's the good of Mercator's North Poles'"** Lewis Carroll, *The Annotated Hunting of the Snark,* ed. Martin Gardner (New York: W. W. Norton, 2006).

249 **"'The Snark is a poem about'"** Gardner, in Carroll, *Annotated Snark,* xxxviii– xxxix.

253 **"Your obligation / is not discharged by"** Seamus Heaney, *Station Island* (New York: Farrar, Straus, and Giroux, 1985), 92–93.

253 **"Take hold of the shaft of"** Ibid., 97.

254–55 **"At the present time, we understand"** Susskind, *The Black Hole War,* 440.

11 Hope Produces Space and Time

274 **"Elementary phenomena are impossible without the"** John Archibald Wheeler, *At Home in the Universe* (Woodbury, NY: AIP Press, 1996), 292.

276 **"The theory of measurement,"** Wigner wrote Eugene Wigner, *Symmetries and Reflections* (Woodbridge, CT: Ox Bow Press, 1967), 179.

277 **"The interpretation of quantum mechanics . . . is"** Hugh Everett III, "The Theory of the Universal Wavefunction," 1955, in *The Many Worlds Interpretation of Quantum Mechanics,* eds. Bryce DeWitt and R. Neill Graham (Princeton, NJ: Princeton University Press, 1973).

277–78 **"does not introduce the idea of"** John Archibald Wheeler, "Assessment of Everett's 'Relative State' Formulation of Quantum Theory," *Reviews of Modern Physics* 29, 3 (July 1957): 464.

279 **"Mice and men and all on"** Wheeler, *At Home in the Universe,* 306.

12 That Hypothetical, Secret Object

293 **"The theory of [quantum] measurement is"** Eugene Wigner, *Symmetries and Reflections* (Woodbridge, CT: Ox Bow Press, 1967), 179.

302 **"The Aleph was probably two or"** Jorge Luis Borges, *Collected Fictions* (New York: Viking, 1998), 283–84.

310 **"The difference between cosmology and the"** J. R. Minkel, "Strung Out on the Universe: Interview with Raphael Bousso," *Scientific American*, April 7, 2003.

311 **"This is just a particularly bad"** Raphael Bousso, "Cosmology and the S-Matrix," *Physical Review D* 71 (2005): 064024; arXiv:hep-th/0412197.

13 Smashing the Glass

330 **"'building physics from scratch'"** Luboš Motl, "Why I Don't Quite Agree with Tom Banks on Eternal Inflation," *The Reference Frame*, October 24, 2011, http://motls.blogspot.com/2011/10/why-i-dont-quite-agree-with-tom-banks.html.

14 Incompleteness

349 **"Complementarity," the authors concluded, "isn't enough."** Ahmed Alm-heiri, Donald Marolf, Joseph Polchinski, and James Sully, "Black Holes: Comple-mentarity or Firewalls?" arXiv:1207.31323[hep-th], July 13, 2012.

351 **Susskind posted a paper on the arXiv** Leonard Susskind, "Complementarity and Firewalls," arXiv:1208.3445[hep-th], August 16, 2012.

360 **"Bousso and Harlow have [advocated] a"** Leonard Susskind, "Black Hole Complementarity and the Harlow-Hayden Conjecture," arXiv:1301.4505v1[hep-th], January 18, 2013.

15 Into the Margin

362–64 **"If different observers give different accounts"** Carlo Rovelli, "Rela-tional Quantum Mechanics," *International Journal of Theoretical Physics* 35 (1996): 1637; arXiv:9609002v2[quant-ph].

367 **Physics World ran an article headlined** Jon Cartwright, "Quantum Physics Says Goodbye to Reality," *Physics World*, April 20, 2007.

368 **"Einstein's reasoning requires the existence of"** Carlo Rovelli and Matteo Smerlak, "Relational EPR," April 2006, arXiv:quant-ph/0604064.

375–76 **"the more general problem that arises"** Raphael Bousso, "Cosmology and the S-Matrix," *Physical Review D* 71 (2005): 064024, arXiv:hep-th/0412197

375 **"no observer can obtain or store"** Thomas Breuer, "The Impossibility of Ac-curate State Self-Measurements," *Philosophy of Science* 62, 2 (June 1995): 197–214.

379 **"Quantum gravity may not admit a"** Raphael Bousso, "Adventures in de Sit-ter Space" in *The Future of Theoretical Physics and Cosmology: Celebrating Stephen Hawking's 60th Birthday*, eds. G. W. Gibbons, E. P. S. Shellard, and S. J. Rankin (Cambridge, UK: Cambridge University Press, 2003), 545.

384 **"The most symmetric phase of the"** Frank Wilczek and Betsy Devine, *Long-ing for the Harmonies* (New York: W. W. Norton, 1987), 275.

385 **"Thanks to this"** Laurent Nottale, "The Principle of Relativity-Emptiness," lec-ture given at Bodhicharya's Ringu Tulku Rinpoche Teachings at La Petite Pierre, France, 2009.

385–86 **"Form is emptiness because"** Ibid.

Suggestions for Further Reading

1 Crashing the Ultimate Reality Party
Science and Ultimate Reality: Quantum Theory, Cosmology and Complexity, edited by John Barrow, Paul Davies, and Charles Harper. Cambridge University Press, 2004.

2 The Perfect Alibi
Three Roads to Quantum Gravity, by Lee Smolin. Basic Books, 2001.

Appearance and Reality: An Introduction to the Philosophy of Physics, by Peter Kosso. Oxford University Press, 1998.

3 Smile!
Coming of Age in the Milky Way, by Timothy Ferris. Perennial, 1988.

4 Delayed Choices
At Home in the Universe, by John Archibald Wheeler. AIP Press, 1996.

5 Schrödinger's Rats
Understanding Philosophy of Science, by James Ladyman. Routledge, 2002.

6 Fictitious Forces
The Force of Symmetry, by Vincent Icke. Cambridge University Press, 1995.

The Comprehensible Cosmos, by Victor Stenger. Prometheus Books, 2006.

Objectivity, Invariance, and Convention: Symmetry in Physical Science, by Talal Debs and Michael Redhead. Harvard University Press, 2007.

The Scientist as Philosopher: Philosophical Consequences of Great Scientific Discoveries, by Friedel Weinert. Springer, 2005.

Symmetries in Physics, edited by Katherine Brading and Elena Castellani. Cambridge University Press, 2003.

Deep Down Things: The Breathtaking Beauty of Particle Physics, by Bruce Schumm. Johns Hopkins University Press, 2004.

7 Carving the World into Pieces
An Introduction to Black Holes, Information and the String Theory Revolution: The Holographic Universe, by Leonard Susskind and James Lindesay. World Scientific, 2005.

The Future of Theoretical Physics and Cosmology, edited by Gary Gibbons, Paul Shellard, and Stuart Rankin. Cambridge University Press, 2003.

A Brief History of Time, by Stephen Hawking. Bantam Books, 1988.

8 Making History

The Elegant Universe, by Brian Greene. Vintage Books, 1999.

The Cosmic Landscape: String Theory and the Illusion of Intelligent Design, by Leonard Susskind. Little, Brown, 2005.

Cosmic Jackpot: Why Our Universe Is Just Right for Life, by Paul Davies. Houghton Mifflin, 2007.

The Grand Design, by Stephen Hawking and Leonard Mlodinow. Bantam Books, 2010.

9 A Hint of How the Universe Is Built

Decoherence and the Quantum-to-Classical Transition, by Maximilian Schlosshauer. Springer, 2010.

10 That Alice-in-Wonderland Shit

The Black Hole War: My Battle with Stephen Hawking to Make the World Safe for Quantum Mechanics, by Leonard Susskind. Back Bay Books, 2008.

11 Hope Produces Space and Time

Geons, Black Holes and Quantum Foam: A Life in Physics, by John Archibald Wheeler with Kenneth Ford. W. W. Norton, 1998.

Gödel's Proof, by Ernest Nagel and James R. Newman. New York University Press, 2001.

Gödel, Escher, Bach: An Eternal Golden Braid, by Douglas Hofstadter. Basic Books, 1979.

Symmetries and Reflections: Scientific Essays, by Eugene Wigner. Ox Bow Press, 1967.

12 That Hypothetical, Secret Object

The Hidden Reality: Parallel Universes and the Deep Laws of the Cosmos, by Brian Greene. Allen Lane, 2011.

Out of This World: Colliding Universes, Branes, Strings, and Other Wild Ideas of Modern Physics, by Stephen Webb. Copernicus Books, 2004.

The Little Book of String Theory, by Steven Gubser. Princeton University Press, 2010.

15 Into the Margin

Quo Vadis Quantum Mechanics? edited by Nancy Kolenda, Avshalom Elitzur, and Shahar Dolev. Springer, 2005.

About the Author

AMANDA GEFTER is a physics and cosmology writer and a consultant for *New Scientist* magazine, where she formerly served as books and arts editor and founded CultureLab. Her writing has been featured in *New Scientist*, *Scientific American*, *Sky and Telescope*, Astronomy.com, and *The Philadelphia Inquirer*. Gefter studied the history and philosophy of science at the London School of Economics and was a 2012–13 Knight Science Journalism Fellow at MIT. She lives in Cambridge, Massachusetts. This is her first book.